BIOTHERMODYNAMICS
Principles and Applications

BIOTHERMODYNAMICS
Principles and Applications

Mustafa Ozilgen • Esra Sorgüven
both of Yeditepe University, Istanbul, Turkey

CRC Press
Taylor & Francis Group
Boca Raton London New York

CRC Press is an imprint of the
Taylor & Francis Group, an **informa** business

MATLAB® is a trademark of The MathWorks, Inc. and is used with permission. The MathWorks does not warrant the accuracy of the text or exercises in this book. This book's use or discussion of MATLAB® software or related products does not constitute endorsement or sponsorship by The MathWorks of a particular pedagogical approach or particular use of the MATLAB® software.

CRC Press
Taylor & Francis Group
6000 Broken Sound Parkway NW, Suite 300
Boca Raton, FL 33487-2742

First issued in paperback 2019

© 2017 by Taylor & Francis Group, LLC
CRC Press is an imprint of Taylor & Francis Group, an Informa business

No claim to original U.S. Government works

ISBN-13: 978-1-4665-8609-3 (hbk)
ISBN-13: 978-0-367-86812-3 (pbk)

Visit the Taylor & Francis Web site at
http://www.taylorandfrancis.com

and the CRC Press Web site at
http://www.crcpress.com

Contents

Preface

This book has emerged from the lecture notes of graduate-level thermodynamics classes. The audience of these classes were students of medicine, genetics, biotechnology, and engineering (mostly mechanical engineers but also chemical, biomedical, and food engineers). Having an audience from different backgrounds enriched in-class discussions and inspired new ideas. Interacting with such a broad range of students and feeling their appreciation and getting their feedback was one of the strongest incentives to this study. We believe that the only way to understand complicated biological phenomena such as aging, muscle efficiency, or cancer is through interdisciplinary team work. This team must involve people with a strong background in engineering thermodynamics, heat transfer, and fluid mechanics and those with a solid background in chemistry, biology, and physiology. We wrote this book in the hope of building a common ground for this interdisciplinary team work. In order to achieve this goal, we kept the language as free of area-specific terminology as possible. We omitted abstract mathematical formulations as much as we could. We have provided numerous examples that demonstrate the biological applications of engineering thermodynamics. The subjects covered in introductory-level chemistry books or the nonbiological applications typically covered in the chemical or mechanical engineering thermodynamics or statistical mechanics books are excluded to keep the focus on "biothermodynamics."

We believe that in education, knowledge is transmitted from the masters to the apprentices. We owe our courage to perform research on a relatively unknown territory of biothermodynamics to the great academicians we had the chance to work with. Mustafa Özilgen got his PhD degree from the University of California at Davis in David F Ollis's group, where he had the opportunity to gain experience in biological systems. Until feeling the need for coauthoring this book, the major teaching material in his classes was the textbook written by Joe Smith (late professor of chemical engineering in the University of California, Davis) and his coauthors. Teaching thermodynamics to the entire faculty in Massey University at New Zealand, where thermodynamic systems are the most important component of the economy, provided him a strong vision to attempt to author such a book. Esra Sorgüven has an MSc degree from Erlangen-Nürnberg University, Germany, where she had the opportunity to work in Franz Durst's group on the simulation of mass, momentum, and energy transfer of chemically reacting flows. During her PhD program, she worked on turbomachinery in the Department of Fluid Machinery, Karlsruhe University (now Karlsruhe Institute of Technology), Germany. As the multidisciplinary environment of the Yeditepe University provided her the chance to work with medical doctors and genetic engineers, her background in chemically reacting flows and turbomachinery led to the development of the analogies between the mechanical and biological systems described in this book. The mechanical engineering vision of this book has deep roots in her education received in Germany.

This is an "engineering thermodynamics" book with biological examples. We believe that teaching based on this book would be most efficient if students prepare a term project as part of the course work. We have observed that most fruitful projects are prepared when the team members have different expertise. If a term project is included, then the first three chapters may be used as the teaching material, and Chapters 4 and 5 may be used as the supplementary material for the projects.

Mustafa Özilgen
Yeditepe University
Istanbul, Turkey

MATLAB® is a registered trademark of The MathWorks, Inc. For product information, please contact:

The MathWorks, Inc.
3 Apple Hill Drive
Natick, MA 01760-2098 USA
Tel: 508-647-7000
Fax: 508-647-7001
E-mail: info@mathworks.com
Web: www.mathworks.com

Acknowledgments

We received substantial help from Ali Bahadır Olcay, Burak Arda Özilgen, Seda Genç, Jale Çatak, Ahmet Çağrı Develi, Bahar Değerli, Kübra Küçük, Bahar Hazal Yalçınkaya, Arda Gürel, Ekin Sönmez, Iltan Aklan, and Sihab Sakib Bayraktar while putting together some of our examples and the related MATLAB codes. We appreciate the author's license from MATLAB (MathWorks Book Program, A#: 1-1276379995).

Acknowledgements

Authors

Mustafa Özilgen is a chemical engineer. He earned a BS and an MS from the Middle East Technical University in Turkey and a PhD from the University of California at Davis. He is an author or coauthor of numerous refereed publications and three other books: *Handbook of Food Process Modeling and Statistical Quality Control* (2nd ed., CRC Press, Boca Raton, Florida), *Information Build-up During Progression of Industrialization* (Arkadas Publishing Co, 2nd ed., Ankara, Turkey, 2011, in Turkish) and *Artistic Narrative of Technology* (Yeditepe University Press, Ankara, Turkey, 2014). Dr. Özilgen has taught numerous classes at the University of California (Davis, California), Middle East Technical University (Ankara, Turkey), and the Massey University at New Zealand. He was a member of the organizing committee and coeditor of the proceedings of CHEMECA 1998, the annual Australian and New Zealander Chemical Engineering conference. He worked for the Marmara Research Center of Scientific and Technological Research Council of Turkey (Gebze, Turkey). He was a recipient of one of the major research awards offered by the Scientific and Technological Research Council of Turkey in 1993. He is working as a professor and chairperson of the Chemical Engineering and Food Engineering Departments at Yeditepe University (Istanbul, Turkey). His current research focuses on biothermodynamics, bioenergetics, and environment-friendly bioprocessing.

Esra Sorgüven received her BSc in chemical engineering from Boğaziçi University, Turkey, and MSc in chemical engineering from Erlangen University, Germany. She has a PhD in mechanical engineering from the University of Karlsruhe (KIT), Germany. She is currently working in the Department of Mechanical Engineering at Yeditepe University, Turkey, as an associate professor. Her research is focused on turbomachinery, computational and experimental fluid dynamics, aeroacoustics, and biothermodynamics. She led about a dozen academic and industrial projects, which involved designing turbomachinery, computational flow simulations, and noise prediction, measurement, and reduction. She authored and coauthored several refereed publications on thermo- and fluid dynamics of mechanical systems and biological applications.

1

Energy, Entropy, and Thermodynamics

1.1 Energy

Energy (e) is the capacity for doing work. It may exist in a variety of forms and may be transformed from one type to another.

Kinetic energy (e_k) refers to the energy associated with the motion. It is proportional to the square of the system's velocity.

Potential energy (e_p) refers to the energy that a system has because of its position or configuration. An object may have the capacity for doing work because of its position in a gravitational field (gravitational potential energy), in an electric field (electric potential energy), or in a magnetic field (magnetic potential energy).

Internal energy (u) refers to the energy associated with the chemical structure of the matter. It includes the energy of the translation, rotation, and vibration of the molecules. It is the energy associated with the static constituents of matter like those of the atoms and their chemical bonds. The internal energy of a matter changes with temperature and the pressure acting on the matter.

Enthalpy (h) is the energy of a fluid in motion. Consider a fluid particle, which is originally at rest and has an internal energy u. If this fluid particle is set to motion, then its internal energy does not change, but its total energy increases because of the flow motion. The energy that the fluid particle possesses to push all the other fluid particles in front of it is called the *flow energy*, and it may be estimated as the multiplication of the particles' pressure and specific volume, pv. The total energy of a flowing fluid particle is then the sum of its internal energy and flow energy. Accordingly, enthalpy is defined as

$$h = u + pv$$

Enthalpy of formation (Δh_f) is the energy required for the formation of 1 mol of a compound from its elements. If all the substances are in their *standard conditions*, then it is called the standard enthalpy of formation, which is denoted with Δh_f^o. The superscript zero indicates that the process has been carried out under standard conditions. While defining the standard conditions, the most common practice is choosing the temperature as 25°C (298.15 K), pressure as 1 bar, and concentration in a solution as 1 mol, pure liquid or solid state at its most stable structure for a substance.

The total specific energy of a system is the sum of all different types of energy that the system possesses:

$$e = u + e_k + e_p + pv$$

The unit of energy in SI is joule (J), which is named after the English physicist James Prescott Joule (1818–1889). He discovered the relationship between heat and the mechanical work, which led to the development of the first law of thermodynamics. Work done by moving an object to a distance of 1 m by applying 1 newton of force is 1 J:

$$J = N\,m$$

Calorie (1 cal = 4.184 J) is a pre-SI metric unit of energy. In some food product labels, kilocalories (kcal) are expressed as Cal, where the first letter is capitalized.

The British thermal unit (BTU) was the traditional unit of energy in England. It is the amount of energy needed to heat 1 pound (454.6 g) of water from 39°F to 40°F (3.8°C to 4.4°C). The BTU unit is still used in the power, steam generation, heating, and air conditioning industries. It may be related with joule as

$$1\ BTU = 1.055\ J$$

Note that the extensive properties depend on the mass of the system. In this book, capital letters are used to denote the extensive properties. For example, the total energy of a system is denoted as E, which has the unit of kJ. The intensive properties are denoted in small letters. For example, the total energy at a point in a system is $e = E/m$ (kJ/kg). Fluxes are represented with a dot. For example, the mass flow rate is $\dot{m}$(kg/s) and the heat flux is $\dot{Q}$ (kJ/s).

1.2 Energy Transfer

Energy can be transferred from one entity to another. In order to study energy transfer, the entity that is to be analyzed has to be defined precisely. In thermodynamics, this entity is defined as a *system*. The precise definition of the system is done by enclosing it with a hypothetical 2-dimensional surface, which is called *system boundaries*. Energy transfer occurs only through the system boundaries.

Energy transfer can occur via mass, work, or heat transfer. Any mass that passes through the system boundaries carries its own energy with itself in or out of the system.

Energy transferred from one system to another by thermal interaction is called *heat*. Amount of heat needed to increase temperature of unit mass of a material by one degree is called the *specific heat*. Parameters c_p and c_v are the specific heat at constant pressure and volume, respectively. Specific heat is expressed in J/mol K or J/g K.

In physics, work is described as the product of the force and the distance through which the force acts:

$$Work = (force)(distance)$$

In the international unit system (SI, *Système international d'unités*), newton (N) is the unit for force. It is named after the British scientist Isaac Newton (1642–1727) and equals to the amount of net force required to accelerate 1 kg mass at a rate of 1 m/s²:

$$N = (kg\,m)/s^2$$

There are different types of work, such as the following:

1. Boundary work ($\delta w = pdv$), which is done by moving a boundary against pressure, like expanding a balloon against the atmospheric pressure.
2. Displacement work ($\delta w = F\,dx$), which is done to move an object a certain distance. For example, if a toy car is pulled with a force F for a distance dx, then the work done for this linear motion is $\delta w = F\,dx$.

Another example for the displacement work can be given from mechanical engineering. In a steam turbine, steam applies a torque on the turbine blades, which causes rotation. The work extracted from the shaft of this steam turbine is $\delta w = \tau\,d\theta$, where τ is torque ($\tau = F/r$) and $d\theta$ is the angle of rotation ($dx = rd\theta$). A schematic drawing of a simple turbine impeller is given in Figure 1.1.

Fluid elements, which are leaving the steam nozzle, have high energy (in the form of enthalpy and kinetic energy). As steam flows through the impeller blades, its energy decreases. The difference in the total energy of the fluid is transferred to the impeller blades. Torque is applied to impeller blades, which cause the rotary motion of the impeller. In the pre-industrial revolution times, water and windmills employed the kinetic energy of water or wind to do the work. We will have a detailed coverage of the work done by an impeller, while we will be learning about the energy cycles in Chapter 3.

Yet another example for the displacement work can be given from biology as the muscle work. During the contraction of a muscle, a tension force is applied and the total length of the muscle bundle is decreased. The work performed by the muscle is $\delta w = Fdl$, where F is the tension and dl is the decrease in the muscle length. Hill's model (Hill, 1938) and Huxley's theory of muscle contraction (1957) laid the foundation for the discussion of the muscle work (Holmes, 2006). The *sliding filament theory* is one of the best established models, which describes the mechanism of the muscle work at the molecular level, in six-step stretching–contracting cycles (Huxley and Hanson, 1954; Huxley and Niedergerke, 1954; Huxley, 2008; Koubassova and Tsaturyan, 2011). Myofibrils are surrounded by calcium-containing sarcoplasmic reticulum. The influx of calcium into the muscle from the sarcoplasmic reticulum triggers the exposure of the binding sites of actin,

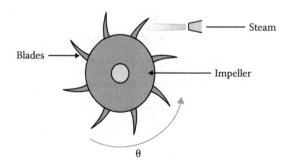

FIGURE 1.1 Schematic drawing of a simple turbine impeller.

and then myosin binds actin. The cross bridge between actin and myosin causes the sliding of the filaments, and the muscle contracts. Then ATP comes in, dissociates into ADP, increases energy of the filaments, and brings the muscle back to the stretched position, while the calcium ions are transported back into the sarcoplasmic reticulum. This cycle repeats itself as long as the muscle work continues. The schematic description of the sliding filament theory is given in Figure 1.2.

The work loop technique is used to evaluate the mechanical work output of muscle contractions (Josephson, 1985). A work loop combines three separate plots: the muscle force versus time, muscle length versus time, and muscle force versus muscle length. The force versus length plot is called the work loop, where each point along the loop corresponds to a force and a length value at a unique point in time.

3. Electrical work ($\delta w = F_E dr$), which is done to move an electrical charge, q, between two points. Here, F_E is the Coulomb force, and dr is distance. The electrical work can also be expressed in terms of voltage difference as $\delta w = qdV$. An example for the electrical work can be given from biology as the neuronal signal transmission. Power is defined as energy utilization or work done in unit time. The SI unit of power is watt (W). One watt is the rate at which work is done when an object's velocity is held constant at 1 m/s against constant opposing 1 N force:

$$W = J/s = (N\ m)/s = (kg\ m^2)/s^3$$

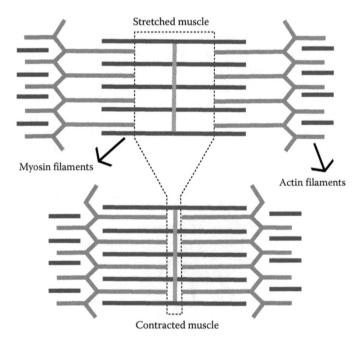

FIGURE 1.2   Schematic description of the sliding filament theory.

Example 1.1: Work Done to Blow Up a Balloon

Calculate the work done to blow up a balloon. The initial volume of the balloon is V_0, and its final volume is V_f.

This example aims to demonstrate the importance of defining the system boundaries precisely. In praxis, it is very common to describe the system only ambiguously. This is partially because the governing equations are traditionally derived with a Lagrangian approach. In Lagrangian mechanics, systems are always solid bodies, for which the system boundary is the same as the body's surface. But as systems involve complicated physical structures or mass transfer through the boundaries, a precise description of the system boundaries is a must. For example, here, it is not enough to define the system as *the balloon*. We have to precisely determine whether the balloon skin is inside or outside the system (Example Figure 1.1.1).

If the system boundary is defined as the inner surface of the balloon skin, then the work done is

$$W_{inner} = p_{inner}(V_f - V_i)$$

If the system boundary is the outer surface of the balloon, then the work done is consumed to push the air molecules in the surroundings. In other words, work is done against a pressure of $P_{surroundings}$:

$$W_{outer} = p_{surroundings}(V_f - V_i)$$

The difference between the two works is the work done to elongate the balloon skin:

$$W_{inner} - W_{outer} = \int_{balloon} Fdl$$

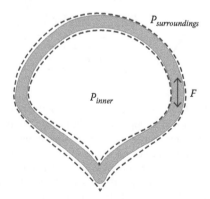

EXAMPLE FIGURE 1.1.1 The alternative choices for the system boundaries of the balloon skin.

Example 1.2: Calculation of the Work Done to Inject the Viral DNA into Host Cell

The lytic cycle is one of the main ways of viral replication, where a phage virus attaches to the surface of a bacterial cell and then injects its own DNA in (Example Figure 1.2.1). The host mistakenly copies the viral DNA and produces capsids, the protein shell of the viral DNA. When the number of viruses inside becomes too much for the cell to hold, the membrane bursts and the viruses are released.

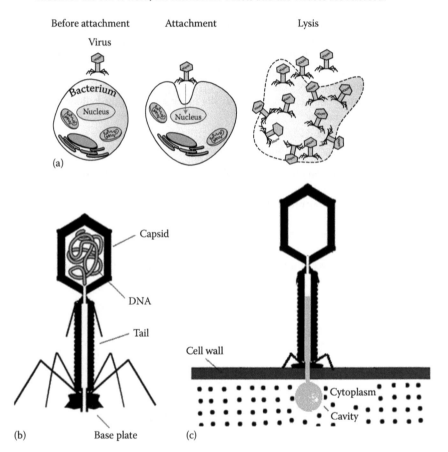

EXAMPLE FIGURE 1.2.1 Schematic drawings of the lytic cycle (a) and the structure of a typical tailed bacteriophage (b). The folded DNA packed into the head (capsid) has high Gibbs free energy. At the beginning of the injection process, the base plate attaches to the receptor on the cell wall (c); the DNA starts unpacking; thus, the Gibbs free energy of the DNA is converted into work (Steven, 1993; Spakowitz and Wang, 2005). The DNA then travels through the tail, makes a cavity in the cytoplasm, while being ejected. The cytoplasm is regarded as a solution of hard spheres. The DNA needs to do work to push the hard coils away while being ejected. The details of the mechanism of the virus-related DNA injection into a bacterial cell are described by Furukawa et al. (1983) and Rossmann et al. (2004).

When the viral DNA pushes the cytoplasmic fluid away to make a cavity to place itself in, the work done against the pressure, p, and the surface tension, σ, is

$$dw = pdv + \sigma ds$$

If the cavity is spherical, we may rewrite this equation as

$$dw = p\left(4\pi r^2\right) dr + \sigma \left(1 - \frac{\delta}{r}\right) 2\pi r dr$$

where the first term $p(4\pi r^2)dr$ represents the work done by moving the spherical surface of area $(4\pi r^2)$ against the cytoplasmic pressure p by distance dr. In the second term, σ is the surface tension of a flat surface; we make it valid for the spherical surface after multiplying with the term $(1-\delta/r)$, where d is referred to as the Tolman length. The term $2\pi r$ is the circumference of the circle of radius r; after multiplying it with dr, we obtain the area of a circular segment $2\pi r dr$. The entire second term of the equation, $\sigma(1-\delta/r)2\pi r dr$, describes the work done to generate the incremental surface area $2\pi r dr$. We may integrate the equation to calculate the total work needed to create a cavity of diameter d_c (Castelnovo et al., 2003; Evilevitch et al., 2004) as

$$\int_0^W dw = \int_0^{d_c/2} p(4\pi r^2)dr + \int_0^{d_c/2} \sigma\left(1 - \frac{\delta}{r}\right) 2\pi r dr$$

or

$$W = \pi\sigma\delta d_c + \frac{\pi\sigma}{4}d_c^2 + \frac{p\pi}{6}d_c^3$$

Pressure, p, applied to the cell wall of *Escherichia coli* was reported to be 3.5×10^5 Pa (Stock et al., 1977) and the surface tension, σ, of the cytoplasm is assumed to be 300 dyn/cm (Koch et al., 1981), and the Tolman length is 0.02 nm (Castelnovo et al., 2003). After substituting the numbers in equation, we will calculate the work as

$$W = \pi\sigma\delta d_c + \frac{\pi\sigma}{4}d_c^2 + \frac{p\pi}{6}d_c^3 = 1.2\times10^{-23} \text{ J}$$

Example 1.3: Estimation of the Flagella Work Done by *Chlamydomonas reinhardtii*

Chlamydomonas reinhardtii is a photosynthetic alga and moves around with the flagella work (Silflow and Lefebvre, 2001). At negligibly small Reynolds numbers after assuming that it has a spherical shape, the drag force F_d on *C. reinhardtii* may be calculated with the help of the Stokes law as

$$F_d = -3\pi\eta dv$$

where η is the viscosity of the medium. Rafai et al. (2010) calculated the drag force F_d on each algal cell according to the Stokes law within the range of 1.1–2.5 mPa s viscosity as 2.510^{-13} N. The drag force when multiplied with the estimated average velocity ($v = 50\,\mu m/s$) gives the estimate of the flagella work performance rate of a single *C. reinhardtii* cell as

$$\dot{W} = (F_d)(v) = (2.5\times10^{-13}\text{ N})(50\,\mu m/s)\left(\frac{1\,m}{10^6\,\mu m}\right) = 1.25\times10^{-17}\text{ J/s}$$

the dot on the symbol $\dot{W}$ tells us that what is calculated here is not work (with unit J) but the work performance rate (with unit J/s).

Example 1.4: Estimation of the Electrical Work Done in Each Opening of a Sodium Gate during Action Potential

There are a lot of examples in the biological systems to the work done while moving a charge against an electric potential difference, such as transmitting the stimulus along the axons (see Example 4.6):

$$\delta w = q_c dV_e$$

where dV_e is the voltage difference in volts across which the charge q in coulombs is traveling. In conventional electricity the charge carriers are usually the electrons, whereas in biological systems Na^+ and K^+ ions may also be the electric charge carriers. The medium charge in conventional electricity is electrical wires; in the biological systems, intracellular and extracellular fluids may serve as the electric carriers (Example Figure 1.4.1).

The nerve and muscle cells are excitable in terms of potentials and currents. At the resting state of the neurons, Na^+ concentration is higher outside and K^+ concentration is higher inside. The K^+ ion leakage across the membrane brings the voltage difference to −70 mV. Normally, the Na^+ gates remain in the closed state. Action potential opens them and the inflow of the Na^+ ions rises the membrane voltage to +50 mV. As the action potential passes in the direction of nerve impulse, K^+ gates open by allowing the K^+ ions to flow out and dropping the voltage again to −70 mV (Hodgkin and Huxley, 1952). In the typical neuron, $10^{+4}Na^+$ ions cross the membrane in millisecond when a single Na^+ gate is open. Since the charge of one electron is known to be about -1.60×10^{-19} C, we may calculate the work done in the electric field during this action as

$$W = \left[\left(|-70| + |+50|\right)\ mV_e\right]\left[(1.60\times10^{-19}\times10^4)C\right] = 1.92\times10^{-18}\text{ J/s}$$

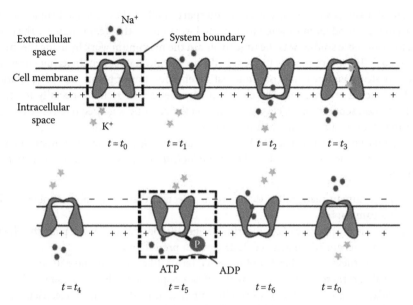

EXAMPLE FIGURE 1.4.1 Schematic description of the ion transport along the membrane.

1.3 The Emergence of the Steam Engine and Thermodynamics

The earliest energy resource employed for doing work was the nutrients used to fuel the muscle work. There were a few major occasions in the history where people switched from one major energy resource to another. The coldest century in the history of Europe was the seventeenth century, where people burned all the firewood for warming, and then all the woods around the human settlements were depleted, which forced people to turn to coal. So, the way to the discovery of the steam engine is opened (Özilgen, 2014). The first transition from one energy source began in the early 1700s with a switch from human and animal muscle work and wood to fossil fuels, which then motivated the discovery of the steam engine and the beginning of the industrial revolution. The steam engine improved industrial production drastically and caused tremendous changes in the prevailing social structure (Özilgen, 2014). The use of the steam engine in locomotives and ships enabled the people to travel to faraway places to set up colonies, which carried the revolution to the second stage, where coal, raw materials, and market for industrial products became highly demandable. The drastic increase in the demand for the raw materials and the market to sell the products led the way to the first and the second world wars, and social changes all around the world, and gave rise to political ideologies like capitalism and communism.

Thermodynamics was the scientific counterpart of all of these events. People started studying thermodynamics when they started working on the steam engines. The driving force of these studies was the urge to obtain the maximum work by using the minimum coal. Thermodynamics first entered into the mechanical engineering curricula to study steam engines, internal combustion engines, power plants, and refrigeration cycles. Later it entered into the chemical engineering curricula especially to study hydrocarbon separation processes. The concept of entropy was used in electrical engineering as a measure of information content.

Thermodynamics was used to be one of the *technological sciences* taught in engineering and science curricula. The *technological sciences*, according to Hansson (2007):

1. Have human-made rather than natural objects as their ultimate study objects, like a steam engine
2. Include the practice of engineering design, like designing a power plant to achieve a predetermined capacity of electric power production
3. Define their study objects in functional terms, like enthalpy and entropy
4. Evaluate these study objects with category-specified value statements, like calculation of coal required to produce predetermined amount of electricity, then relating these numbers with the efficiency of the plant
5. Employ less far-reaching idealizations than the natural sciences, like using block diagrams of the equipment, rather than detailed sketches
6. Do not need an exact mathematical solution when a sufficiently close approximation is available

"The development that meets the needs of the present generations without compromising the ability of the future generations to meet their own needs" is referred to as the sustainable development (Borland et al., 1987). In the assessment of new energy sources, sustainability, that is, renewability, is an important concern. Sustainability depends both on the energy content of the input materials and the process pathways (Thamsiriroj and Murphy, 2009). A successful energy source has to be sustainable in economic, social, environmental, and thermodynamic aspects, including—among other criteria—minimization of the net carbon dioxide emission, and to be complementary with the food supply and waste management programs (Cramer et al., 2006; Worldwatch Institute, 2007; Goldemberg et al., 2008). Since the fossil fuels are not renewable, industrial development that is based on their consumption is not sustainable.

Steam engine was not built by one person; it was built by many contributors in many centuries (Figure 1.3). The inputs to the steam engine are coal and air, and the outputs are carbon dioxide, ash, waste heat, and work. People have met with the concept of industrial environmental pollution after the invention of the steam engine. Carbon dioxide and the other greenhouse gases, waste heat, and ash are among the major environmental pollutants of the twenty-first century. With the buildup of the scientific knowledge, the field of application of thermodynamics expanded, and after the early 2000s thermodynamics started to go beyond the limits of the technological science,

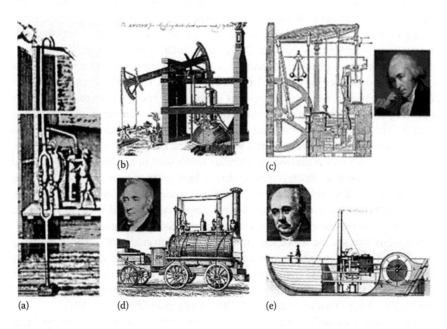

FIGURE 1.3   Steam engine was not built by one person; it was built by many contributors in many centuries. (a) Thomas Savery built the first commercial steam engine *Miner's friend* to drain the flood waters from the coal mines in 1702. (Courtesy of Savery, T., *The Miners Friend or an Engine to Raise Water by Fire*, London, 1702, available at http://himedo.net/ TheHopkinThomasProject/TimeLine/Wales/Steam/URochesterCollection/Savery/index. htm, accessed September 9, 2016.) (b) In 1711, Thomas Newcomen developed a steam engine in England. It could do work equivalent to those of 500 horses, which led to the definition of *horse power* (hp) as a power unit. Newcomen's engine required very large amounts of coal to operate. (Courtesy of Black, N.H. and Davis, H.N., *Practical Physics*. MacMillan, New York, 1914.) (c) James Watt used insulation to reduce the heat loss and increased the fuel efficiency of the steam engine by 75%. Power unit, *watt* (W), was named after James Watt. Although horse power is regarded as a historic power unit, one hp equals to 746 W or 746 J/s. We will frequently need to express the power in 10^3, 10^6, etc., Joules. We will refer to the prefix described in Table A1.1 while doing that. (From Thurston, R.H., *A History of the Growth of the Steam Engine*, Stevens Institute of Technology, Hoboken, NJ, 1878; James Watt's photo by Carl Frederik von Breda.) (d) George Stevenson (1781–1848) built a steam-powered locomotive in 1814, which laid the foundation for modern railways. (Courtesy of Keeling, A.E., *Great Britain and Her Queen*, Project Gutenberg eBook, http://www.gutenberg.org/etext/13103, accessed September 9, 2016.) (e) William Symington (1764–1831) mounted the steam engine to a ship successfully. Steam technology served as a *school* for the later technologies throughout the world. (Courtesy of Bowie R., Scottish Shipbuilding Innovations, 1883, available at http://www.martinfrost.ws/, accessed September 9, 2016.)

and associated with the biological sciences (Petela, 2008), and papers suggesting application of thermodynamics while studying response of the human body to the environmental changes (Tokunaga and Shukuya, 2011), health sciences and nutrition (Fienman and Fine, 2004; Lusting, 2006), aging (Silva and Annamalai, 2008, 2009), and ecology (Jorgensen and Nielsen, 2007; Smith, 2008) appeared in the journals.

Terminology used in thermodynamics is adapted from old Greek and Latin. The word thermodynamics was derived from Greek words *thermos, heat* and *dynamis, power*. It was spelled as thermo-dynamics in the 1850s and 1860s to refer to a newly appearing science about the operation of the steam engine, and then the hyphen was omitted. The words *temperature* and *pressure* originate from Latin words *tempatura* and *pressura*. The term *volume* is related with Greek and Latin words *volumen, bulk, mass, quantity*. The term *heat* is derived from similar words used in medieval English and old German. The symbol q, which is used to denote heat, is taken from the French version of heat *calorique*. The terms *energy* and *enthalpy* originate from Greek words *energeia, activity and operation* and *enqalpein, to heat*, respectively. Rudolf Clausius combined the Greek words *en, contents* and *tropein, transformation* implying *contents in transformation* while introducing the concept of entropy (Battino et al., 1997). The term exergy comes from the Greek words *ex, from* and *ergon, work* (Dinçer and Çengel, 2001). Exergy is the measure of the useful energy of a thermodynamic system, with respect to a datum that is usually the dead state after being corrected for the entropy effects (Dinçer and Çengel, 2001; Sorgüven and Özilgen, 2010, 2012; Özilgen and Sorgüven, 2011).

Steam engine was inefficient in terms of the ratio of the work output to energy input and massive in size and therefore replaced with the internal combustion engine during the twentieth century. Growth of the automotive and the oil industries reinforced each other. Henry Ford and his contemporaries transformed the automobile industry from being a craft to mass production in the 1920s. Oil embargo, which was imposed by OPEC members against the United States, the United Kingdom, and some other countries, convinced the developed nations to turn onto the renewable energy. Natural resources with the ability of being replaced through biological or other natural processes with the passage of time are called renewable resources. They are part of our natural environment and form our ecosystem. Brazil turned onto bioethanol, and France chose the nuclear power, while the search continued for renewable alternatives (Solomon and Krishna, 2011; Özilgen, 2014).

Petroleum became a valuable product in the 1860s. Its price was high at the very beginning as it was used as a fuel for street lights. Oil prices reached a relatively low level after the 1900s. The major consumption was for gas lighting in the 1900s (British Petroleum Company, 2011). Diesel engines ran on peanut oil when it was first manufactured in the early 1900s, but it was replaced with petroleum later due to the lower cost of petroleum. Ford Model T was the first mass-produced car in the 1920s.Oil prices followed a reasonably smooth course until 1973. Elvis Presley's Pink Cadillac, which was big enough for two people to perform a dance show on, may be regarded as the symbol of the days when the idea of energy savings was not prevalent (Özilgen, 2011, 2014). After the Arab–Israeli War, oil prices quadrupled between October 1973 and January 1974 (Park, 1992). People started to look for ways to improve the energy efficiency of the process and later for renewable alternatives to fossil fuels after this hike. Biodiesel became one of the most reasonable alternatives to fossil fuels at the beginning of the twenty-first century. Vehicles such as soybean bus, i.e., the bus running on biodiesel produced from soybean, became popular in the early twenty-first century.

1.4 The Source, Storage, and Utilization of Energy in Biological Systems

Solar energy is the major energy source on Earth. Plants grow by utilizing solar energy for photosynthesis, where CO_2 is the starting chemical. Animals eat the plants to grow. Plant materials like starch and cellulose and fat stored in the animal cells are among the chemicals that contain high-energy bonds. If the plants and animals die and entrapped under the ground, they may be converted into the fossil fuels. The other alternative to fuel production is converting plant and animal material with chemical reactions into biofuels in a factory (Figure 1.4). Regardless of whether we use a fossil or biofuel, when we burn it, we produce CO_2, the starting molecule of the cycle.

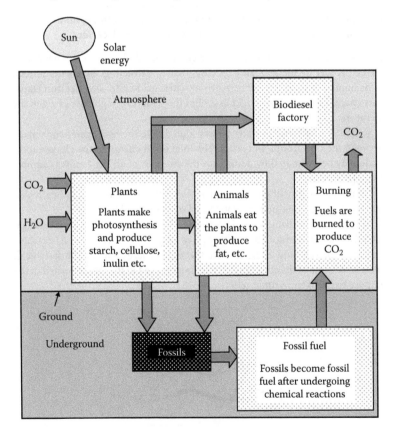

FIGURE 1.4 The cycle of carbon dioxide release in energy-utilizing processes and its recapture via photosynthesis.

TABLE 1.1   Energies of Some Biologically Important Bonds

Bond	C≡C	C=O	C=C	O—H	H—H	C—H	N—H	C—C	C—O
Bond energy (kJ/mol)	828	723	606	456	431	410	385	334	330

Source: Adapted from Laidler, K.J., *Physical Chemistry with Biological Applications*, Benjamin/ Cummings Publishing Co., Menlo Park, CA, 1980, p. 167.

The biological compounds, as explained in Figure 1.4 describing the formation of the bio or fossil fuels, store the chemical energy in their bonds. The *bond energy* is the measure of the strength of a chemical bond and defined as the energy required for breaking it (Benson, 1965; Gronert, 2006). The bond energies are usually affected by the chemical structure of the neighboring bonds; therefore, the bond energies given in Table 1.1 are approximate values.

Example 1.5: Comparison of the Bond Energies of Two Important Metabolites—Glucose and the Oleic Acid Tail of a Triglyceride

Glucose is the starting molecule of the energy metabolism. Energy demand of the cell is met by metabolizing glucose, that is, extracting the energy hidden in its bonds, through glycolytic pathway, citric acid cycle, and electron transport chain. Example Figure 1.5.1 describes the structure of glucose given in the literature.

Solubility of glucose in water is very high due to the hydrogen bonds made between the OH groups and water. The total bond energy of the glucose molecule may be estimated by adding up the energies of all the bonds (Example Table 1.5.1).

Total bond energy of glucose implies that if the molecule should be divided into its atoms, approximately 9230 kJ/mol of energy will be released.

In the animal cells, glucose is converted into fats to store internal energy. A triglyceride is produced by reacting three fatty acids with a glycerol (Example Figure 1.5.2). The tails, R groups, are the major energy source of a fatty acid. Oleic acid is one of the common fatty acids employed in the fats deposited in the adipose tissue of the animal cells.

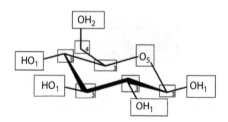

EXAMPLE FIGURE 1.5.1   Molecular structure of glucose.

EXAMPLE TABLE 1.5.1 Details of the Procedure of the Estimation of the Total Bond Energies of Glucose

Bond	Bond Energy (kJ/mol)	Number of Bonds/ Molecule	Total Bond Energy/mol (kJ/mol)
C–H	410	7	2870
C–O	330	7	2310
C–C	334	5	1670
O–H	456	5	2280
H bonds	20	5	100
		Total bond energy	9230

EXAMPLE FIGURE 1.5.2 The triglyceride synthesis reaction.

EXAMPLE TABLE 1.5.2 Details of the Procedure of the Estimation of the Total Bond Energies of Oleic Acid

Bond	Bond Energy (kJ/mol)	Number of Bonds/ Molecule	Total Bond Energy/mol (kJ/mol)
C=O	723	1	723
C=C	606	1	606
O–H	456	1	456
C–H	410	33	13,530
C–C	334	16	5,344
C–O	330	1	330
		Total	20,989

Estimation of the bond energy of 1 mol of oleic acid $CH_3(CH_2)_7CH=CH(CH_2)_7COOH$ is described in Example Table 1.5.2:

$$(20,989 \text{ kJ/mol})\left(\frac{1}{282.5 \text{ g/mol}}\right) = 74.3 \text{ kJ/g}$$

Bond energy of 1 g of glucose is (9230 kJ/mol)(1 mol/180 g) = 51.3 kJ/g; therefore, bond energy of 1 g of oleic acid is 1.4 times of that of glucose.

Utilization of chemical energy stored in the bonds of the carbohydrates and fats powers the biological processes in every organism. Organisms are designed by nature to store an energy reserve to be used in case of emergency. In the human body, any disruption or disorder in this process signals for a health problem (Beloukas et al., 2013; Villa et al. 2013). Short polymers of glucose like glycogen are employed as the short-term energy reserve for organisms since they are simple to metabolize. Fatty acids, triglycerides, and other lipids are commonly used for long-term energy storage by the animal cells. The hydrophobic character of lipids requires them to get into a compact form away from the water. Plant cells utilize starch, a long- and multichain polymer of glucose to store the chemical energy. The strong affinity of carbohydrates for water makes storage of large quantities difficult due to the large molecular weight of the solvated water–carbohydrate complex. Since the plants do not move around, it is feasible for them to store starch as the energy source, but this is not feasible for the animals, since the energy cost of moving around with the large volumes of the carbohydrate–water complex would be very high. Triglycerides coming from the foods are split into glycerol and fatty acids in the intestine, carried with the blood to the adipose tissue, rebuilt there, and stored as fat. When the body requires fatty acids as an energy source, the hormone glucagon signals the breakdown of the triglycerides known as lipolysis. The free fatty acids released by lipolysis enter into the bloodstream and circulate throughout the body. They are broken down in the mitochondria to generate acetyl-CoA. In the energy metabolism, the electron transport chain follows the citric acid cycle. Acetyl-CoA enters into the citric acid cycle. Oxygen enters into the oxidative phosphorylation process, where carbon dioxide and water are the final products. Converting the carbohydrates into lipids for storage and then reconverting them to the intermediaries of the energy metabolism are a costly process when compared with storing them as the polymers of glucose (Özilgen and Sorgüven, 2013); this makes starch a preferred storage material for the photosynthetic organisms. Processing of the high-energy storage carbohydrates in the energy metabolism with the purpose of ATP production is depicted in Figure 1.5.

Brewing is a typical example of such a metabolic process (Figure 1.5). In a typical brewing process, malted barley is almost always the main source of starch and enzymes. In the malting stage, starch is converted into fermentable sugars. The next step is the mashing step during which malt is hydrolyzed and its soluble fractions are extracted. Mashing is followed by the separation of the nonsoluble components and boiling with hops to incorporate desirable flavor and aroma to the beer. In the fermentation stage, fermentable sugars are mainly converted into ethanol. During the downstream processing stage, the beer is filtered, stabilized, and bottled (Linko et al., 1998). Most cancer cells predominantly produce energy by a high rate of glycolysis followed by lactic acid fermentation in the cytosol, rather than by a comparatively low rate of glycolysis followed by oxidation of pyruvate in mitochondria as in most normal cells (Bao et al., 2013). This is called the Warburg effect in oncology (Bensinger et al., 2012). Skipping the glycolytic pathways and providing energy input

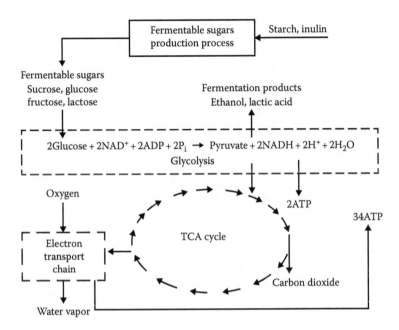

FIGURE 1.5 Schematic description of the major pathways of the energy metabolism: Glycolytic pathway is anaerobic, whereas TCA cycle and electron transport chain are aerobic pathways. Pyruvate is produced in the glycolytic pathway, then either converted into ethanol, lactic acid, etc., or sent to the aerobic metabolic pathways.

to the other stages of the energy metabolism may be a promising treatment method. Similar views were also expressed by Seyfried et al. (2011) while reviewing brain cancer management.

1.5 First Law of Thermodynamics

The principle of the conservation of energy is known as the *first law of thermodynamics*. The first law is usually summarized as *energy can neither be generated nor destroyed but may be converted into other forms of energy, heat or work* as described in Figure 1.6.

Present form of the first law of thermodynamics has been structured with the contributions made by numerous scientists (Keenan and Shapiro, 1947; Bejan, 2006) as

$$\sum_{in}\left[\dot{m}\left(h+e_p+e_k\right)\right]_{in} - \sum_{out}\left[\dot{m}\left(h+e_p+e_k\right)\right]_{out} + \sum_i \dot{Q}_i - \dot{W} = \frac{d\left[m\left(u+e_p+e_k\right)\right]_{system}}{dt}$$

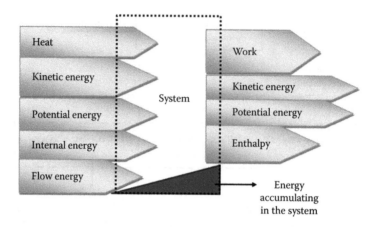

FIGURE 1.6 First law of thermodynamics gives the accounting of the conversion of different forms of energy into any other form of energy, heat, or work. This figure represents a case where kinetic, potential, and internal and flow energies are entering a system, heat, work, and enthalpy are leaving. The same law is applicable to the cases where any of these arrows are reversed.

The left-hand side of the equation (de/dt) represents the energy accumulation within the system boundaries. This term can be positive (or negative), if there is more (or less) energy flowing into the system than flowing out of the system. Under steady-state conditions, that is, if nothing changes with respect to time, then $de/dt = 0$. The right-hand side of this equation represents the net energy flow into the system. Energy flow through the system boundaries may occur via mass, heat, or work transfer. Since thermodynamics was born from analysis of steam engines, heat going into and work going out of the system are regarded as positive quantities. Each stream entering into a system carries its energy in (Figure 1.6). The total energy of a stream contains of its

- Internal energy, for example, energy associated with the atomic and molecular structure, u
- Kinetic energy, e_k
- Potential energy, e_p
- *Flow energy*, pv, for example, the work done on a molecule in the flow system by the other molecules that are pushing it forward

In chemical (Balzhiser et al., 1972; Smith et al., 2005) and mechanical (Çengel and Boles, 2007) engineering texts, work is usually related with the expansion of the system boundaries against a pressure; Bejan (2006) refers to *muscle work* briefly and without going into biological details while discussing the *constructal theory of running, swimming and flight*. Hill's model of muscle contraction (Hill, 1938) and Huxley's theory of muscle contraction (1957) laid the foundation for the discussion of the muscle work (Holmes, 2006). The *sliding filament theory* suggests that the myofibrils in a muscle are surrounded by calcium-containing sarcoplasmic reticulum. The influx of calcium into the muscle from the sarcoplasmic reticulum triggers the exposure of the binding sites of actin, and then myosin binds actin. The cross bridge between actin and myosin causes the sliding of the filaments, and the muscle contracts. Then ATP comes in, dissociates into ADP, increases energy of the filaments, and brings the muscle back to the stretched position, while the calcium ions are transported back into the sarcoplasmic reticulum. This cycle repeats itself as long as the muscle work continues (Huxley and Hanson, 1954; Huxley and Niedergerke, 1954; Huxley, 2008). The work loop technique is used to evaluate the mechanical work output of muscle contractions (Josephson, 1985). The experimental data are usually collected as the muscle force versus time, and the muscle length versus time, and then these data are combined in the plot of the muscle force versus the muscle length, which is called the work loop, where each point along the loop corresponds to a force and a length value at a unique point in time. Human walking may be simulated with the *inverted pendulum-like model of movement* (Griffin et al., 2004; O'Neill and Schmitt, 2012), while human running and hopping and galloping of the animals may be studied with *the bouncing model* (Geyer et al., 2005; Legramandi et al., 2013).

While performing muscle work heat, $\dot{q}$, is released. Resting-state heat release accompanying ATP production in the energy metabolism, excluding the muscular activity, is described in detail by Genç et al. (2013a,b). There are detailed molecular level studies in the literature as reviewed by Ketzer and de Meis (2008) concerning the heat produced during the muscular activity. There are also studies in the literature relating the heat released by the body to the diet (Tataranni, 1995; Gougeon et al., 2005). After reviewing 50 studies, Granata and Brandon (2002) reported that the methodological variations in the protocols of the studies aiming the determination of the thermic effects of foods make it very difficult to make a sound conclusion. The first law of thermodynamics has also been the topic of the studies related with dieting or obesity (Buchholz and Schoeller, 2004; Feinman and Fine, 2004; Fine and Feinman, 2004; Lusting, 2006). Among the second law studies of biological concern, we may cite the ones on human body exergy metabolism (Mady et al., 2012; Mady de Olivera, 2013) and muscle work efficiency (Çatak et al., 2015; Sorgüven and Özilgen, 2015) of significant concern.

For a steady flow system with one inlet and one outlet and no heat or work transfer across the system boundaries, the energy balance can be rearranged as

$$\dot{m}(u_{p,in} - u_{p,out}) + \dot{m}(pv_{in} - pv_{out}) + \dot{m}(e_{p,in} - e_{p,out}) + \dot{m}(e_{k,in} - e_{k,out}) = 0$$

For incompressible fluids, this equation can be further simplified as

$$\dot{m}(p_{in} - p_{out})v + \dot{m}(e_{p,in} - e_{p,out}) + \dot{m}(e_{k,in} - e_{k,out}) = 0$$

By dividing the equation over the mass flow rate and v, we can obtain the energy balance between two points (point 1 and point 2) as

$$(p_1 - p_2) + \rho g_a (h_{1,1} - h_{1,2}) + \frac{1}{2}\rho(v_1^2 - v_2^2) = 0$$

In engineering, this equation is referred to as the *Bernoulli equation*. The Bernoulli equation shows that the total energy of a fluid remains constant at each point in a flow field, if heat or work transfer does not occur.

Example 1.6: Work Done by the Heart

The detailed depiction of the cardiovascular system is available in physiology books such as Tortola and Grabowski (1993) and Shier et al. (2015). Heart is the organ that performs the work to circulate the blood in the body. The solutions offered to the heart problems developed in proportion with the information obtained about the energy metabolism (Beloukas et al., 2013). Blood is circulated through the body with the pumping action of the heart. The schematic description of the circulatory system is given in Example Figure 1.6.1, where *venae cavae* bring O_2-poor blood to the right atrium, and then the blood passes through a valve and flows to the right ventricle, then through the pulmonary arteries to the lungs, where O_2-poor blood becomes O_2 rich after the exchange of CO_2 and O_2. The pulmonary vein delivers the O_2-rich blood to the left atrium, and then it flows into the left ventricle. Heart muscles contract and pump blood out through the aortic valve into the aorta. Blood flowing through the aorta is circulated through the arteriovenous system of the body.

In this example, we will calculate the total work done by the heart. We are going to consider the right and left sides of the heart as systems 1 and 2, respectively. The system boundaries are shown in Example Figure 1.6.1. Note that the system boundaries involve the blood filled inside the heart cavity. The heart muscles are not included in the system. The work done by the contracting heart muscles is transferred through the system boundaries as work. This work is used to increase the total energy of the blood. Since blood is an incompressible fluid undergoing an isothermal process, increasing its total energy means to increase its pressure.

In this example, we used cardiovascular data of an average adult man, who has a heart rate of 65 beats per minute with a stroke volume of 70 mL. Stroke volume is defined as the volume of blood discharged from the ventricle to the aorta with each heartbeat. The pressure difference between the outlet and the inlet of system 1 (right side of the heart) is given as $\Delta p = 15$ mmHg and system 2 (left side of the heart) as $\Delta p = 91$ mmHg.

The volume and the mass of the blood pumped in 1 min are

$$\dot{V}_{out} = (70 \text{ mL/beat})(65 \text{ beats/min}) = 4550 \text{ mL/min}$$

corresponding to a mass flow rate of

$$\dot{m}_{out} = (\dot{V}_{out})(\rho_{blood}) = (4,550 \text{ mL/min})(1.056 \text{ g/mL})(60 \text{ min/h}) = 288,288 \text{ g/h}$$

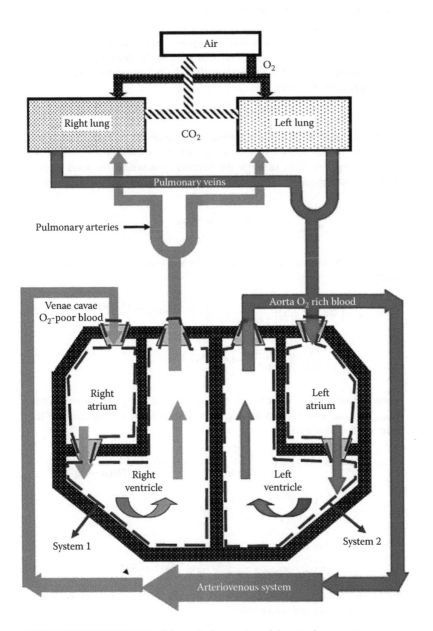

EXAMPLE FIGURE 1.6.1 Schematic description of the circulatory system.

The first law of thermodynamics requires

$$\left[\dot{m}(u+e_p+e_k+pv)\right]_{in} - \left[\dot{m}(u+e_p+e_k+pv)\right]_{out} + \dot{q}-\dot{w} = \frac{d}{dt}\left[m(u+e_p+e_k)\right]_{acc}$$

There is no internal energy, potential energy, or kinetic energy change involved in the process. There is no energy accumulation in the heart and also there is no heat transfer involved in the process; then the expression for the first law of thermodynamics becomes

$$\dot{m}_{in}(pv)_{in} - \dot{m}_{out}(pv)_{out} - \dot{w}_{leftside} = 0$$

when we consider the left side of the heart, $\dot{m}_{in}$ stands for the mass of the blood coming to the heart through venae cavae, and $\dot{m}_{out}$ stands for the mass of the blood leaving the heart through the pulmonary arteries. The work is performed via contraction and relaxation of the heart muscles. When we substitute $v = 1/\rho_{blood}$, then

$$W_{leftside} = \dot{m}_{leftside}\frac{p_{pulmanarvein} - p_{aorta}}{\rho_{blood}} = \dot{m}_{leftside}\frac{-\Delta p}{\rho_{blood}}$$

When $\Delta p = 91$ mmHg, we can calculate the work done by the left side of the heart as

$$W_{leftside} = \left(288,288 \text{ g/h}\right)\left(\frac{1}{1.056 \text{ g/mL}}\right)\left(-91 \text{ mmHg}\frac{1.01325\times10^5 \text{ N/m}^2}{760 \text{ mmHg}}\right)$$

$$\times\left(\frac{1 \text{ J}}{1 \text{ N m}}\right)\left(\frac{1 \text{ m}^3}{1\times10^6 \text{ mL}}\right) = -3,312 \text{ J}$$

where the minus sign indicates that the work is entering the system. When $\dot{m}_{leftside} = \dot{m}_{rightside}$ and $\Delta p = 15$ mmHg, then work done by the right side of the heart will be calculated as

$$W_{rightside} = \left(288,288 \text{ g/h}\right)\left(\frac{1}{1.056 \text{ g/mL}}\right)\left(-15 \text{ mmHg}\frac{1.01325\times10^5 \text{ N/m}^2}{760 \text{ mmHg}}\right)$$

$$\times\left(\frac{1 \text{ J}}{1 \text{ N m}}\right)\left(\frac{1 \text{ m}^3}{1\times10^6 \text{ mL}}\right) = -546 \text{ J}$$

Total work done by the heart is

$$W_{heart} = W_{leftside} + W_{righside} = -546 - 3312 = -3858 \text{ J}$$

MATLAB® code E1.6 calculates the work done by the heart muscles as a function of beating rate and stroke volume.

MATLAB CODE E1.6

```
Command Window
clear all
close all
```

```
% enter the data
rho=1.056; % density of the blood (g/mL)
Ppul_ve=6; % mmHg
Paort=97; % mmHg
Ppul_art=15; % mmHg
Pven_cav=0; % mmHg
z=(50:5:100); % mL of blood per beat
b=(50:5:100)'; % beats per minute
v=b*z; % volumetric flow rate of blood
m=rho*v; % mass flow rate of the blood (g/min)

% compute the work done
wRight=m*60*(Pven_cav-Ppul_art)/rho*1.01325*10^5/760/10^6;
wLeft=m*60*(Ppul_ve-Paort)/rho*1.01325*10^5/760/10^6;
wTotal=wRight+wLeft; % total work done by the heart

% plot the data
surf(b,z,wTotal);
xlabel('Beating Rate (beats/min)')
ylabel('Blood Volume/Beat (ml/beat)')
zlabel('Work Done by the Heart (Joule)')
title('Work done by the heart at various heartbeat rates')
```

When we run the code EXAMPLE Figure 1.6.2 will appear in the screen:

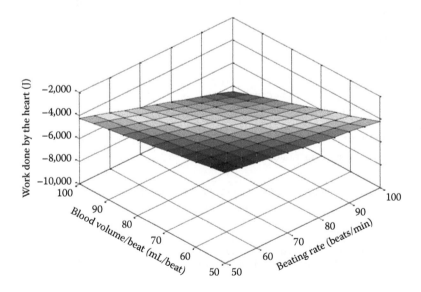

EXAMPLE FIGURE 1.6.2   Variation of the work done by heart with the volume of the blood pumped and the beating rate of the heart.

Example 1.7: Calculation of the Muscle Work from a Work Loop

The work loop concept is used in many applications to calculate the muscle work (Paphangkorakit et al., 2008). Example Figure 1.7.1 demonstrates how a work loop graph is sketched. Usually muscle tension and length are recorded simultaneously. In open literature, both in vivo and in vitro measurements for different muscle groups can be found. Then these two graphs are combined and force versus length graph is sketched. The area of this graph shows the work performed by the muscle.

Energy for the muscle work is provided by dissociation of ATP to ADP:

$$ATP + H_2O \rightarrow ADP + P_i$$

This reaction releases about 22.3 kJ/mol of energy. ATP is produced in the cellular energy metabolism by catabolizing nutrients.

Example Table 1.7.1 presents the muscle force versus the muscle length data measured during pedaling.

MATLAB code E1.7 calculates the work done in one cycle by using the expression $W = \oint F dl$:

MATLAB CODE E1.7
Command Window

```
clear all
close all

% enter the data
L =[0.0638 0.0675 0.0750 0.0825 0.0900 0.0975 0.1050 0.1125 0.1163
   0.1125 0.1050 0.0975 0.0900 0.0825 0.0750 0.0675 0.0638]; %
   muscle length (m)
F= [50 40 20 10 5 3 3 1 0 200 500 510 400 350 200 50 50]; %
   Force (N)

hold all
box on
plot(L,F,'r-', 'LineWidth',2.85)
xlabel('L (m)')
ylabel('F (N)')
ylim([-200 800]); % limit of the range of the y axis

% calculate the muscle work

for i=2:length(F)
  deltaW(i)=(F(i)*(L(i)-L(i-1))); % incremental work
end

w= sum(deltaW)
```

When we run the code the following lines and Example Figure 1.7.2 will appear in the screen:

```
w =
  -15.5570
```

The work done in one cycle is calculated as $W = 15.6$ N m or 15.6 J via the MATLAB code E1.7.

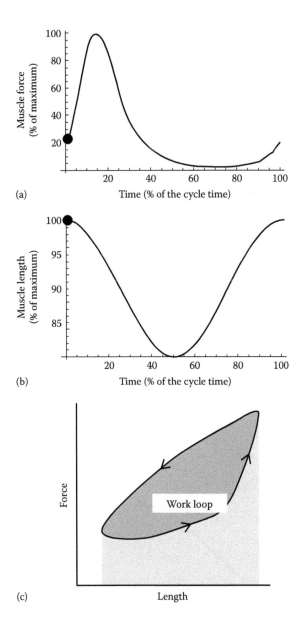

EXAMPLE FIGURE 1.7.1 Variation of the muscle force (a) and the muscle length (b) during the cycle time. (c) Arrows indicate the time course of the loop. The area in the loop refers to the muscle work, area below the loop describes the work lost to passive and active resistance, and their sum refers to the total work.

EXAMPLE TABLE 1.7.1   Muscle Force versus Muscle Length Data Obtained during Pedaling

Muscle Length (m)	Muscle Force (N)
0.0638	50
0.0675	40
0.0750	20
0.0825	10
0.0900	5
0.0975	3
0.1050	3
0.1125	1
0.1163	0
0.1125	200
0.1050	500
0.0975	510
0.0900	400
0.0825	350
0.0750	200
0.0675	50
0.0638	50

Source: Adapted from Neptune, R.R. and Kautz, S.A., *Exerc. Sport Sci. Rev.,* 29, 76, 2001.

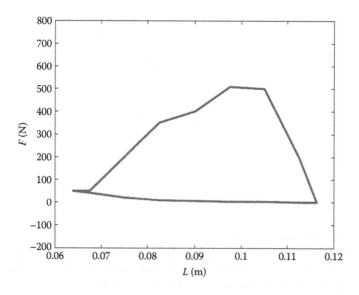

EXAMPLE FIGURE 1.7.2   Printout of the MATLAB® code describing the work loop.

Example 1.8: The Cycling Work

Neptune and van der Bogert (1998) measured the power output of each muscle instrumentally in a biking experiment. The pedaling power, depicted in Example Table 1.8.1, is the average of power output measured at the pedal (Example Figure 1.8.1).

The average muscle work done in one cycle of pedaling may be calculated by using the pedaling power data as

$$w_{cycle} = t_{cycle}\left[\frac{1}{n}\sum_{i=1}^{n}|\dot{w}_i|\right] = (1.5\ \text{s/cycle})(158\ \text{J/s/cycle}) = 237\ \text{J}$$

where

$$\left[\frac{1}{n}\sum_{i=1}^{n}|\dot{w}_i|\right] = 214\ \text{J/s/cycle is the average power output of the muscles in the cycle}$$

$t_{cycle} = 1.5$ s/cycle is the time needed to complete one cycle

EXAMPLE TABLE 1.8.1 Experimental Data Collected during One Cycle of Pedaling

Crank angle (degrees)	0	30	60	90	120	150	180	210	240	270	300	330	360
Power output ($\dot{w}$) of the iliacus psoas (inner hip) muscles (J/s)	−20	−23	−27	−33	−26	−23	−15	7	23	30	23	−15	−20
Pedaling power (J/s)	10	190	380	520	380	190	40	−70	−100	−120	−100	−40	10

Source: Adapted from Neptune, R.R. and van der Bogert, A.J., *J. Biomech.*, 31, 239, 1998.

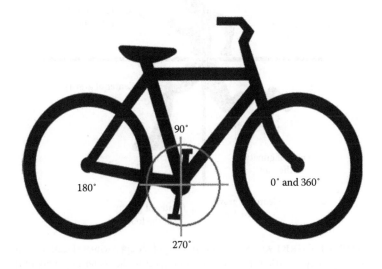

EXAMPLE FIGURE 1.8.1 The angles of measurement.

Contribution of the iliacus psoas muscles to the power output is

$$w_{PSOAS} = t_{cycle}\left[\frac{1}{n}\sum_{i=1}^{n}|\dot{w}_i|\right] = (1.5 \text{ s/cycle})(22 \text{ J/s/cycle}) = 33 \text{ J}$$

Example 1.9: Work for Locomotion

There are two fundamental mechanisms for the terrestrial locomotion. The first mechanism is the pendulum-like model of movement, which applies to human walking. The second one is the bouncing model, which applies to human running and hopping, as well as galloping of the animals.

Willems et al. (1995) suggested a model to describe the locomotion of human body, where the entire body is divided into n segments (Example Figure 1.9.1). Accordingly, the total energy of the body is calculated as the sum of the potential and kinetic energies of each segment:

$$e_{Tot} = \sum_{i=1}^{n} e_{p,i} + e_{k,i} = \sum_{i=1}^{n}\left(m_i g_a h_{l,i} + \frac{1}{2}m_i v_i^2 + \frac{1}{2}m_i \kappa_i^2 \omega_i^2\right)$$

where
m_i is the mass at the center of mass of the ith segment of the body
h_i is the height of the center of the mass, above the ground level

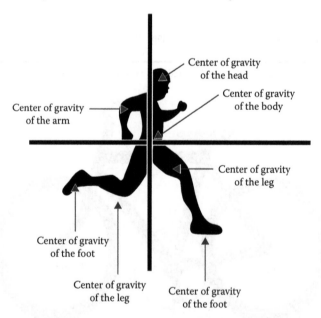

EXAMPLE FIGURE 1.9.1 Schematic drawing of the approximate locations of the center of gravity of various body parts with respect to the center of gravity of the entire body during running.

The terms $(1/2)m_i v_i^2$ and $(1/2)m_i \omega_i^2$ refer to the linear and the rotational kinetic energies of the center of the mass. The terms in this equation can be evaluated either via numerical simulations (with multibody system dynamics software) or via experimental measurements. Parameters v_i and ω_i refer to the linear and the rotational velocities of the center of the mass, respectively, and κ is the radius of rotation. During locomotion, the velocity of each body segment v_i may be expressed as

$$v_i = v_c + v_{r,i}$$

where

v_c is the velocity of the center of the mass of the moving body
$v_{r,i}$ is the velocity of the ith body part, with respect to the center of the mass of the moving body

When we substitute h_c as the height of the center of mass of the body, we will obtain an approximation as (Willems et al., 1995)

$$e_{tot} = m_{tot} h_c g + \frac{1}{2} m_{tot} v_c^2 + e_{arms\ and\ legs}$$

where

$$e_{arms\ and\ legs} = \sum_{i=1}^{n} \left(\frac{1}{2} m_i v_{i,r}^2 + \frac{1}{2} m_i \kappa_i^2 \omega_i^2 \right)$$

Willems et al. (1995) recorded walking and running of several subjects with camera and performed image processing to calculate the parameters in the energy equations. Their calculations show that when the subjects are walking at an average speed of 7 km/h, then $(1/2)\, m_{tot} v_c^2 = 0.4$ J/kg-m, $m_{tot} h_{l,c} g_a = 1.1$ J/kg-m, and $e_{arms\ and\ legs} = 0.65$ J/kg-m. When the subjects are running at an average speed of 27 km/h, then $(1/2)m_{tot} v_c^2 = 0.9$ J/kg-m, $m_{tot} h_{l,c} g_a = 0.3$ J/kg-m, and $e_{arms\ and\ legs} = 1.6$ J/kg-m.

Let us define the efficiency of locomotion as the ratio of the kinetic energy to the total consumed energy. The purpose of running is to move the entire body in forward direction. Vertical motion and movement of the arms and legs accompanies the forward locomotion with energy expenditure. Therefore, we may calculate the fraction of the energy expenditure to maintain forward movement as

$$\eta = \frac{\dfrac{1}{2} m_{tot} v_c^2}{m_{tot} h_{l,c} g_a + \dfrac{1}{2} m_{tot} v_c^2 + \displaystyle\sum_{i=1}^{n} \left(\dfrac{1}{2} m_i v_{i,r}^2 + \dfrac{1}{2} m_i \kappa_i^2 \omega_i^2 \right)}$$

We may calculate the efficiency for gaining inertia for locomotion in forward direction after substituting the numbers as $\eta = 0.19$ while walking and $\eta = 0.32$ while running.

Example 1.10: Thermodynamic Limits of the Work Production in Muscles

Muscles produce work by repositioning the myosin head on the actin filament. Example Figure 1.10.1 shows schematically how ATP hydrolysis enables the fibers to contract and perform work. One mole of ATP is required to engage one myosin head in the muscle contraction.

Muscle cells produce ATP from the energy metabolism, which involves a series of biochemical reactions occurring both in the cytoplasm and mitochondria. Details of this will be explained in Chapter 4. Here, for the sake of simplicity, let us summarize the energy metabolism with one overall reaction:

$$C_6H_{12}O_6 + 6O_2 \rightarrow 6CO_2 + 6H_2O \qquad\qquad \text{(E1.10.1)}$$

Depending on the preferred metabolic pathway and the physiological conditions, 30–38 mol of ATP may be produced accompanying reaction (E1.10.1).

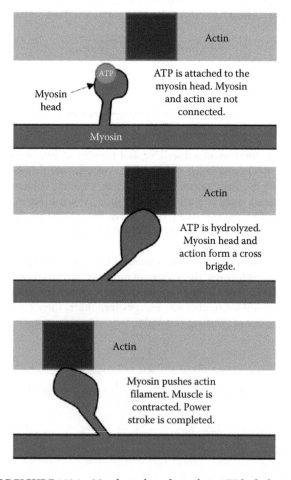

EXAMPLE FIGURE 1.10.1   Muscle work performed via ATP hydrolysis.

When we consider a muscle cell catabolizing 1 mol of glucose to perform contraction work, the maximum work, which may be done after oxidizing 1 mol of glucose in the energy metabolism, would be equal to the Gibbs free energy released with reaction (E1.10.1):

$$w_{max} = \Delta g_{rxn} = \Delta h_{rxn} - TDs_{rxn} \qquad \text{(E1.10.2)}$$

Genç et al. (2013a) calculated the enthalpy and Gibbs free energy of this reaction, under physiological conditions as $\Delta h = -4586$ kJ/mol and $\Delta g = -3862$ kJ/mol of glucose. This implies that the maximum work is $w_{max} = -3862$ kJ/mol of glucose. Detailed discussion of the thermodynamic limits of the work production by a muscle cell is provided by Sorgüven et al. (2015).

Example 1.11: Electric Shock Production by Eel

One of the unique energy conversion systems is the electric shock production by an eel (*Electrophorus electricus*). Electric eels inhabiting in the Amazons have long cylindrical body reaching up to 2.5 m in length and 20 kg in weight. Their electric power–producing cells, for example, the electrocytes, are located in the columns parallel to their spinal cords (Example Figure 1.11.1). Detailed information regarding the electrocyte physiology (Markham, 2013) and neural mechanisms (Zakon, 2003; Rose, 2004) of electric fish are available in the literature.

Eels produce and transmit electricity with the sodium–potassium pumps, like the ones found in all animal nerve cells. We will solve this example by referring to this similarity. In the brain, as the action potential travels down an axon, polarity changes across its membrane. The impulse travels in the axon toward its terminal, where the signal is transmitted to the other neurons (Example Figure 1.11.2).

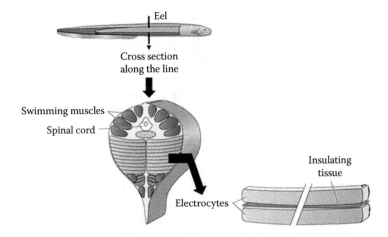

EXAMPLE FIGURE 1.11.1   A simple anatomical sketch of the electric organs of eel.

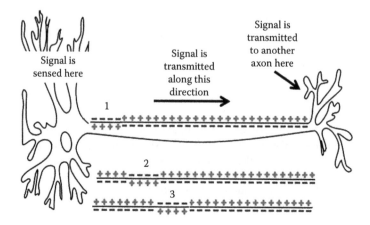

EXAMPLE FIGURE 1.11.2 Propagation of the potential change along an axon during the action: 1 beginning of the action, 2 and 3 are the forthcoming stages.

At the beginning of an action, the Na^+ channels are opened to let the Na^+ move into the axon, and then the K^+ channels are opened to move K^+ out of the axon to change the polarity between the outside and the inside of the cell. Example Figure 1.11.3 describes a case with +60 mV peak potential. The continued leave of the K^+ ions from the cell causes hyperpolarization; at this stage, the

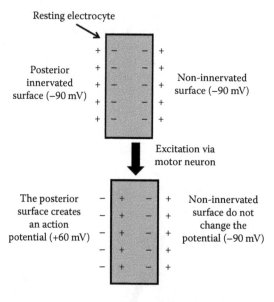

EXAMPLE FIGURE 1.11.3 Schematic diagram of the development of the action potential: the potential difference at the outer surfaces of the electrocyte before the excitation is –90 mV. The posterior surface creates an action potential of +60 mV upon excitation, and then the potential difference between two surfaces becomes + 60 mV – (–90 mV) = 150 mV = 0.15 V.

K^+ channels are closed and Na^+ starts entering into the cell through leak channels to restitute the resting potential. In all of these processes, all-or-none law is valid, for example, either action potential is seen because the voltage threshold is passed over or it is not seen because the voltage threshold is not passed over.

The following reactions occur in each cycle of the action (Tinoco et al., 2013):

$$3Na_{in}^+ \rightarrow 3Na_{out}^+ \text{ with } Dg_R = 13.56 \text{ kJ/mol}$$

$$2K_{out}^+ \rightarrow 2K_{in}^+ \text{ with } Dg_R = 0.97 \text{ kJ/mol}$$

$$ATP + H_2O \rightarrow ADP + P_i \text{ with } Dg_R = -49.1 \text{ kJ/mol}$$

The total Gibbs free energy change of these reactions is

$$Dg_{R,total} = 3(13.56) + 2(0.97) + (-49.1) = -6.5 \text{ kJ/mol}$$

For the production of an electric shock of 600 J/s (potential) = 600 V, current = 1 A, duration = 2 ms, power (600 J/s) $(2 \times 10^{-3} \text{ s}) = 1.2$ J) by the electric eel, we need $(1.2 \text{ J})/(6.5 \text{ kJ/mol}) = 1.85 \times 10^{-4}$ mol of ATP. Since 6.02×10^{23} molecules or ions make 1 mol, we need $(6.02 \times 10^{23} \text{ molecules/mol})$ $(1.85 \times 10^{-4} \text{ mol}) = 1.1 \times 10^{20}$ molecules of ATP dissociated and $3(1.1 \times 10^{20} \text{ mol}) = 3.3 \times 10^{20}$ ions of Na and $2(1.1 \times 10^{20} \text{ mol}) = 2.2 \times 10^{20}$ ions of K transported to support this current. Note that we assumed a thermodynamic efficiency of 100%. A major fraction of the electric power is lost to water as heat; therefore, the real cost of the process is substantially higher.

1.6 Irreversibility

The basic tendency of all the matter in the universe is achieving a lower energy level and a maximum disorder. If there is a lower energy configuration available, matter tries to achieve it. For example, wood is made of cellulose, which has high-energy bonds. Under certain environmental conditions, fires may start spontaneously in search of a lower energy level; heat released when the wood is burned is indeed the difference of the bond energies of the final products and those of the reactants.

The second law of thermodynamics is formulated as a result of observations regarding irreversibility of processes, such as cooling of a cup of hot coffee when left in a room at a lower temperature and atmospheric pressure. The coffee cools down, until a thermal equilibrium between the air and the coffee is achieved. Coffee cannot get spontaneously hot again by cooling the surrounding air. We need to heat the coffee to bring it back to its original temperature.

In Figure 1.7, we have a rock at rest at elevation h, at time $t = 0$. When we disturb it to fall down, its potential energy is converted into kinetic energy between $0 < t < t_f$, where t_f is the time when the rock hits the ground. When the rock is about to touch the ground, for example, $t \approx t_f$ it will have only one form of energy, that is, the kinetic energy. And finally, as it hits the ground, for example, $t \geq t_f$ all of its kinetic energy will be converted

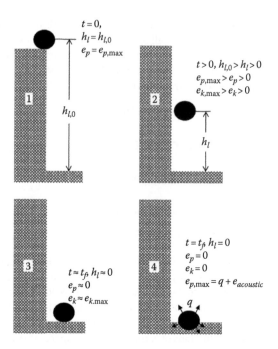

FIGURE 1.7 Schematic description of the conversion of the potential energy of a rock as it falls down the cliff. Numbers 1–4 refer to the progress of falling, $t = 0$ is the beginning and $t = t_f$ is the end.

into heat and acoustic energy. We know from experience that this is an irreversible process that may occur only in this direction. It is not possible to elevate the rock back to its original position by making noise or heating the rock. It is not possible to convert internal energy into potential energy spontaneously.

There are many other examples such as combustion of fuels, free expansion of gases, and diffusion of fluids where a process prefers to proceed in one direction only. The observations show that heat transfer, free expansion of fluid, chemical reactions, mixing, and friction are among the major reasons of irreversibility. These observations cannot be explained with the first law. According to the first law, there are different types of energy, and these can be converted to each other, provided that total energy is conserved. In other words, first law would be satisfied if a rock could gain potential energy by absorbing internal energy of the surroundings. All the actual processes are irreversible. To make things simple, to be able to model, we define reversibility, so we may attempt to solve the sophisticated problems by following a relatively simple path.

In a cylinder where work is done by moving the piston against a force, if all the molecules should change their thermodynamic properties at the same rate, and the thermodynamic properties would be uniform inside the cylinder, at any given instant, then there would be no relative motion between the molecules and no energy would be needed to overcome turbulence. Additionally, if there is no friction between the piston and the cylinder, all the initial energy of the fluid would be used to increase the occupied space, that is, to expand. Maximum work can be produced from a reversible expansion (Figure 1.8).

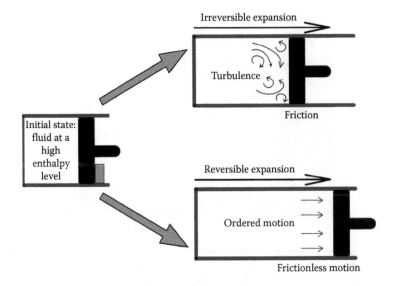

FIGURE 1.8 Schematic description of irreversible and reversible expansion.

Analogous to the expansion process, a compression process would require the minimum work if it occurs reversibly, that is, without turbulence and friction.

Hydraulic turbines use the potential energy difference to produce work (Figure 1.9). Water flowing down from a height h can be used to operate a hydraulic turbine. Irreversibilities like turbulence, friction, and noise generation occur in the flow upstream and downstream the turbine.

Part of the potential energy of water at the height h will be transformed into these dissipative forms of energy. This part of the energy is not useful to produce work. As h increases, the initial potential energy increases, but flow becomes more instable and the part of the energy in dissipative forms increases as well. As h decreases, flow becomes more ordered. If the height difference is small, then fluid molecules move together and a lesser percent of the initial potential energy is transformed into dissipative scales. In the limit as h approaches to zero, energy transformed to dissipative forms diminishes. Then the ideal hydraulic turbine would let the fluid fall in infinitesimally small height increments (dh_l) and convert all the potential energy into work:

$$\lim_{dh \to 0} de_{\text{dissipative forms}} = 0$$

$$dw = \rho g_a dh_l$$

The driving force for the potential energy is the height difference, and if the driving force approaches zero, then irreversibilities become zero too. Analogous to that, the driving force for heat transfer is the temperature difference, and if the driving force (i.e., the temperature difference) approaches zero, then irreversibilities become zero too.

With the pioneering work of Carnot and the formulation of the second law of thermodynamics, a measure for the irreversibilities is defined as the *entropy generation*.

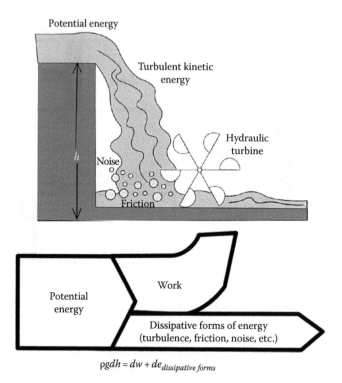

$$\rho g dh = dw + de_{dissipative\ forms}$$

FIGURE 1.9 Schematic description of conversion of potential energy into work by a hydraulic turbine in the presence of friction and noise generation.

1.7 Entropy

Entropy is a thermodynamic property that provides a measure for the *randomness*. In statistical mechanics, entropy (which is a macroscopic thermodynamic property) is defined based on microdynamics. Consider a system that involves N particles. The exact position and momentum of each particle define its microstate (Figure 1.10).

At a given moment, the macroscopic state (T, p, u, …) of the system is defined by the microscopic states of the particles. If the probability of having particles in the ith microstate is p_i, then the entropy of the system is

$$s = -k_B \sum_i p_i \ln p_i$$

This equation is known as the Boltzmann equation, and $k_B = 1.38 \times 10^{-23}$ J/K is the Boltzmann constant. In statistical mechanics, a specific microscopic configuration of a system may occur with a certain probability in the course of its thermal fluctuations. Note that molecular fluctuations affect the intensity of the random motion of

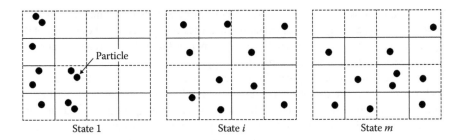

FIGURE 1.10 Schematic description of the microstates.

the particles. The number of positions that the particles can occupy increases with the molecular velocity.

As temperature decreases, fluctuations diminish. If a gas is cooled down, then its entropy decreases. The decrease of the entropy continues with condensation and solidification. In the limit, as the absolute zero temperature is reached, all substances would form a crystalline structure and their molecular velocity would be zero. Consequently, each atomic particle would occupy one certain position, which would not change with respect to time. Then p becomes 1 and entropy becomes $s = 0$ at 0 K.

The Boltzmann equation shows clearly that entropy is a thermodynamic property, just like enthalpy or internal energy. In other words, if we can determine the thermodynamic state of a system, then we can calculate its entropy. In the following, a few examples from our everyday experiences, engineering applications, and biological phenomena are given to demonstrate the use of entropy.

Consider two cups of water, one at 80°C and the other at 10°C. Molecules of the water at 80°C will have a higher molecular velocity than liquid water molecules at 10°C. So, the number of possibilities for an atomic particle to occupy a certain space with a certain momentum is larger for the water at 80°C. Thus, it has a higher entropy.

Figure 1.11 shows several stages of the development of a chicken embryo. At day 0, the egg contains mainly disorganized organic material. A few hours after fertilization, cell

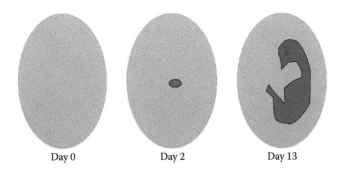

Day 0   Day 2   Day 13

FIGURE 1.11 Schematic description of the stages of the development of a chicken embryo.

division begins. Molecules are organized to form a vascular system and the heart. At day 2, heart begins to beat. Degree of organization increases as days pass by. The entropy of the egg decreases as the embryo grows. Entropy at day 13 is lower than the entropy at day 2, which is even lower than the egg's initial entropy.

Another example can be given from mechanical engineering. Turbines are fluid machineries that make use of the energy of the fluid particles to generate mechanical work. Consider a turbine that is fed with steam at a certain temperature T_{in} and pressure P_{in}. Since entropy is a property of the macroscopic state, the entropy of the steam at T_{in} and P_{in} can be determined. Similarly, at the outlet of the turbine, the steam possesses a unique thermodynamic state T_{out}, p_{out}, and s_{out}.

Even though entropy is a thermodynamic property, *entropy generation* is not. Entropy generation depends on the path that the process follows, just like heat or work. As systems undergo processes, irreversible phenomena like friction or noise generation may occur. These irreversibilities generate randomness, that is, entropy. To exemplify the entropy generation, let us consider the steam that enters the turbine (Figure 1.12). The randomness of the microstates of the particles that make up the steam at the turbine inlet is measured as $s_{in}(T_{in}, p_{in})$. As the inflowing steam passes through the turbine blades, it transfers part of its energy to the turbine blades and performs work. In an ideal turbine, the steam would follow the blades without any flow separation, turbulence, viscous dissipation, or friction. The microstates of the particles would not be affected by the flow and the entropy of the steam would remain constant. In an actual turbine, all of the aforementioned irreversible phenomena may occur: fluid particles adjacent to solid walls are decelerated, secondary flows appear, vortices are formed, etc. Part of the fluids energy shifts into a dissipative form. Particles are mixed and their microstates achieve a new equilibrium. Thus, the outflowing steam possesses a higher entropy. Entropy is generated inside the turbine. The amount of the entropy generation increases with the extent of the irreversibilities.

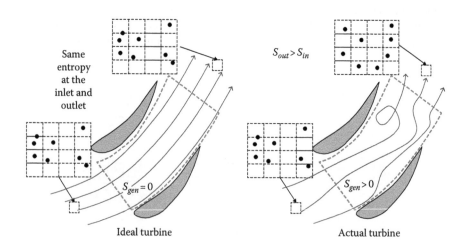

FIGURE 1.12   Schematic description of the entropy in an actual turbine.

The misuse of the terms *entropy generation* and *entropy change* is a common mistake. Entropy generation refers to the extent of the irreversibilities (losses) in a process. For actual processes, it is always positive. For an ideal system that undergoes a totally reversible process, entropy generation is zero. However, entropy generation can never be negative. Entropy change, on the other hand, refers to the difference between the final and initial entropies of a system. Entropy of a system may

- Remain constant (if the system undergoes a steady-state process)
- Increase (if the system gets *disordered*)
- Decrease (if the system achieves a higher ordered state)

For example, if we leave a cup of water at a temperature of 80°C in a room, where the ambient temperature is 20°C, the water will cool down until it reaches a thermal equilibrium with the ambient air. This is an irreversible process; thus, it must generate entropy ($s_{gen} > 0$). At the same time, the water inside the cup reaches a more organized state, so that its entropy is decreased ($s_{final} - s_{initial} < 0$).

1.8 Second Law of Thermodynamics

The second law of thermodynamics is formulated to differentiate between the useful and dissipative forms of energy. Schematic description of the entropy flow in an open system is described in Figure 1.13. Every real heat transfer increases the entropy (disorder) of the universe. In most energy transformations, ordered forms of energy are converted at least in part to heat. This explains why energy is conserved (as heat is a form of energy), yet heat transfer increases randomness. An organism takes matter and energy from its surroundings and converts them into less ordered forms. Animals obtain starch, proteins, and complex molecules from food (stored chemical energy), catabolic processes break them down and release as simpler molecules (CO_2 and H_2O) with less chemical energy, and the remainder of the energy is released as heat. Since disorder of the structure of the matter increases with temperature, we may say that the matter should have the most orderly structure at absolute zero, for example, 0 K.

The second law of thermodynamics describes how entropy in a system changes. Entropy of a system can change with respect to time either due to a net entropy transfer or an entropy generation within the system boundaries. Observations show that entropy

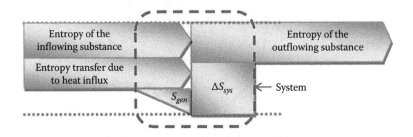

FIGURE 1.13 Schematic description of the entropy flow in an open system.

can be transferred either with the transferred mass or with heat. The mathematical formulation of the second law is

$$\frac{dS_{system}}{dt} = \sum_{in} (\dot{m}s)_{in} - \sum_{out} (\dot{m}s)_{out} + \sum_i \frac{\delta \dot{Q}_i}{T_{b,i}} + \dot{S}_{gen}$$

where T_b is the temperature at the boundary.

Example 1.12: Embryonic Development of a Turkey

A series of biochemical reactions occur as a chicken embryo is developing inside an egg. These chemical reactions cause the egg to consume oxygen and to lose water, as well as to release metabolic heat. Dietz (1995) reports that the metabolic heat released by an 88 g turkey egg is 174 mW. The incubator temperature is constant at 37.5°C and the temperature inside the egg is 0.9°C larger than the incubator temperature (French, 1997). Determine the entropy transfer due to heat transfer.

The entropy transferred with heat can be written as $\sum_i (\delta \dot{Q}_i / T_{b,i}) = \delta \dot{Q}/T_b$, since in this case there is only one thermal reservoir (i.e., the incubator) that is in contact with the system. Here, the system boundary is chosen a few millimeters far from the egg surface, so that on the system boundary temperature is equal to the incubator temperature (Example Figure 1.12.1).

The heat released per gram of egg is

$$\frac{\delta \dot{Q}}{T_b} = \frac{-174 \, \text{mW/88 g}}{(273.15 + 37.5)\text{K}} = -6.36 \, \mu\text{W/g K}$$

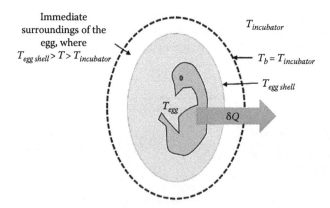

EXAMPLE FIGURE 1.12.1   Schematic description of the embryonic development of turkey.

Remark 1: If the system boundary would be drawn exactly on the egg shell, then the boundary temperature would be higher than the incubator temperature. The entropy that passes through the egg shell is $\delta\dot{Q}/T_{egg\ shell} < \delta\dot{Q}/T_b$. This means that a lower amount of entropy is transferred out of the egg shell, and then the amount of the heat transferred out of the system boundary. This shows that the heat transfer through a finite temperature difference ($T_{egg\ shell}-T_b$) causes an entropy generation of

$$S_{gen\ in\ the\ immediate\ surroundings\ of\ the\ egg} = \frac{\delta\dot{Q}}{T_b} - \frac{\delta\dot{Q}}{T_{egg\ shell}}$$

Remark 2: This system can be regarded as a closed system, if the entropy of the O_2 entering in and H_2O leaving are neglected. Then the entropy equation is

$$\frac{dS_{system}}{dt} = \frac{\delta\dot{Q}}{T_b} + \dot{S}_{gen}$$

If all the processes occur reversibly, then the total entropy within the egg has to decrease at a rate of

$$\frac{dS_{system}}{dt} = \frac{\delta\dot{Q}}{T_b} = -6.36\ \mu W/g\ K$$

Irreversibilities that may occur would cause a positive entropy generation ($s_{gen} > 0$). Then for any actual egg, the rate of entropy decrease is

$$\frac{dS_{system}}{dt} > -6.36\ \mu W/g\ K$$

We may speculate further that a healthier egg (in which biochemical processes occur nearly reversibly) would have a larger entropy decrease. Another conclusion that can be drawn is that if we could measure the entropy decrease rate in a living organism, then we could quantify the organism's health level.

Example 1.13: Estimation of the Entropy of Helium with the Boltzmann Equation

Under standard temperature and pressure (25°C and 1 atm), 1 mol ($N_A = 6.02 \times 10^{23}$) of ideal gas occupies 22.4 L of volume, where N_A is the Avogadro number. Each atom may be assumed to occupy 0.05 nm × 0.05 nm × 0.05 nm of volume (0.05 nm is the Bohr radius of an atom). Then the total number of the microstates available for helium to occupy will be $\Lambda = (22.4 \times 10^{-3} m^3)/(5 \times 10^{-11} m)^3 = 1.8 \times 10^{29}$, $N_A = 6.02 \times 10^{23}$ atoms may arrange themselves in $\Lambda = 1.8 \times 10^{29}$ microstates in $\left(\Lambda/N_A\right)^{N_A}$ different ways. Then the entropy of 1 mol of helium under the standard temperature and pressure may be calculated with the Boltzmann equation as

$$s = k_B \ln\left(\frac{\Lambda}{N_A}\right)^{N_A} = k_B N_A \ln\left(\frac{\Lambda}{N_A}\right)$$

$$= \left(1.38\times10^{-23}\ J/K\right)\left(6.02\times10^{23}\right)\ln\left(\frac{1.8\times10^{29}}{6.02\times10^{23}}\right) = 104.7\ J/K$$

If we should double the volume of the gas by keeping the number of moles of the gas and temperature the same, we will have $\Lambda = (2\times22.4\times10^{-3}\ m^3)/(5\times10^{-11}\ m)^3 = 3.6\times10^{29}$; then the entropy of 1 mol of helium will be

$$s = k_B N_A \ln\left(\frac{\Lambda}{N_A}\right) = \left(1.38\times10^{-23}\ J/K\right)\left(6.02\times10^{23}\right)\ln\left(\frac{3.6\times10^{29}}{6.02\times10^{23}}\right) = 110.5\ J/K$$

There is a general tendency in the nature that the matter always seeks ways to increase their entropy; as we see in this example, entropy of 1 mol of helium increased from 104.7 to 110.5 J/K; therefore, it fills all the available volume.

1.9 Exergy

After the 1973 Arab–Israeli War, the notion of energy has changed, and the energy sources were started to be regarded as valuable commodities, and research for their conservation has started. Energy intensity index is a measure of the energy utilized to produce a unit mass of a commodity. Figure 1.14 shows how energy intensity indices in the chemical and the paper and pulp industries have changed in Japan, for example, an industrial country with very limited energy sources, between the years of 1993 and 2002. This figure shows that the energy intensity indices attained their minimum values in the early 1990s and then kept on fluctuating.

The first law of thermodynamics states that energy is conserved; that is energy can neither be created nor destroyed. But energy can be transferred from one system to another via heat, work of mass transfer. First law does not differentiate between the different modes of energy transfer. The second law of thermodynamics makes this differentiation by defining *entropy*, which is a measure for randomness and increases due to losses involved in the processes. It provides insight on the irreversibilities and helps to quantify the energy losses and proposes measures for minimization of the loss. Exergy is a combination of these two laws of thermodynamics. The exergy balance equation can be derived by multiplying the entropy balance equation with T_0 and subtracting it from the energy balance equation.

The total energy of a system is not always available for performing work. For example, when a chemical reaction occurs at a constant temperature T, the enthalpy change caused by the reaction, Δh, would be accompanied with an increase in the randomness in the structure of the matter, as expressed by the term *entropy*, Δs. Under these

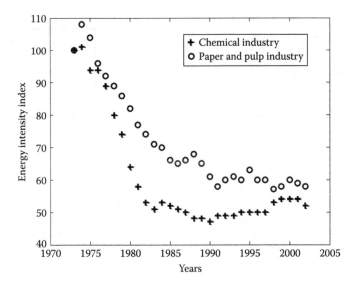

FIGURE 1.14 Variation of the energy intensity index in the chemical and pulp and paper industries in Japan between the years of 1973 and 2002. The energy intensity index had the value of 100 in the base year 1973; it is based on the sum of production weight-averaged by share of industrial production items revised by the change of the monetary value. (Data adapted from *EDCM Handbook of Energy and Economic Statistics in Japan*, the Energy Conservation Center, Tokyo, Japan 2004; Nuibe, T., *Energy Intensity in Industrial Sub Sectors*, the Energy Conservation Center, Tokyo, Japan, 2007, http://www.asiaeec-col.eccj.or.jp/eas/01/pdf/02.pdf.)

circumstances, $T\Delta s$ would be the amount of the energy utilized by the atoms or the molecules of the matter to vibrate, rotate, or travel around. Energy, which remains after deducting the amount consumed to increase the internal randomness, is called the Gibbs free energy, which may be used to do work:

$$\Delta g = \Delta h - T\Delta s$$

Exergy (also called *availability*) is defined as the useful work potential. Exergy of a system is the maximum work that this system can produce if it is brought to thermal, mechanical, and chemical equilibrium with its environment via reversible processes (Figure 1.15). In other words, exergy is the maximum work that can be extracted from a system without violating the laws of thermodynamics. The exergy of a system that is in equilibrium with its environment is zero. The state of the environment is defined as the dead state. Thermomechanical equilibrium is reached at the *restricted* dead state. The thermomechanical exergy, ex_{th}, of a pure substance is defined as

$$ex_{th} = (h - h_0) - T_0(s - s_0)$$

The total exergy of a mixture would be smaller than the sum of the exergies of the species that make up the mixture, because mixing is an irreversible process. The work potential

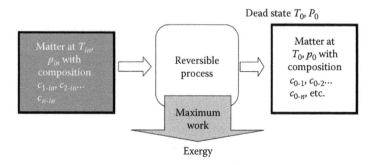

FIGURE 1.15 Exergy of a system is the maximum work that this system can produce if it is brought to thermal, mechanical, and chemical equilibrium with its surroundings via reversible processes. The dead state is at T_0 and p_0, and $c_{0-1}, c_{0-2} \dots c_{0-n}$ are in equilibrium with the surroundings. Exergy is also defined as the minimum work, required to produce a matter at T_{in}, p_{in} with composition $c_{1-in}, c_{2-in} \dots c_{n-in}$ after starting with the matter at T_0, p_0 with composition $c_{0-1}, c_{0-2} \dots c_{0-r}$.

lost due to mixing decreases the absolute value of the exergy of a mixture. The specific exergy of a mixture of *n*-species is

$$ex_{th,mixture} = (h - h_0) - T_0(s - s_0) - \sum_{i=1}^{n}(\mu_i - \mu_{0,i})x_i$$

where
   μ is the chemical potential
   x is the concentration

In mechanical engineering literature, the last term is sometimes referred to as the chemical exergy. In this book, we will regard this term as a part of the thermomechanical exergy, since mixing is a physical process and does not change the intramolecular structure of the species. We define the chemical exergy in relation with the chemical reactions. Chemical equilibrium is achieved at the *true* dead state, when all the components in the system are oxidized or reduced in a reversible way, so that the only components that remain in the system are the components in the environment. The chemical exergy of a substance, which has a molecular structure made up from a number of *k* elements, is defined as

$$ex_{chem} = \Delta g_f^0 + \sum_{i=1}^{k} n_i ex_i$$

where
   Δg_f^0 is the standard Gibbs free energy of formation
   ex_i is the chemical exergy of the *i*th element

The total exergy of a substance is the sum of its thermomechanical and chemical exergies:

$$ex = ex_{th} + ex_{chem}$$

Environment is defined based on the surroundings of the system. For example, in thermodynamic analyses concerning combustion in aircraft engines, environment is taken as the atmospheric air. Accordingly, the restricted dead state is chosen as $T_0 = 298.15$ K and $p_0 = 1$ atm, and the true dead state is defined with the composition of the atmospheric air as 20.35% O_2, 0.03% CO_2, 75.67% N_2, and 3.03% H_2O. Similarly, for processes occurring in the Earth's crust, temperature, pressure, and the composition of the solid components in the external layer of the Earth's crust define the true dead state. For the thermodynamic analyses of systems in the sea, seawater is defined as the environment. If a thermodynamic analysis is performed for a cell, then a convenient choice for the environment is extracellular fluid. Then the true dead state would be defined with $T_0 = 330.15$ Kw, $p_0 = 1$ atm, and $c_{i,0} = c_{i,extracellular\,fluid}$.

The specific exergy of an ideal gas mixture is defined as (Uche et al., 2013)

$$ex = c_p \left[T - T_0 - T_0 \ln\left(\frac{T}{T_0} \right) \right]$$

$$+ \Delta g_f + \sum_{e=1}^{k} n_e ex_e - RT_0 \sum_{i=1}^{n} x_i \ln\left(\frac{x_i}{x_{0,i}} \right) + \frac{1}{2}\left(v^2 - v_0^2 \right) + g\left[h_l - h_{l,0} \right]$$

Each term in this expression refers to the work performance via the following driving forces:

$c_p \left[T - T_0 - T_0 \ln\left(\frac{T}{T_0} \right) \right]$: Thermal exergy, work capacity of the system when it goes from temperature T to temperature of the dead state T_0

$\Delta g_f + \sum_{e=1}^{k} n_e ex_e$: Chemical exergy, work capacity of the system when its present chemical structure is reduced to the chemical structure of the true dead state

$RT_0 \sum_i x_i \ln\left(\frac{x_i}{x_{0,i}} \right)$: Concentration exergy, work capacity of the system when it goes from its current chemical composition to the chemical composition of the true dead state

$\frac{1}{2}\left(v^2 - v_0^2 \right)$: Kinetic energy exergy, work capacity of the system when it goes from its present velocity v to the dead state velocity v_0

$g[h_l - h_{l,0}]$: Potential energy exergy, work capacity of the system when it goes from its present elevation level h_l to the dead state elevation level $h_{l,0}$

In the last few decades, numerous studies are published on exergy analysis (Ayres, 1998, 1999; Caton, 2000; Rakopolous et al., 2006; Talens et al., 2007). Especially in the assessment of renewable energy sources, where we need to weigh various processes and fuels with respect to their ability to produce useful work and to identify their impact on environment, exergy analysis provides a fair tool for comparison. Figure 1.15 implies that with the available technology no further saving could be made in energy utilization, after attaining the minimum energy intensity, but exergy analysis may actually show the stages in a production process where considerable amounts of exergy are destroyed, and then new technology may be developed to improve the exergy efficiency of those stages, to reduce the energy intensity further by employing the newer technology.

The exergetic efficiency η_e of a real process is defined as

$$\eta_e = \frac{\text{Total exergy of the products}}{\text{Total exergy of the inputs}}$$

Exergetic efficiency of ammonia and urea production and raw materials for nitrogenous chemical fertilizers is reported to be about 44% (Hinderink, 1996). Exergetic efficiency of nitric acid is 4%, implying that 25 times of the exergy of nitric acid in the form of those of the raw materials, fuels, electricity, etc., is consumed while producing it (Hinderink et al., 1996). Exergetic efficiency of the biological and the industrial bioprocesses is discussed in the forthcoming chapters.

Energy and exergy are always expressed with respect to a reference level. The specific (implying per unit mass of the object) potential energy is calculated as $e_p = gh$; the reference level of this equation is the location where $h = 0$. The specific kinetic energy is calculated as $e_k = (1/2)v^2$; the reference level of this equation is the state where $v = 0$. The specific internal energy and enthalpy are always expressed with respect to a reference temperature T_{ref}, for example, $u(T) = \int_{T_{ref}}^{T} c_v dT$ and $h(T) = \int_{T_{ref}}^{T} c_p dT$. The reference letter is referred to as the dead state in the exergy calculations. Szargut et al. (2005) provide comprehensive discussion about how the calculated values of the chemical exergies differ as the reference state changes.

1.10 Governing Equations of Mass, Energy, Entropy, and Exergy

Mass balance and the laws of thermodynamics (Table 1.2) are the mathematical relations that need to be satisfied to make activities of the organisms *feasible*. The eighteenth-century chemists, for example, Lavoisier (1743–1794), found out that while matter could be transferred from one form to another, it could be neither created nor destroyed. They always ended up with the same amount of material as in the beginning. This experimental observation gave rise to the concept of *the law of conservation of mass*, concisely the *mass balance* (Whitwell and Toner, 1969). The expression of the mass balance equation establishes basis for numerous feeding and dieting schemes. In the integral biological systems like the circulatory system, which involve many organs, they are expected to assure steady state to avoid health problems. Any disturbance in one of the terms of either mass, energy, entropy, or exergy balances would disturb the others due to their interrelation. Ventura-Clapier et al. (2003), Damman et al. (2007), and Ormerod et al. (2008) reviewed the entire literature to establish relations between the terms of the

TABLE 1.2 Governing Equations of Mass, Energy, Entropy, and Exergy

$$\sum_{in} \dot{m}_{in} - \sum_{out} \dot{m}_{out} = \frac{dm_{system}}{dt}$$

$$\sum_{in} \left[\dot{m}\left(h + e_p + e_k\right)\right]_{in} - \sum_{out}\left[\dot{m}\left(h + e_p + e_k\right)\right]_{out} + \sum_{i} \dot{Q}_i - \dot{W} = \frac{d\left[m\left(u + e_p + e_k\right)\right]_{system}}{dt}$$

$$\sum_{in}\left[\dot{m}s\right]_{in} - \sum_{out}\left[\dot{m}s\right]_{out} + \sum_{i}\frac{\dot{Q}_i}{T_{b,i}} + \dot{S}_{gen} = \frac{d\left[m\,s\right]_{system}}{dt}$$

$$\sum_{in}\left[\dot{m}ex\right]_{in} - \sum_{out}\left[\dot{m}ex\right]_{out} + \sum_{i}\left(1 - \frac{T_0}{T_{b,i}}\right)\dot{Q}_i - W - \dot{E}x_{destr} = \frac{d\left[m\,ex\right]_{system}}{dt}$$

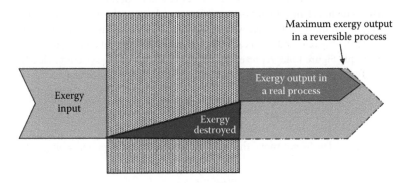

FIGURE 1.16 In a real process a fraction of the exergy input is destroyed due to the irreversibilities.

equations of the laws of thermodynamics. The laws of thermodynamics are indeed a sound basis for the systematic analysis and may be used to improve our understanding of the biological systems and consequently the longevity of the steady stage, for example, the healthy period in the life span (Figure 1.16).

The first step of a thermodynamic analysis is choosing the system boundaries as shown in Examples 1.14 through 1.18, which is followed by performing mass, energy, entropy, or exergy balances (Sorgüven and Özilgen, 2010, 2011, 2012). The *system* is referred to as the *closed system* when no mass crosses through the *system boundary*, for example, $\sum_i \dot{m}_{i,in} = \sum_j \dot{m}_{j,out} = 0$ in the mass balance equation; $\sum_i \dot{m}_{i,in}(u_i + e_p + e_k + pv) = \sum_j \dot{m}_{j,out}(u_j + e_p + e_k + pv) = 0$ in the energy balance equation; $\sum_i \dot{m}_{i,in}s_{i,in} = \sum_j \dot{m}_{j,out}s_{j,out} = 0$ in the entropy balance equation; and $\left(\sum_{k=1}^{N} \dot{m}_k ex_k\right)_{in} = \left(\sum_{j=1}^{M} \dot{m}_j ex_j\right)_{out} = 0$ in the exergy balance equation. A *closed system* is referred to as an *isolated system* if it is impermeable to *heat* and *work* transfer in addition to being *impermeable to mass transfer*, which leads to assigning $\dot{W} = \dot{Q} = dQ/dt = 0$ in the energy, entropy, and exergy balance equations.

Example 1.14: Work Done in the Ear during the Process of Hearing

Sound travels through the air as vibrations of air pressure. The outer part of the ear catches and directs these vibrations into the ear canal. Generally, the sound waves generate extremely small force changes on the eardrum, but even the minimal forces make it to move a good distance. The vibrations are amplified by a group of tiny bones, malleus, incus, and stapes, in the middle ear and translated into electrical signals and transmitted to the brain in the forthcoming sections of the ear.

The system (Example Figure 1.14.1) consists of eardrum, 3 bones (malleus, incus, and stapes) found in the middle ear, and oval window (separates middle ear from inner ear). The sound wave is creating pressure P_{in} in the eardrum. The vibrations travel along after hitting the eardrum, transmitted to the bones, and reach the oval window where it is converted into the waves again as P_{out}. The sound is transmitted with a series of cluster and rebound processes, and air particles do not change their positions. The sound makes the molecules constituting the air move and collide with each other and form cluster (compression). After that, they rebound from each other, like ball hitting and coming back.

Example Figure 1.14.2 represents a simplified sketch of the system to clearly identify the energy transfer modes involved in the process of hearing:

$$\sum_i m_{i,in} - \sum_j m_{j,out} = \left(\frac{dm}{dt}\right)_{system}$$

As air particles do not move through the system boundaries, $m_{in} = m_{out} = 0$; hence,

$$m_{final} = m_{initial}$$

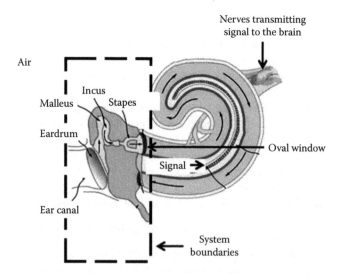

EXAMPLE FIGURE 1.14.1 The system and its boundaries.

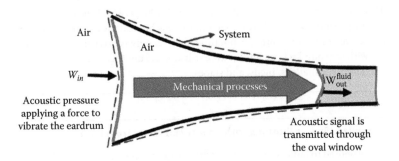

EXAMPLE FIGURE 1.14.2 The simplified sketch of the system and its boundaries.

For this steady system, there is no energy transfer via mass transfer and there is no accumulation. Thus, the energy balance is

$$-\dot{w}_{in} + \dot{w}_{out} + \dot{q} = 0$$

Sound is sensed as vibrations of the eardrum, which are transmitted via mechanical processes through the oval window. Acoustic pressure waves transfer work by vibrating the eardrum (w_{in}). Some of this work is dissipated due to friction (q) and the rest is transferred out of the system boundaries through the oval window (w_{out}). Work related to acoustic pressure fluctuations can be written as

$$\dot{w} = F\frac{dx}{dt} = pA\frac{dx}{dt} = pAc$$

where
  p is the acoustic pressure
  A is the area
  c is the speed of the sound wave

In this case, let us assume that the acoustic signal reaching the ear has an acoustic pressure of $p_{in} = 1$ Pa. This acoustic pressure corresponds to the threshold of human hearing, which is roughly the sound of a mosquito flying 3 m away. Let us estimate the geometrical dimensions of the ear based on an adult man as $A_{eardrum} = 5.0 \times 10^{-5}$ m^2 (the area of an average human eardrum) and $A_{oval\ window} = 2.0 \times 10^{-6}$ m^2 (the area of an average human oval window is 25 times smaller than that of the eardrum) and the acoustic pressure at the oval window as $p_{out} = 1.3$ Pa. The speed of sound in air at 20°C is approximately $c_{sound,in} = 343$ m/s. The outgoing acoustic signal travels in the ear fluid, which can be approximated as water with $c_{sound,out} = 1482$ m/s (velocity of the sound waves in water at 37°C):

$$\dot{w}_{in} = 1\,Pa \cdot 5.0 \cdot 10^{-5}\ m^2 \cdot 343\ m/s = 1.7 \cdot 10^{-2}\ J/s$$

$$\dot{w}_{out} = 1.3\,Pa \cdot 2.0 \cdot 10^{-6}\ m^2 \cdot 1482\ m/s = 3.8 \cdot 10^{-3}\ J/s$$

This result implies that $(3.8 \times 10^{-3})/(1.7 \times 10^{-2})$ (100) = 78% of the energy of the incoming sound in the air is utilized in the mechanical process of amplifying of the pressure of the sound. A fraction of this energy is carried away from the ear with the sound reflected from the eardrum, and another fraction is lost with friction of the malleus, incus, and stapes.

Example 1.15: Car Running with Energy Harvested from Rain

It is proposed to build a car running with energy harvested from rain. To assess the proposal,

(a) Draw a picture of a raindrop, choose appropriate system boundaries, and apply the first law of thermodynamics to calculate the velocity of the droplet when it hits the car. The schematic drawing of the raindrop and the system boundaries is described in Example Figure 1.15.1.

Water balance equation around the raindrop is

$$m_{water,in} - m_{water,out} = m_{water,acc}$$

We will analyze each term of the general conservation equation individually:

$m_{water,in}$ = 0 (no water enters in through the system boundaries)

$m_{water,out}$ = 0 (no water leaves out through the system boundaries, raindrop does not evaporate, vapor do not go out through the system boundaries)

$m_{water,acc}$ = 0

Energy balance around the drop (system: raindrop)
We will use the first law of thermodynamics as

$$\left[\dot{m}(u+e_p+e_k+pv)\right]_{in} - \left[\dot{m}(u+e_p+e_k+pv)\right]_{out} + \dot{q} - \dot{w} = \frac{d}{dt}\left[m(u+e_p+e_k)\right]_{acc}$$

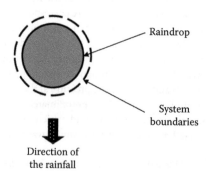

Raindrop

System boundaries

Direction of the rainfall

EXAMPLE FIGURE 1.15.1   System boundaries of the falling droplet.

where $\dot{m}$ refers to the mass rate of change of the droplet. We will analyze each term individually to simplify the equation:

$$\left[\dot{m}(u+e_p+e_k+pv)\right]_{in} = 0 \,(\text{no water enters into the system to bring in additional}$$

internal, potential, or kinetic energy, or pv work)

$$\left[\dot{m}(u+e_p+e_k+pv)\right]_{out} = 0 \,(\text{no water leaves the system to take out internal,}$$

potential, or kinetic energy, or pv work)

$$Q = 0 \,(\text{there is no heat transfer to the system, heating by sun, etc., is neglected})$$

$$W = 0 \,(\text{system is not doing any work, effect of friction of air, etc., are neglected})$$

After substituting numerical values of $[\dot{m}(u+e_p+e_k+pv)]_{in}$, $[\dot{m}(u+e_p+e_k+pv)]_{out}$, Q, and W in the first law of thermodynamics, we will end up with $d/dt\left[m(u+e_p+e_k)\right]_{acc} = 0$, which may be rewritten as

$$\frac{d}{dt}\left[m\,(u+e_p+e_k)\right]_{acc} = \frac{\left[m\,(u+e_p+e_k)\right]_{final} - \left[m\,(u+e_p+e_k)\right]_0}{t_{final}-t_0} = 0$$

or

$$\left[m\,(u+e_p+e_k)\right]_{final} - \left[m\,(u+e_p+e_k)\right]_{initial} = 0$$

Since $m_{final} = m_{inital}$, $u_{final} = u_{inital}$, and $e_{p,final} = e_{k,initial} = 0$, we may determine the final value of the kinetic energy of the drop as $e_{k,final} = e_{p,initial}$, or $1/2v_{final}^2 = g_a h_{clouds}$, where g is the gravitational acceleration and h_{clouds} is the initial height of the raindrop from the ground level. Then we may calculate the velocity of the raindrop when it hits the ground as

$$v_{final} = \sqrt{2g_a h_{clouds}}$$

(b) We will choose the impeller as the system (Example Figure 1.15.2), apply the first law of thermodynamics, and then obtain a relation between the rainfall rate, the height of the clouds, and the work done by the impeller:

$$\left[\dot{m}(u+e_p+e_k+pv)\right]_{in} - \left[\dot{m}(u+e_p+e_k+pv)\right]_{out} + \dot{Q} - \dot{W} = \frac{d}{dt}\left[m\,(u+e_p+e_k)\right]_{acc}$$

We may assume that the steady state is prevailing and neglect any heat losses in the impeller. Then the work produced by the impeller has to be equal to the energy difference between the inlet and the outlet:

$$\dot{W} = \left[\dot{m}(u+e_p+e_k+pv)\right]_{in} - \left[\dot{m}(u+e_p+e_k+pv)\right]_{out}$$

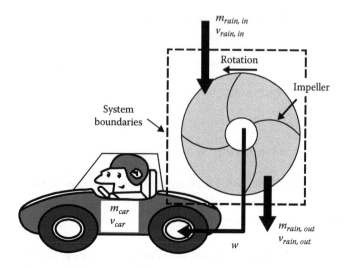

EXAMPLE FIGURE 1.15.2 System boundaries around the impeller, where the droplets are falling and work is produced to move the car.

Here, only kinetic energy is relevant:

$$\frac{\dot{w}}{\dot{m}} = \left[e_k\right]_{in} - \left[e_k\right]_{out}$$

Theoretically, the velocity of water at the outlet of the turbine can be approximated to zero by increasing the area of flow and all of the kinetic energy can be converted into mechanical work. So, theoretically,

$$\frac{\dot{w}_{max}}{\dot{m}} = \frac{1}{2}v_{final}^2 = g_a h_{clouds}$$

Note that decelerating a fluid is hydrodynamically a very difficult process. Therefore, in actual turbines, the efficiency of converting kinetic energy into work is limited.

Example 1.16: Fat-Consuming Runner

(a) A girl with a body mass of m runs for 1000 m with a speed of 5 km/h (Example Figure 1.16.1). The final destination is 100 m above the starting point. While the girl is running, heat is also released from her muscles. Assume that the runner's metabolism operates so that all of the energy for running comes from catabolizing fat. Neglect the energy requirement for metabolic activities and derive an expression to describe how the runner's internal energy is converted into other forms of energy.

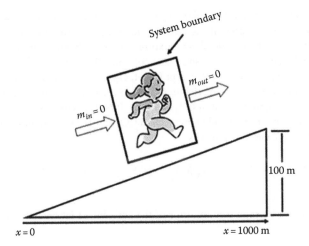

EXAMPLE FIGURE 1.16.1 The fat-consuming girl running uphill.

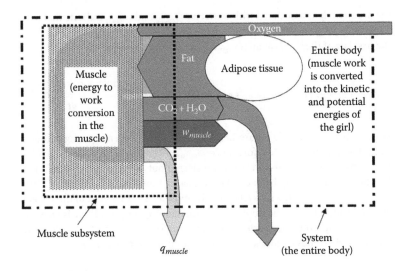

EXAMPLE FIGURE 1.16.2 Schematic description of the fat burning and muscle work done by the girl to increase the kinetic and potential energy of her body.

Energy conversion in this process is described in Example Figure 1.16.2, where only $q_{released}$ crosses through the system boundaries. It should be noticed that the muscle work is produced in the muscle subsystem, which is then employed to make the entire body run.

Let us first analyze the muscle subsystem with the energy balance:

$$\left[\dot{m}\left(u+e_p+e_k+pv\right)\right]_{in} - \left[\dot{m}\left(u+e_p+e_k+pv\right)\right]_{out} - \dot{Q}_{muscle} - \dot{W}_{muscle} = 0$$

In this expression, the first, $[\dot{m}(u+e_p+e_k+pv)]_{in}$, and the second, $[\dot{m}(u+e_p+e_k+pv)]_{out}$, terms represent the rates of the energy entering into and leaving the muscle with blood. Blood is a mixture of several chemical species. In this simplified analysis, we will assume that concentrations of all of the chemicals in the blood remain constant, except for fat, O_2, CO_2, and H_2O. Blood flowing into the muscle subsystem is rich in fat and oxygen. Inside muscle subsystem, fat is oxidized and reduced to CO_2 and H_2O. Blood flowing out of the muscle subsystem is poor in fat and O_2 but rich in CO_2 and H_2O. Thus, for the muscle subsystem, the energy balance is

$$\left[\left(\dot{m}h\right)_{fat}+\left(\dot{m}h\right)_{O_2}\right]_{in}-\left[\left(\dot{m}h\right)_{CO_2}+\left(\dot{m}h\right)_{H_2O}\right]_{out}-\dot{Q}_{muscle}-\dot{W}_{muscle}=0$$

This can also be expressed in terms of the enthalpy of fat oxidation reaction (Δh_{rxn}):

$$\dot{m}_{fat}\Delta h_{rxn}-\dot{Q}_{muscle}-\dot{W}_{muscle}=0$$

The negative signs associated with the terms $\dot{Q}_{muscle}$ and $\dot{w}_{muscle}$ indicate that heat is released and work is done by the muscle subsystem.

The first law of thermodynamics regarding the entire system is

$$\left[\dot{m}\left(u+e_p+e_k+pv\right)\right]_{in}-\left[\dot{m}\left(u+e_p+e_k+pv\right)\right]_{out}-\dot{Q}-\dot{W}=\frac{d\left[m\left(u+e_p+e_k\right)\right]_{entire\,body}}{dt}$$

As described in Example Figure 1.16.2, there is only oxygen entering through the system boundaries; therefore, $\dot{m}_{in}=\dot{m}_{in,oxygen}$. The products of the fat oxidation (carbon dioxide and water) are transferred out of the system boundaries; therefore, in the energy balance, their enthalpy must be accounted for. Heat, which is produced in the subsystem by the muscle, is transferred to the system and then transferred to the environment; therefore, $\dot{Q}=\dot{Q}_{muscle}$. There is no work passing through the boundaries of the entire system, that is, $\dot{W}=0$. Then the first law of thermodynamics as applied to the entire body becomes

$$\left(\dot{m}h\right)_{O_2}-\left[\left(\dot{m}h\right)_{CO_2}+\left(\dot{m}h\right)_{H_2O}\right]_{out}-\dot{q}_{muscle}=\frac{d\left[m\left(u+e_p+e_k\right)\right]_{entire\,body}}{dt}$$

We may rearrange this equation as

$$\left(\dot{m}h\right)_{O_2}-\left[\left(\dot{m}h\right)_{CO_2}+\left(\dot{m}h\right)_{H_2O}\right]_{out}-\dot{q}_{muscle}dt=d\left(mu\right)_{entire\,body}$$
$$+d\left(me_p\right)_{entire\,body}+d\left(me_k\right)_{entire\,body}$$

The internal energy change of the entire body equals to those of the fat consumed:

$$d(mu)_{entire\ body} = d(mu)_{fat}$$

The total amount of fat consumed by the girl during running is

$$d(mu)_{entire\ body} = (\dot{m}h)_{O_2} - \left[(\dot{m}h)_{CO_2} + (\dot{m}h)_{H_2O}\right]_{out}$$

$$-\dot{q}_{muscle}dt - d(me_p)_{entire\ body} - d(me_k)_{entire\ body}$$

The total amount of fat consumed is

$$m_{fat} = \frac{1}{u_{fat}}\left[q_{muscle} + \int_{(me_p)\text{at }t=0}^{(me_p)\text{ at }t=t_{final}} d(me_p)_{entire\ body} - (\dot{m}h)_{O_2} + \left[(\dot{m}h)_{CO_2} + (\dot{m}h)_{H_2O}\right]_{out} \right]$$

where

$$\int_{(me_p)\text{at }t=0}^{(me_p)\text{ at }t=t_{final}} d(me_p)_{entire\ body} = mg_a\Delta h_l$$

Δh is the elevation the girl reached after running

While analyzing the thermodynamics of running, Myers (1982) reported that the work done by the muscles is identified as the sum of three terms: work done to overcome gravity (e.g., generating potential energy), work done against the inertial forces while accelerating (e.g., kinetic energy generation), and work done to overcome wind resistance, which is neglected in this example.
You may also work on the following cases to calculate

(b) How much fatty acids will be consumed to provide the utilized internal energy.

(c) How much water will be needed for perspiration to remove the heat released.

(d) When the girl runs on the treadmill, it needs to perform 20% more work than the total energy consumed by the girl to overcome the friction and other losses. Calculate the total amount of CO_2 released in this process, attributable to respiration of the girl plus the electric power consumption by the treadmill.
Data: coefficient = 0.14 kg CO_2 is emitted/MJ of electric power utilization, J = kg/m^2/s^2, and g = 9.81 m/s^2. You may assume that 1 kg of fat is equivalent to 900 g of stearic acid ($C_{18}H_{36}O_2$), and stearic acid is converted to CO_2 and H_2O.

(e) You may also work on the case when the girl described in Example Figure 1.16.1 is running downhill.

Example 1.17: Hydrodynamic Changes in the Heart due to Amputation or Limb Transplantation

The rates of the heart failure are high among the people who have limb amputations (Hrubec and Ryder, 1980; Modan et al., 1998). Amputation implies reduction of the resistance to the blood flow; therefore, the steady state achieved in a healthy person's body is disturbed after amputation. The heart needs to attain a new level of work performing, to establish a new level of steady state. Limb transplantation, in the simplest thermodynamic sense, means higher level of workload for the heart. An amputee who underwent the world's first would-be quadruple limb transplant died in Turkey, a few days after surgery, on February 27, 2012, after his heart and vascular system failed to sustain the new limbs (*Hürriyet Daily News*, 2012). In the biological systems, an organ or metabolic pathway adapts itself to the existing conditions, to minimize the energy utilization. Therefore, the amputee's heart might have adapted itself to pump blood to the remaining part of the body, when he lost his both arms and legs after an accident that had occurred years ago. The additional work that the heart expected to perform after the transplantation is assessed in this example.

For a healthy adult man, the volumetric flow rate of the blood through the left ventricle ranges between 4 and 7 L/min. Assuming a flow rate of 5 L/min, corresponds to a mass flow rate of

$$\dot{m} = \dot{v}\rho = \left(5\,\text{L/min}\right)\left(\frac{1\,\text{min}}{60\,\text{s}}\right)\left(1.060\,\text{g/mL}\right)\left(1000\,\text{mL/L}\right)\left(\frac{1\,\text{kg}}{1000\,\text{g}}\right) = 0.0883\,\text{kg/s}$$

Pressure is the highest in the aorta at the exit of the heart and decreases as the blood travels along the circulatory system. Small arterioles and arteries have the greatest resistance and thus produce the highest pressure loss through their length.

A cardiac cycle has four phases (Guyton and Hall, 2016):

1. *Filling*: Blood flows into the ventricle from the left atrium and the volume increases.
2. *Constant-volume contraction*: The volume of the ventricle remains constant since all the valves are closed, and the pressure increases.
3. *Ejection*: The aortic valve is opened, blood flows into the aorta, and the volume decreases.
4. *Constant-volume relaxation*: The aortic valve closes, so the pressure increases back to the diastolic pressure while the volume remains unchanged.

The numerical values employed in this example are extracted from the intraventricular pressure versus the volume plot given by Rubenstein et al. (2011) for a healthy man. Work done by the left ventricle is (Morley et al., 2007)

$$W_{Left\ ventricle} = \int_{v_0}^{v_f} p\,dv \qquad\qquad (\text{E1.17.1})$$

Pressure drop Δp along each blood vessel is calculated as (Çengel and Cimbala, 2006)

$$\Delta p = \left(\frac{fL}{d}\right)\left(\frac{\rho v^2}{2}\right) \qquad \text{(E1.17.2)}$$

where

f is the Darcy friction factor
L is the characteristic length
d is the characteristic diameter of the blood vessel
ρ and v are the density and the velocity of the blood, respectively

In the laminar flow regime, for example, $Re = (\rho v d)/\mu < 2300$, where Re is referred to as the Reynolds number, the Darcy friction factor is

$$f = \frac{64}{Re} \qquad \text{(E1.17.3)}$$

Mass flow rate in each vessel is calculated as

$$m^{\bullet} = \frac{\pi d^2 \rho v}{4} \qquad \text{(E1.17.4)}$$

After taking the density and the viscosity of the blood as $\rho = 1.060$ g/mL and $\mu = 3.5 \times 10^{-3}$ kg/s m (Cho and Kensey, 1991) and employing Equations E1.17.1 through E1.17.4, the following table is constructed. It should be noted here that it was discovered almost a century ago that the viscosity of the blood is not a constant quantity, but dependent on the diameter of the tube, for example, the vein. Below a critical point at a diameter of about 0.3 mm, the viscosity decreases strongly with reduced diameter of the tube. As a consequence, the resistance to the flow of blood in the arterioles (and in the small veins) is considerably less than would be the case if the streaming of the blood followed the law of Poiseuille, that is, behaved as a fluid with regard to its viscosity (Fahraeus and Lindqvist, 1931), therefore assuming a constant blood velocity here may be an oversimplification in a research.

Example Table 1.17.1 shows that the mass flow rates of the blood through each arm and each leg are calculated as 6.18×10^{-3} and 7.42×10^{-3} kg/s, respectively. These values are deducted from the blood flow rate through the left ventricle of the healthy man to calculate the blood flow rates of an amputee (Example Table 1.17.2).

Work done by left ventricle of a healthy man when one beat pumps 5 L blood per minute is calculated by integrating the pressure versus the ventricle volume data (Rubenstein et al., 2011) as given in the MATLAB code. The reduction caused by amputation in the blood volume and pressure is introduced into calculations:

MATLAB CODE E1.17

Command Window

```
clc
clear all
close all
format compact
```

EXAMPLE TABLE 1.17.1 Pressure Drop in Each Arm and Leg

Body Part	Volumetric Flow Rate (L/min)	$\dot{m}$(kg/s)	d (mm)	v (m/s)	Re	f	Δp (mmHg)
Arm	0.35 (Voliantis and Secher 2002)	6.18×10^{-3}	4 (brachial artery)	0.46	557	0.1	12.82
Leg	0.42 (Jorfeldt and Wahren 1971)	7.42×10^{-3} kg/s	4 (profunda artery)	0.56	678	0.09	22.44

Source: Shoemaker, J.K. et al., *Cardiovasc. Res.*, 35, 125, 1997; Radegran, G. and Saitin, B., *AJP-Heart*, 278, H162, 1999.

Note: The calculations are based on the brachial artery and the profunda artery, since they are the arteries with the largest diameter, 4 mm, in the arm and the leg, respectively.

EXAMPLE TABLE 1.17.2 The Blood Flow Rates through the Left Ventricle of the Healthy Man and the Amputee

	Δp across the Left Ventricle (mmHg)	$\dot{m}$ across the Left Ventricle (kg/s)	% Decrease of $\dot{m}$ through the Left Ventricle ($f_m \times 100$)	Decrease of Δp through the Left Ventricle (mmHg)	Work Performed by the Heart (Calculated by the Code)
No amputation	117.95	0.0883	0	0	0.958
One leg amputated	95.51	0.0808	9	22.44	0.689
Two legs amputated	73.07	0.0734	17	44.88	0.458
Two legs and an arm are amputated	60.25	0.0673	24	67.32	0.265
Two legs and two arms are amputated	47.43	0.0611	31	70.52	0.221

```
% ENTER THE BLOOD PRESSURE AND VOLUME DATA AFTER AMPUTATION
fm = 0.31; % fraction of the reduction in blood mass due to
  the amputation
dP_mercury =70.52; % decrease in the blood pressure after
  amputation
rm= (1-fm)


% ENTER THE FILLING PHASE DATA
vol_DA =rm*[5.01019E-05 5.11776E-05 5.35578E-05 5.62996E-05
  5.94771E-05 6.16439E-05 6.46777E-05 6.69890E-05 7.00233E-05
  7.36355E-05 7.75372E-05 8.05723E-05 8.33184E-05 8.64979E-05
  8.98220E-05 9.38699E-05 9.76284E-05 1.02400E-04 1.05726E-04
  1.08618E-04 1.11367E-04 1.14838E-04 1.16863E-04 1.19466E-04]';
  % LV volume (m3) transpose the vector to save the space

p_DA = [1327.58708 962.55537 774.67575 605.35981 529.37566
  491.16747 452.63504 414.37272 394.53902 374.48920
  373.02999 390.59388 408.26589 425.77578 443.23155
  516.51352 571.20476 700.31294 792.56388 884.97695
```

```
    1052.23919 1163.13501 1237.17345 1329.69464]'; % Pressure
    (Pa) transpose the vector to save the space

% ENTER THE EJECTION PHASE DATA
vol_BC = rm*[1.18097E-04 1.16520E-04 1.14652E-04 1.12784E-04
    1.10189E-04 1.08319E-04 1.06448E-04 1.05153E-04 1.02991E-
    04 1.00974E-04 9.82348E-05 9.56387E-05 9.37643E-05
    9.13114E-05 8.77025E-05 8.49589E-05 8.20704E-05 7.87472E-
    05 7.60015E-05 7.25325E-05 6.94962E-05 6.67493E-05
    6.44347E-05 6.06737E-05 5.80697E-05 5.50312E-05 5.30027E-
    05 5.14069E-05 5.06797E-05 5.02383E-05]'; % LV volume (m3)
    transpose the vector to save the space

P_BC = [11203.24994 11783.50688 12251.67918 12701.15274
    12982.60733 13394.68317 13713.26524 13956.83587 14238.12834
    14481.96922 14782.17666 15007.53493 15195.22536 15364.43317
    15534.07336 15628.59411 15704.47014 15724.41183 15725.43868
    15689.33819 15615.67784 15560.60837 15449.28033 15282.39632
    15152.47754 14985.32329 14742.99562 14463.10824 14257.69180
    13902.57700]'; % pressure (Pa), transpose the vector to
    save the space

n = 4; % set the degree of the polynomials

% unit conversions
vol_DA_ml = vol_DA.*10^6; % unit conversion from m3 to mm3
p_DA_mmHg = p_DA.*(760/101325); % unit conversion from Pa to
    mmHg
vol_BC_ml = vol_BC.*10^6;
P_BC_mmHg = P_BC.*(760/101325); % unit conversion from Pa to
    mmHg
dP = dP_mercury*101325/760; % conversion from mmHg to Pa

% FILLING PHASE CALCULATIONS
f = polyfit(vol_DA,p_DA,n); % fit a polynomial (SI units)
fit = polyval(f,vol_DA); % in SI units

% Work done during filling of left ventricle (J)
int_fit = polyint(f); % integrate polynomial analytically
a = vol_DA(1);
l = length(vol_DA);
b = vol_DA(l);
W_filling = abs(polyval(int_fit,b) - polyval(int_fit,a));

% EJECTION PHASE CALCULATIONS
f1 = polyfit(vol_BC,P_BC,n); % fit a polynomial (SI units)
f1_unit = polyfit( vol_BC_ml, P_BC_mmHg,n);
fit1 = polyval(f1,vol_BC);
fit1_unit = polyval(f1_unit,vol_BC_ml,n);

% Work during ejection of left ventricle in joules
int_fit1 = polyint(f1);
a1 = vol_BC(1);
l1 = length(vol_BC);
b1 = vol_BC(l1);
W1 = abs(polyval(int_fit1,b1) - polyval(int_fit1,a1));
```

```
%Pressure decrease - work calculation for amputees

for i =1:11
  P_BC_amputee_trans (i) = P_BC(i)-dP; % in Pa
  P_BC_amputee_trans_unit(i) = P_BC_mmHg(i)-dP_mercury; %in
  mmHg
end

  P_BC_amputee = P_BC_amputee_trans';
  P_BC_amputee_unit = P_BC_amputee_trans_unit';

f2 = polyfit(vol_BC,P_BC_amputee,n);
f2_unit = polyfit(vol_BC_ml, P_BC_amputee_unit,n);
fit2 = polyval(f2,vol_BC);
fit2_unit = polyval(f2_unit,vol_BC_ml);

% work done by heart during ejection in amputee (area under
  the BC curve)
int_fit2 = polyint(f2);
a2 = vol_BC(1);
l2 = length(vol_BC);
b2 = vol_BC(l2);
W_ejection = abs(polyval(int_fit2,b2) - polyval(int_fit2,a2));

% Calculate the net work done by heart in 1 cycle (J)
Work = W_ejection-W_filling

% plot the P versus V data
hold all
f_unit = polyfit(vol_DA_ml,p_DA_mmHg,n); % with standard units
fit_unit = polyval(f_unit,vol_DA_ml); % with standard units
plot(vol_DA_ml,p_DA_mmHg,'r-')
plot(vol_BC_ml,fit2_unit,'b'), grid
title ('The Cardiac cycle')
xlabel('Left Ventricular volume(ml)')
ylabel('LV pressure mmHg')
```

When we run this code for a healthy man with fm= 0 and
  dP_mercury =0 mmHg of pressure drop reduction due to
  amputation, we will have the following lines and Example
  Figure 1.17.1 appears in the screen:

```
Work =
  0.9580
```

When we run this code for an amputee with fm= 0.31 and
  70.12 mmHg of pressure drop reduction due to amputation,
  we will have Example Figure 1.17.2 and the following lines
  will appear in the screen:

```
Work =
  0.2208
```

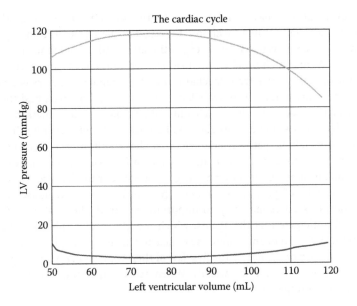

EXAMPLE FIGURE 1.17.1 The cardiac cycle of a healthy man with $f_m = 0$ and $=0$ mmHg of pressure drop reduction due to amputation.

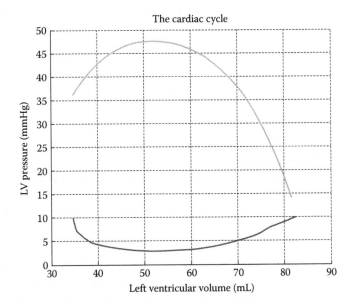

EXAMPLE FIGURE 1.17.2 The cardiac cycle of an amputee with a $f_m = 0.31$ and dP_mercury = 70.12 mmHg of pressure drop reduction due to amputation.

The energy balance equation relates the energy in and outflow with the work performed in a system. When we choose the heart as the system, the energy balance equation may be employed to relate energy utilization and the pump work performed by the heart. A mismatch between the load applied to the heart and energy needed to meet the load may arise from mechanical and/or metabolic factors is among the causes of the heart failure (Ventura-Clapier et al., 2003). The mass and the energy balance equations describe the match implied in this sentence. The same authors also state that factors that lead to abnormal contraction and relaxation in a failing heart include metabolic pathway abnormalities that result in decreased energy production, energy transfer, and energy utilization. Patients suffering from heart failure always complain about early muscular fatigue and exercise intolerance. The decreases in mitochondrial ATP production and energy transfer to the heart muscles through phosphotransfer kinases may play an important role in the heart attacks. Ormerod et al. (2008) anticipate that the agents, which modify cardiac substrate utilization, may have the potential to ameliorate the deficiency in energy supply and increase the pumping efficiency of the heart. The first law of thermodynamics makes it possible to study these phenomena mathematically. When heart is operating under steady-state conditions, for example, when

$$\left(\frac{d\left[m\left(u+e_p+e_k\right)\right]}{dt} \right)_{system} = 0$$

after setting $\dot{m}_{in} = \dot{m}_{out} = \dot{m}$, we may express the first law of thermodynamics as

$$\dot{m}\left(u_{in}-u_{out}\right)+\dot{m}\left(e_{p,in}-e_{p,out}\right)+\dot{m}\left(e_{k,in}-k_{k,out}\right)$$

$$+\dot{m}\left(\frac{p_{in}}{\rho_{in}}-\frac{p_{out}}{\rho_{out}}\right)-\dot{m}\int pd\dot{v}-\sum \dot{W}_{shaft}-\sum \dot{f}=0$$

where
  $\dot{m}$ is the constant flow rate of the blood through heart
  the term $\dot{m}\int pd\dot{v}$ represents the work done by expanding the system boundaries against constant pressure, P
  the term $\sum \dot{f}$ is the energy consumption to overcome the total friction in the circulatory system

The left side of the heart pumps blood coming from lungs via pulmonary vein to the aorta, to travel around the entire circulatory system of the body. The right side of the heart pumps the blood coming from the lower and the upper parts of the body via vena cava to lungs via pulmonary artery. There is no shaft work in the circulatory system, implying that $\sum \dot{W}_{shaft} = 0$; the potential energy term, for example, $\dot{m}\left(e_{p,in} - e_{p,out}\right)$, is neglected when the engineering Bernoulli equation is used in association with the circulatory system; then we may write that

$$\dot{m}\left(u_{in} - u_{out}\right) = \dot{m}\left(e_{k,out} - k_{k,in}\right) + \dot{m}p\left(\frac{p_{out}}{\rho_{out}} - \frac{p_{in}}{\rho_{in}}\right) + \dot{m}\int p d\dot{v} + \sum \dot{f}$$

where $\dot{m}\left(u_{in} - u_{out}\right)$ describes the energy cost of the heart work, where carbohydrates are employed in the energy metabolism to produce ATP, which is then employed in the heart to do the muscle work. The theory of Ventura-Clapier et al. (2003) states that heart attack will be inevitable, if the internal energy supply to the heart should not meet the energy demand. In the circulatory system internal energy is used to compensate the potential, for example, $\dot{m}\left(\left(p_{out}/\rho_{out}\right) - \left(p_{in}/\rho_{in}\right)\right)$, and the kinetic energy, for example, $\dot{m}\left(e_{k,in} - e_{k,out}\right)$, changes, the muscle work, for example, $\dot{m}\int p \, d\dot{v}$, and the friction, for example, $\sum \dot{f}$ losses. In a healthy person, the heart needs to be operating under the steady-state conditions; any change in P_{in} or P_{out} indicates a blood pressure problem, for example, hypertension; since the velocity of the blood in the veins is related with the cross-sectional area of the veins, any change in $\dot{m}\left(e_{k,in} - e_{k,out}\right)$ may indicate problem in the veins, such as narrowing due to cholesterol accumulation, which also causes a change in $\sum \dot{f}$. Skhiri et al. (2010) reported pulmonary hypertension, for example, increase in P_{out}, for example, pressure in the pulmonary artery would increase $\dot{m}\left(\left(p_{out}/\rho_{out}\right) - \left(p_{in}/\rho_{in}\right)\right)$, and obstruction of the ventricular outflow track would increase $\sum \dot{f}$, and may be among the reasons of the heart failure. It is also known that the pulse pressure affects the artery function (Hayman et al., 2012). Among the other reasons that Skhiri et al. (2010) reported as the cause of the right heart failure were volume overload, caused by tricuspid regurgitation, pulmonary regurgitation, atrial septal defect, total or partial anomalous pulmonary return, and carcinoid syndrome, caused by a narrowing or constriction of the diameter of a bodily passage or orifice, and inflow limitation, caused by tricuspid stenosis and vena cava stenosis, all of which indicate disturbances in the mass balance equation, making it almost impossible to attain a steady-state value. Pericardial disease, for example, constrictive pericarditis, may disturb $\dot{m}\int p d\dot{v}$, creating an unbalance between $\dot{m}\left(u_{in} - u_{out}\right)$ and $\dot{m}\left(e_{k,out} - k_{k,in}\right) + \dot{m}p\left(\left(p_{out}/\rho_{out}\right) - \left(p_{in}/\rho_{in}\right)\right) + \dot{m}\int p d\dot{v} + \sum \dot{f}$ is also stated among the causes of the right heart failure by Skhiri et al. (2010). Damman et al. (2007) draw attention to the point that worsening renal function is another predictor of clinical outcome in heart failure, both of which may be attributed to a common cause (Pearce et al., 2005). It should be noticed that the increased blood pressure implies a disturbance in $\dot{m}\left(\left(p_{out}/\rho_{out}\right) - \left(p_{in}/\rho_{in}\right)\right)$, which subsequently upsets the steady state imposed by the engineering Bernoulli equation and causes the heart failure.

Example 1.18: Mass and Energy Balances around the Human Body

In this example, the human body will be modeled as a combination of two subsystems. The first subsystem includes all of the respiratory and circulatory activities. The other subsystem is a cell, where the cellular metabolism occurs (Example Figure 1.18.1). The first subsystem delivers the necessary nutrients and oxygen to the cell (subsystem 2) and removes the waste of the energy metabolism (i.e., CO_2, H_2O).

Assuming that the main nutrient of the cell is glucose, then the cellular metabolism occurring in the subsystem 2 can be summarized with the following reaction:

$$C_6H_{12}O_6 + 6O_2 \rightarrow 6CO_2 + 6H_2O \qquad (E1.18.1)$$

In steady state, the enthalpy of this reaction is converted into heat and work. Work can be electrical work, if subsystem 2 is a neuronal cell, or displacement work, if subsystem 2 is a muscle cell. A cell also performs intracellular activities that require work consumption. Examples of such activities are enzyme expression, transportation of chemicals across negative concentration gradients, etc. If work is performed only for intracellular activities and no work passes through the system boundaries, then the enthalpy of

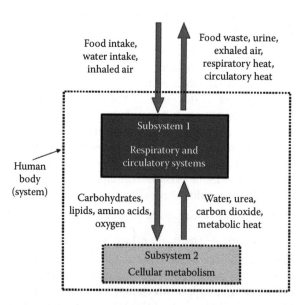

EXAMPLE FIGURE 1.18.1 Schematic description of the human body as a combination of the respiratory and the circulatory and metabolic activity subsystems.

reaction is equal to the heat release. Under such circumstances, the heat release is called the metabolic heat, q_{EM}:

$$q_{EM} = 6h_{CO_2} + 6h_{H_2O} - h_{C_6H_{12}O_6} - h_{O_2} \tag{E1.18.2}$$

If the main nutrient is a lipid, then lipid metabolism occurs. Here, the lipid is simulated as triglycerides of palmitic acid:

$$C_{16}H_{32}O_2 + 23O_2 \rightarrow 16CO_2 + 16H_2O \tag{E1.18.3}$$

The metabolic heat related to the lipid metabolism, q_{LM}, is defined as

$$q_{LM} = 16h_{CO_2} + 16h_{H_2O} - h_{C_{16}H_{32}O_2} - 23h_{O_2} \tag{E1.18.4}$$

A third chemical reaction is considered to model the conversion of amino acids to urea before excretion:

$$C_6H_{12}N_2O_4S_2 + 7.5O_2 \rightarrow 5CO_2 + 4H_2O + CH_4N_2O + 2SO_2 \tag{E1.18.5}$$

The metabolic heat related with reaction E.1.18.5, q_{AAM}, is defined as

$$q_{AAM} = 5h_{CO_2} + 4h_{H_2O} + h_{CH_4N_2O} + 2h_{SO_2} - h_{C_6H_{12}N_2O_4S_2} - 7.5h_{O_2} \tag{E1.18.6}$$

The enthalpy of the metabolites occurring in reactions (E1.18.1), (E1.18.3), and (E1.18.5) is listed in Example Table 1.18.1.

MATLAB code E1.18 calculates the energy requirement of the runner to achieve a certain kinetic energy, the amount of oxygen and carbon dioxide, which accompany the metabolic activities, and the heat released from the runner's body (Example Figures 1.18.2 through 1.18.5).

EXAMPLE TABLE 1.18.1 Enthalpy of the Metabolites Contributing to Reactions (E1.18.1), (E1.18.3), and (E1.18.5) under Physiological Conditions

Metabolite	MW (g/mol)	h (kJ/mol)	Reference
Glucose $C_6H_{12}O_6$ (aq)	180	−1267.40	Genç et al. (2013a)
Oxygen O_2 (aq)	32	−11.70	Genç et al. (2013a)
Carbon dioxide CO_2 (aq)	44	−700.59	Genç et al. (2013a)
Water vapor H_2O (v)	18	−241.6	Genç et al. (2013a)
Palmitic acid $C_{16}H_{32}O_2$ (l)	256	−436.9	Domalski (1972)
Amino acid (L-cystine) $C_6H_{12}N_2O_4S_2$ (c)	240	−525.8	Domalski (1972)
Urea CH_4N_2O (aq)	60	−317.47	Laidler (1980)
Sulfur dioxide SO_2 (aq)	64	−466.35	Johnson and Ambrose (1963)

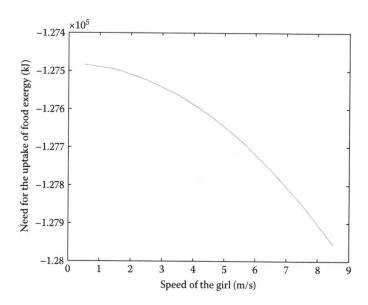

EXAMPLE FIGURE 1.18.2   Estimation of the variation of the need for the food with the speed of the girl while running.

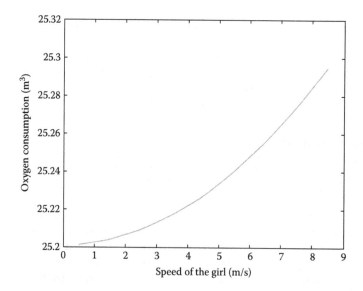

EXAMPLE FIGURE 1.18.3   Estimation of the variation of the oxygen consumption with the speed of the girl while running.

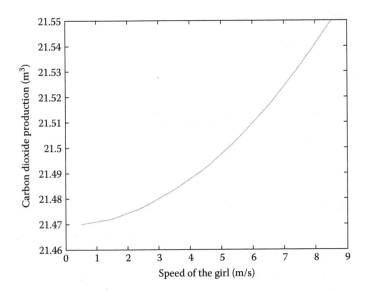

EXAMPLE FIGURE 1.18.4 Estimation of the variation of the carbon dioxide production with the speed of the girl while running.

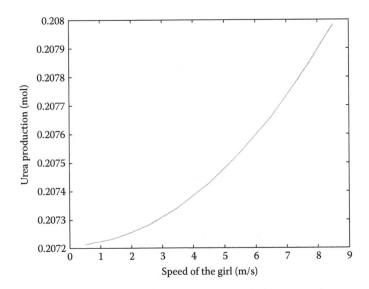

EXAMPLE FIGURE 1.18.5 Estimation of the variation of the urea production with the speed of the girl while running.

MATLAB CODE E1.18

Command Window

```
clear all
close all

% Enthalpy of formation of the nutrients
hCarbohydrate= - 1267.4; % glucose (kJ/mol)
hLipid= - 436.9; % palmitic acid (kJ/mol)
hAminoAcid= - 525.8; % L-cystine (kJ/mol)


% Molecular weights of the substances
MWglucose=0.180; % molecular weight of glucose (kg/mol)
MWpalmiticAcid=0.256; % molecular weight of palmitic acid
  (kg/mol)
MWAminoAcid=1.940; % molecular weight of amino acid (kg/mol)


% Metabolic heat release
qEM= - 4315.5; % Heat released in energy metabolism (kJ/mol
  glucose)
qLM= - 1020.5; % Heat released in lipid metabolism (kJ/mol
  palmitic acid)
qAAM= - 5106.0; % Heat released in amino acid metabolism
  (kJ/mol amino acid)


% Exergy of formation of the nutrients (from Cortassa et al.
  (2002) and Hayne (2008))
exCarbohydrate= - 16511; % exergy of the carbohydrate (kJ/kg
  glucose)
exLipid= - 39182; % exergy of the lipid (kJ/kg palmitic acid)
exAminoAcid= - 19059.5; % exergy of the amino acid (kJ/kg
  amino acid)


% enter the composition of the food
pCarbohydrate=input('Enter the percentage of carbohydrate in
  food: ');
pLipid=input('Enter the percentage of lipid: ');
pAminoAcid=input('Enter the percentage of amino acid: ');


% mole fraction of the components the food
fCarbohydrate= pCarbohydrate/ MWglucose;
fLipid= pLipid/ MWpalmiticAcid;
fAminoAcid= pAminoAcid/MWAminoAcid;


% enthalpy of formation of the food (kJ/kg)
hFood=fCarbohydrate*hCarbohydrate+fLipid
  *hLipid+fAminoAcid*hAminoAcid


% exergy of formation of the food (kJ/kg)
exFood=fCarbohydrate*exCarbohydrate+fLipid*
  exLipid+fAminoAcid*exAminoAcid
```

```
% heat generated (kJ/kg)
qGenerated=fCarbohydrate*qEM+fLipid*qLM+fAminoAcid*qAAM %
  (kJ/kg food)


% data regarding the person
mGirl=input('Enter the weight of the girl(kg): ');
vGirl=input('Enter the speed of the girl(m/s): ');
% food needed to provide the kinetic energy gain for running
mFoodForKineticEnergy=((1/2)*mGirl*(vGirl^2)/ hFood) /1000
  % kJ


% heat generation associated with food consumption to
  provide the kinetic energy
mFoodForHeatRelease= (mFoodForKineticEnergy)*(hFood) %


% food needed to provide the kinetic energy gain for running
  plus heat generation
mFood= ((mFoodForKineticEnergy)+(mFoodForHeatRelease))*1000
  % g food


disp(['total weight of food required is '...
  num2str(mFood) ' g to maintain ' num2str(vGirl) ' m/s of
  speed while running.'])


% estimation of the exergy of the food uptake (kJ/kg food)
vGirl=0.5:1.0:9.0; % speed of the girl while running
mFood=((1/2)*mGirl*(vGirl.^2)+abs(qGenerated))/abs(exFood)
ex=(mFood)*(pCarbohydrate*exCarbohydrate+pLipid*
  exLipid+pAminoAcid*exAminoAcid)


% plot the exergy input needed to keep the girl running
figure % start a new figure
plot(vGirl,ex);
xlabel('Speed of the girl (m/s)')
ylabel('need for the uptake of food exergy (kJ)')


% plot the cubic meters of oxygen needed to keep the girl
  running
O2=mFood*((fCarbohydrate/MWglucose)*6+(fLipid/
  MWpalmiticAcid)*23)*22.4e-3;
figure,plot(vGirl,O2);xlabel('Speed of the girl
  (m/s)');ylabel('oxygen consumption (m^3)')


% plot the cubic meters of carbon dioxide released while the
  girl is running
CO2=mFood*((fCarbohydrate/MWglucose)*6+(fLipid/
  MWpalmiticAcid)*16)*22.4e-3;
figure % start a new figure
plot(vGirl,CO2);xlabel('Speed of the girl
  (m/s)');ylabel('carbon dioxide production (m^3)')
```

```
% plot the amounts of urea produced while the girl is running
Urea=mFood*((fAminoAcid/MWAminoAcid));
figure % start a new figure
plot(vGirl,Urea);xlabel('Speed of the girl
  (m/s)');ylabel('urea production (mol)')
```

When we run the code commands will appear in the screen.
  When we enter the requested numbers the code will plot
  Figures E.1.18.2 through E.1.18.6 as exemplified

```
Enter the percentage of carbohydrate in food: 20
Enter the percentage of lipid: 10
Enter the percentage of amino acid: 5

hFood =
   -1.5924e+05

exFood =
   -3.4142e+06

qGenerated =
   -5.3252e+05

Enter the weight of the girl(kg): 55
Enter the speed of the girl(m/s): 0.9

mFoodForKineticEnergy =
   -1.3988e-07

mFoodForHeatRelease =
    0.0223

mFood =
   22.2749

total weight of food required is 22.2749 g to maintain 0.9
  m/s of speed while running.
mFood =
  0.1560   0.1560   0.1560   0.1561   0.1561   0.1562   0.1563
  0.1564   0.1566

ex =
   1.0e+05 *
  -1.2748 -1.2750 -1.2752 -1.2756 -1.2761 -1.2768 -1.2776
  -1.2785 -1.2796
```

The metabolic heats released as a result of reactions (E1.18.1), (E1.18.3), and (E1.18.5) are calculated from Equations E.1.18.2, E.1.18.4, and E.1.18.6 are q_{EM} = −4,315.5 kJ/mol, q_{LM} = −14,369.2 kJ/mol, and q_{AAM} = −5,106.0 kJ/mol.

1.11 Questions for Discussion

Q1.1 The device that is pictured in Question Figure 1.1.1 was used in the eighteenth century as a hearing aid. If the cross-sectional area of this device at the entrance of the sound is A_1, and that of the eardrum is A_2, use the first law of thermodynamics to calculate the amplification of the sound.

Q1.2 In the example regarding the car running with the energy harvested from rain, two droplets collide while dropping down from the cloud (Question Figure 1.2.1). Initially both the larger and the smaller droplets were at a distance h_1 above the ground, the larger one started falling down some time later than the smaller one. The droplets collided at a distance h_2 from the ground level. The two droplets started traveling together after the collision and arrived the car at the same time. Obtain an expression to describe the velocity of the droplet just before arriving the car if

(a) No energy is lost in collision.

(b) 10% of the kinetic energy of the larger droplet is released as heat.

(c) Calculate the entropy generation during this process. If the droplet is at the same temperature with the environment T_{env}.

Q1.3 Woodchucks (*Marmota monax*) are large squirrels (Question Figure 1.3.1). They dig the ground to make burrows, where they sleep, rear young, and hibernate. In the summer, they eat wild grasses and other vegetation, hydrate through eating leafy plants rather than drinking water, and accumulate fat. The fat is utilized as an energy source in winter while they hibernate.

(a) By using the system boundaries given in Question Figure 1.1.1, make a fat balance around a woodchuck to describe its accumulation in the woodchuck's body during the summer. Then use the first law of thermodynamics to express the accumulation of internal energy in its body.

(b) Make material balance around the woodchuck to describe the depletion of fat in the winter and then use the first law of thermodynamics to describe the depletion of the internal energy while hibernating.

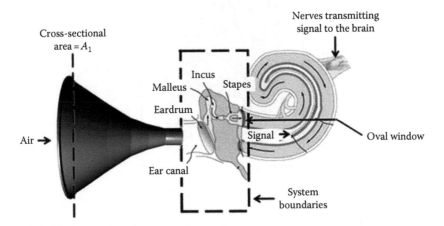

QUESTION FIGURE 1.1.1   Schematic description of an eighteenth-century hearing device.

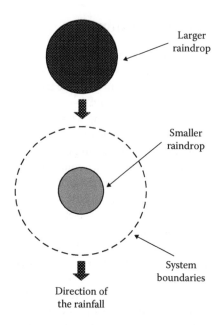

Larger raindrop

Smaller raindrop

System boundaries

Direction of the rainfall

QUESTION FIGURE 1.2.1   Schematic description of the collision process.

$m_{fat,\ in}$

$m_{fat,\ out}$

$m_{fat,\ consumption}$

$m_{fat,\ generation}$

System boundaries

QUESTION FIGURE 1.3.1   The system, for example, woodchuck and the system boundaries.

(c) It was reported by Fenn et al. (2009) that the free-living South Carolina wood-chucks, which had 2340 g average body weight, lost 32.4% of their body weight upon hibernation for 2 months. Their average oxygen consumption rate was 1.43 mL of O_2 per gram of body mass per day. We may summarize lipid consumption within the tissues as

$$-\overset{\overset{\displaystyle H}{|}}{\underset{\underset{\displaystyle H}{|}}{C}}- \; + \; 3/2\,O_2 \longrightarrow CO_2 + H_2O$$

Calculate the percentage of the weight loss of the woodchucks caused by con-sumption of the lipids in the energy metabolism.

Q1.4 A man falls into a frozen sea while walking on the ice due to a fracture on the surface (Question Figure 1.4.1). His body burns fat to reestablish his body tem-perature, but it keeps on declining. Write down the first law of thermodynamics to explain these phenomena.

Q1.5 Life cycle of the Stuart salmon is described in Question Figure 1.5.1:

Fraser River at British Columbia in Canada is the natural habitat of *Oncorhynchus nerka*, for example, early Stuart sockeye salmon. The life cycle of a sockeye salmon begins in the fall when an adult female returns to her natal fresh-water stream to lay up her eggs in a nest at about 2000 m elevation from the ocean. In the late winter, the eggs hatch into alevins, which remain hidden in the nest and feed from the nutrient-rich yolk sac until it is completely absorbed. In about 2 years, the alevins grow to become fry, which migrate to a freshwater lake and

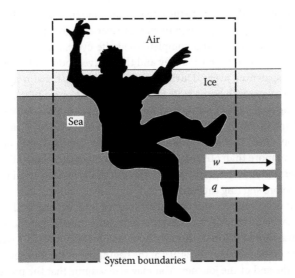

QUESTION FIGURE 1.4.1 The man in the frozen sea, the system boundaries, and the direction of the heat and work flow.

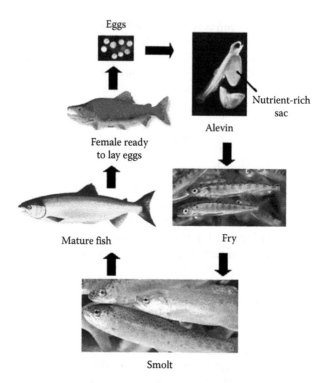

Eggs

Nutrient-rich
sac

Alevin

Female ready
to lay eggs

Mature fish

Fry

Smolt

QUESTION FIGURE 1.5.1   Life cycle of early Stuart sockeye salmon.

become smolts, and then go to the open ocean and remain there for 2–3 years, and then follow their path up the mainstream river and go back to where they were born in order to spawn and complete their life cycle. Early Stuart sockeye stop eating before starting their journey and then travel about 1500 km to reach their headwater spawning areas. During the journey, the fish swim against the water flow, which may reach 6 m/s at its fastest point (Hich et al., 1996).

Kiessling et al. (2004), while studying the changes in the composition of the fat and proteins in the muscles at various locations of the body of early Stuart sockeye salmon, noted that after 1408 km of travel, the fish consumes up to 95% of the total fat and more than 60% of the protein content of their body. By using the first law of thermodynamics, estimate the amount of the fat they need to store in their muscles before starting the journey.

Data: It is estimated from the data presented by Kiessling et al. (2004) that the average flow rate of the river against the swimming direction of the fish was 1.5 m/s. The fish traveled 1408 km of distance by nonstop swimming in 7 weeks. The average weight of the females was 2419 g at the beginning of the journey, which became 2088 g at the end of the journey. You may also assume that the protein to fat ratio of the salmon is the same as that of fillets, for example, 22 protein/12 fats. Energy obtained after metabolizing the muscle fat and protein reserve is the same as the fillet, for example, 18 kJ/g proteins and 38.8 kJ/g fats (Nutritiondata.self.com, 2012).

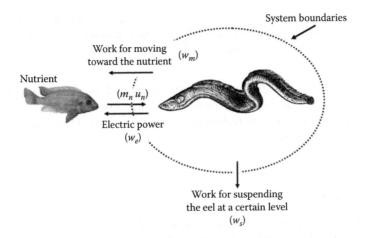

QUESTION FIGURE 1.6.1 Schematic description of the eel while swimming toward its pray and trying to suspend its body in water and producing electric power.

Q1.6 Work performed by an eel to swim toward its pray, for example, nutrient, is w_m; at the same time, it tries to suspend its body in water by performing work, w_s, and the work done to produce electric power is w_e (Question Figure 1.6.1). Write down the first law of thermodynamics to explain these phenomena.

If should define the energy efficiency of this process as the total energy gained from the nutrient divided by the total work done for hunting, write an expression for the energy efficiency of the process.

Q1.7 The hydraulic turbine shown in Question Figure 1.7.1 is being used to produce turbine work ($w_{turbine}$), it produces heat with friction ($Q_{friction}$), and a fraction

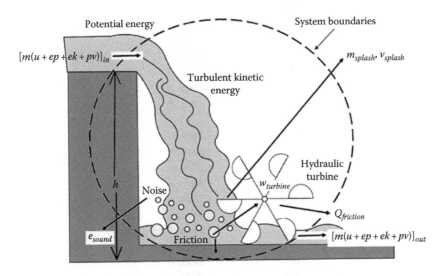

QUESTION FIGURE 1.7.1 The hydraulic turbine and the system boundaries.

of the incoming mass of water is splashed out (m_{splash}) and leaves with velocity v_{splash}, and some energy leaves the system as sound (e_{sound}). Write down the first law of thermodynamics to calculate the turbine work ($w_{turbine}$) and also show that

(a) As the amount of the droplets splashing around and their velocity increase, the amount of the work done by the turbine decreases.
(b) As the friction of the turbine increases, the amount of the work done by the turbine decreases.

1A Appendix

Prefix and Symbols for Powers of 10 and the Greek Equivalents of the Latin Letters

Prefix and symbols for powers of 10 and the Greek equivalents of the Latin letters are used frequently in this book and presented in Tables A1.1 and A1.2, respectively.

TABLE A1.1   Prefix and Symbols for Powers of 10

	Prefix and Symbol
10^{24}	Yotta, Y
10^{21}	Zetta, Z
10^{18}	Exa, E
10^{15}	Peta, P
10^{12}	Tera, T
10^{9}	Giga, G
10^{6}	Mega, M
10^{3}	Kilo, k
10^{2}	Hecto, h
10^{1}	Deca, da
10^{0}	
10^{-1}	Deci, d
10^{-2}	Centi, c
10^{-3}	Milli, m
10^{-6}	Micro, μ
10^{-9}	Nano, n
10^{-12}	Pico, p
10^{-15}	Femto, f
10^{-18}	Atto, a
10^{-21}	Zepto, z
10^{-24}	Yocto, y

TABLE A1.2   The Greek Equivalents of the Latin Letters

Latin Letter	Greek Letter	Name of the Greek Letter
A a	A α	Alpha
B b	B β	Beta
C c	X χ	Chi
D d	Δ δ	Delta
E e	E ε	Epsilon
F f	Φ φ	Phi
G g	Γ γ	Gamma
H h	H η	Eta
I i	I ι	Iota
J j	ϑ φ	This is not a Greek letter, but a keyboard equivalent
K k	K κ	Kappa
L l	Λ λ	Lambda
M m	M μ	Mu
N n	N ν	Nu
O o	O o	Omicron
P p	Π π	Pi
Q q	Θ θ	Theta
R r	P ρ	Rho
S s	Σ σ	Sigma
T t	T τ	Tau
U u	Υ υ	Upsilon
X x	Ξ ξ	Xi
W w	Ω ϖ	Omega
V v	ς ϖ	This is not a Greek letter, but a keyboard equivalent
Y y	Ψ ψ	Psi
Z z	Z ζ	Zeta

Example A1.1: Greek and Latin Letter Equivalents of a Word

Write down the following word with Greek letters and their Latin equivalents:

Greek letters of the word: Beta, iota, omicron, tau, eta, epsilon, rho, mu, delta, psi, nu, alpha, mu, chi, and sigma

We will find each Greek letter from Table A1.2 and write them side by side as Βιοτηερμοδψναμιχσ. We will convert this string of letters into Latin letters by referring to the same table and end up with the word Biothermodynamics.

2

Estimation of Thermodynamic Properties

In the previous chapter, the governing equations of thermodynamics (mass, energy, entropy, and exergy balance equations) are derived. In order to solve these equations, numerical values of the thermodynamic properties are required. Thermodynamic properties can either be experimentally measured or numerically estimated. This chapter aims to clarify how the thermodynamic properties are determined. The most abundant chemicals in the earth are air and water. Both also play important roles in biochemical reactions. So, first, we will discuss the determination of the thermodynamic properties of air and water and define ideal gas and steam table. For most of the engineeringly relevant substances, thermodynamic properties can be found in the literature. Since experimental data are more reliable than the estimated values based on the correlations, we have to use the experimentally measured data wherever possible. Databases like NIST chemistry webbook (http://webbook.nist.gov/) contain thermochemical data for thousands of organic and inorganic pure substances in gas, liquid, or solid phases. However, literature is limited for biochemical species. To estimate properties of such substances, numerical techniques such as group contribution method will be explained. Once the background information on the properties of simple substances is given, we will deepen our discussion by analyzing mixtures. Almost all real systems (such as blood, extracellular fluid, seawater, fuels) are mixtures. Toward the end of this chapter, we will provide equations to estimate the properties of mixtures based on the properties of the pure substances that make up the mixture.

2.1 How Many Thermodynamic Properties May Be Fixed Externally?

Thermodynamic properties (p, T, h...) are interrelated to each other. Modifying one of them will affect the whole system and change the values of the remaining properties. In a thermodynamic system where the number of the phases is π and the number of the chemicals is c, the phase rule $f = 2 - \pi + c$ tells us the number of the thermodynamic properties that may be adjusted externally. For example, when $f = 2$, it means that two properties define the thermodynamic state, in other words, if we set two properties (e.g., temperature and pressure), the system will set all the other thermodynamic properties itself. In such a system, we have the freedom of setting only two parameters as we wish.

In a single-phase ($\pi = 1$) mixture of n chemicals ($c = n$), $f = 2-1 + n = n + 1$, implying that $n + 1$ properties are needed to define the state. These properties may be chosen arbitrarily as

$$u(T,p,c_1,c_2,c_3,...,c_{n-1})$$

or

$$u(T,p,v,c_1,c_2,c_3,...,c_{n-2})$$

or

$$u(T,v,c_1,c_2,c_3,...,c_{n-1})$$

where c_i is the concentration of the ith species.

Example 2.1: Degree of Freedom in an Ethanol Water Mixture

How many of the intensive properties of an ethanol water mixture (Example Figure 2.1.1) define the thermodynamic state?

$$f = 2 - \pi + c$$

where
    $\pi = 2$ (vapor and liquid phases)
    $c = 2$ (ethanol and water)

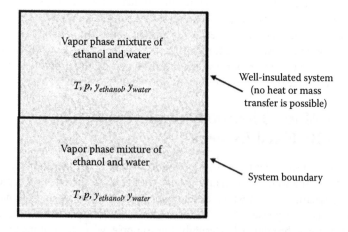

EXAMPLE FIGURE 2.1.1 Schematic description of a system of two phases and two chemicals at equilibrium.

After substituting the numbers, we will calculate $f = 2$. The extensive properties are T, π, x_{water}, $x_{ethanol}$, y_{water}, and $y_{ethanol}$. We may fix any two of these properties as we wish, and then all the others will be established by the equilibrium, for example, we may assign values for T and $y_{ethanol}$, and then p, x_{water}, $x_{ethanol}$, y_{water}, etc., will be established by the equilibrium. If we should attempt to fix one more parameter, such as x_{water}, we will not be able to keep the previously set values of T and $y_{ethanol}$.

On the labels of the wine bottle, we usually see recommendation like *drink at room temperature* or *refrigerate before drinking* etc. If we should consider wine as a mixture of water and ethanol, by neglecting the others, and remember that $x_{ethanol}$ has already set by the wine maker, we can easily understand that by setting the second parameter, T, we actually set all the others too. The parameters x_1 and x_2 contribute to the taste, and the parameters y_1 and y_2 contribute to the aroma. Therefore, drinking the wine at the recommended temperature determines its sensory properties.

2.2 Interrelation between the Thermodynamic Properties of Ideal and Real Gases

The ideal gas law and the law of the corresponding states are useful in relating the gas-phase thermodynamic properties. The ideal gas law sums up the result of the centuries-long research described by Boyle's (1627–1691), Charles' (1746–1823), and Gay-Lusac's (1778–1850) laws in a very short expression as (Mahan, 1969):

$$pV = nRT$$

Furthermore, for an ideal gas,

$$du = c_v dT$$

$$dh = c_p dT$$

$$ds = c_p \frac{dT}{T} - R \frac{dP}{P}$$

The application of this relation has been extended to the real gases when one more parameter, z, which is also called the compressibility, is added to the ideal gas law:

$$pV = nzRT$$

The theorem of the corresponding law indicates that all gases, when compared at the same reduced temperature and reduced pressure, have approximately the same compressibility factor and all deviate from ideal gas behavior to about the same degree. At high temperature and low pressures, if gas molecules do not interact, then $z = 1$. Under these conditions, the gas is referred to as the ideal gas. As the gas deviates from

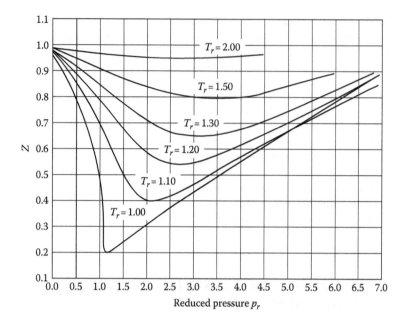

FIGURE 2.1   Generalized compressibility factor diagram.

the ideal behavior, value of the compressibility factor Z deviates from one. The compressibility factor z may be evaluated by using the generalized compressibility factor diagram (Figure 2.1), as a function of $T_r = T/T_c$ and $p_r = p/p_c$ where T_r is the reduced temperature, p_r is the reduced pressure, T_c is the critical temperature, and P_c is the critical pressure (Table 2.1).

At low temperatures and high pressures, the gas molecules come close to each other and start interacting. This is the reason for the deviation from the ideal gas behavior.

TABLE 2.1   Molar Mass, Critical Temperature, and Critical Pressure of Some Biologically Important Chemicals

	Molar Mass	T_c (K)	p_c (Pa)
Acetic acid	60.1	592.0	57.9×10^5
Ammonia	17.0	405.7	112.8×10^5
Carbon dioxide	44.0	304.2	73.8×10^5
Ethanol	46.1	513.9	61.5×10^5
Methanol	30.0	512.6	81.0×10^5
n-Butyric acid	88.1	615.7	40.6×10^5
Nitrogen	28.0	126.2	24.0×10^5
Oxygen	32.0	154.6	50.4×10^5
Water	18.0	647.1	220.6×10^5

Source: Smith, J.M. et al., *Introduction to Chemical Engineering Thermodynamics*, 7th ed., McGraw-Hill, Singapore, 2005.

Example 2.2: Calculation of the Molar Volumes of Carbon Dioxide and Acetic Acid When $T = 225°C$ (498 K) and $p = 20$ atm ($p = 2026$ kPa) by Using the Compressibility Factor

(a) The critical temperature and the critical pressure for carbon dioxide are $T_c = 304.2$ K and 73.8×10^5 Pa. The reduced temperature and the reduced pressure are

$$T_r = \frac{T}{T_c} = \frac{498}{304.2} = 1.64$$

and

$$p_r = \frac{p}{p_c} = \frac{20.3 \times 10^5 \text{ Pa}}{73.8 \times 10^5 \text{ Pa}} = 0.274$$

The compressibility factor is read from Figure 2.1 as $z = 0.98$. The gas law with the compressibility factor is

$$v = \frac{V}{n} = \frac{zRT}{p} = \frac{(0.98)(8.31 \text{ Pa/m}^3 \text{ mol K})(498 \text{ K})}{(2026 \text{ kPa})(1000 \text{ Pa/1 kPa})} = 2.00 \times 10^{-3} \text{ m}^3/\text{mol}$$

(b) The critical temperature and the critical pressure for acetic acid are $T_c = 592.0$ K and $P_c = 57.9 \times 10^5$ Pa. The reduced temperature and the reduced pressure are

$$T_r = \frac{T}{T_c} = \frac{498 \text{ K}}{592 \text{ K}} = 0.84$$

and

$$p_r = \frac{p}{p_c} = \frac{20.3 \times 10^5 \text{ Pa}}{57.9 \times 10^5 \text{ Pa}} = 0.35$$

We may read the compressibility factor from Figure 2.1 with extrapolation as $z = 0.12$. The gas law with the compressibility factor is

$$v = \frac{V}{n} = \frac{zRT}{p} = \frac{(0.12)(8.31 \text{ Pa m}^3/\text{mol K})(498 \text{ K})}{(2026 \text{ kPa})(1000 \text{ Pa/1 kPa})} = 2.4 \times 10^{-3} \text{ m}^3/\text{mol}$$

Example 2.3: Calculation of the Volume of a Real Gas from the Number of Moles Data

A fermentation process consumes 1000 mol of O_2/day. How many m³ of gas needs to be supplied when $T = 25°C$ and $p = 10 \times 10^5$ Pa? Solve the problem by using the corresponding states theory.

It is given in the problem statement that $\dot{n} = 1000\,\text{g mol/day}$, $T = 25°C = 298\,\text{K}$, and $p = 10 \times 10^5$ Pa. We can see from Table 2.1 that $T_c = 154.6\,\text{K}$ and $p_c = 50.43\,10^5$ Pa; therefore, $T_r = T/T_c = 1.93$ and $p_r = p/p_c = 0.2$. From Figure 2.1, it is shown that $z = 0.99$. The corresponding theory states that $V = znRT/p$; after substituting values of the parameters in equation, we will calculate
$\dot{V} = (0.99)(1000\,\text{g mol})(8.42\,\text{kPa m}^3/\text{kg mol K})(298\,\text{K})/(100\,\text{kPa})(1\,\text{kg mol}/1000\,\text{g mol})$
$= 24.8\,\text{m}^3/\text{day}$.

Heat needed to increase the temperature of a material by one degree is called specific heat. The specific heat also referred to as the heat capacity and may be defined as

$$c_p = \left(\frac{dh}{dT}\right)_{p=constant}$$

and

$$c_v = \left(\frac{du}{dT}\right)_{v=constant}.$$

For ideal gases, heat capacities can be estimated based on the kinetic theory of gases and the ideal gas equation. The kinetic theory of gases suggests that the monatomic gases move around with an energy of $u = (3/2)RT$. When the temperature of this gas increases at constant volume by ΔT, its energy increases by Δu, implying that $\Delta u = (3/2)R\Delta T$, and then the c_v of the gas will be

$$c_v = \frac{\Delta u}{\Delta T} = \frac{3}{2}R$$

The enthalpy is defined as $h = u + pv$. For an ideal gas, we can substitute $pv = RT$. Accordingly, the heat capacity of an ideal gas under constant pressure is

$$c_p = \frac{dh}{dT} = \frac{d(u+pv)}{dT} = \frac{du}{dT} + \frac{d(pv)}{dT} = c_v + \frac{d(RT)}{dT} = c_v + R = \frac{3}{2}R + R = \frac{5}{2}R$$

These expressions lead us to calculate $c_p/c_v = 1.67$; this ratio varies between 1.64 and 1.68 for most monatomic gases including He, Ne, Ar, Kr, and Xe (Mahan, 1969). For air at 1 atm and 20°C, this ratio is $c_p/c_v = 1.4$.

Specific heat depends strongly on temperature. Measurements of c_p and c_v with respect to temperature can be evaluated to determine an empirical equation for c_p or c_v as a function of temperature. In literature, such empirical equations are usually given in the form of either a second-order or a fourth-order polynomial, like

$$c_p = a + bT + cT^2$$

where
   T is temperature in K
   a, b, and c are empirical constants given in Table 2.2

TABLE 2.2 Constants of the Empirical Equation $c_p = a + bT + cT^2$ for Biologically Important Gases When c_p Is Expressed in J/mol K

Gas	a	b	c
N_2	26.990	5.806×10^{-3}	-2.884×10^{-6}
O_2	25.569	13.2×10^{-3}	-4.201×10^{-6}
H_2O	28.829	11.0×10^{-3}	0.1919×10^{-6}
CO_2	26.497	42.4×10^{-3}	-14.275×10^{-6}
Air	26.694	7.4×10^{-3}	-1.110×10^{-6}
NH_3	24.746	37.5×10^{-3}	-7.374×10^{-6}

Source: Adapted from Whitwell, J.C. and Toner, R.K., *Conservation of Mass and Energy*, McGraw-Hill, New York, 1973.

Numerical values of thermodynamic properties are evaluated based on a reference state. At the reference state, the absolute value of the property is set as zero (e.g., $u_{ref} = 0$). Different databases may take different thermodynamic states as reference states. Therefore, if thermodynamic data from different databases are collected, then one should assure that the same reference state is used for each of the employed databases. If the reference states of the databases are different, then we must correct the data accordingly.

2.3 Interrelation between the Thermodynamic Properties of Water

Thermodynamic properties of steam are tabulated in steam tables. These tables make it possible to estimate the values of the unknown thermodynamic properties from those of the known. In 1920s and 1930s, engineers working in the rapidly growing electric power industry needed internationally accepted tables listing properties of water and steam. Such tables would provide a common basis for designs, contractual specifications, and evaluation of the performance of purchased equipment. They would also help to avoid redundant calculations in an age when there were no calculators or computers and to make them readily available for everybody. After several meetings, the international community in 1934 agreed on steam tables for volume and enthalpy over a range of temperature and pressure. These tables were the basis for the widely used steam tables of Keenan and Keyes (1936), which became the backbone for engineering calculations for many years. Those international discussions gave birth to the organization called the International Association for the Properties of Water and Steam [IAPWS].

If we put a mass of pure water into a piston-cylinder system (inset in Figure 2.2) and heat it, then its temperature and volume will change, but its pressure will remain constant since the piston will adjust its position to keep the pressure on both sides equal. Let us assume that we have a piston-cylinder device filled with a liquid and we heat it. If we would measure the volume and the temperature continuously during the heating process and plot temperature versus volume, then we would obtain the constant pressure line as shown with a dash-dot-line in Figure 2.2.

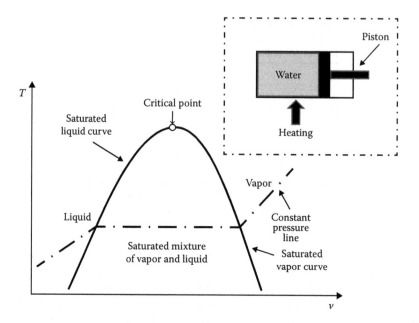

FIGURE 2.2   The piston-cylinder system (inset) and the T versus v diagram for water.

The constant pressure line remains horizontal, when water is a mixture of vapor and liquid phases. The two ends of the horizontal section represent the saturated liquid and saturated vapor points. If we repeat this procedure at different pressures and connect all the points representing saturated liquid and saturated vapor, then we obtain the saturation curve (the bold curve in Figure 2.2). The maximum point of the saturation curve is defined as the critical point. There are three regions in this diagram—the liquid, the saturated mixture, and the vapor.

Saturated mixtures contain one chemical ($c = 1$) in two states ($p = 2$). The phase rule let us calculate the degree of freedom as $f = 2-2 + 1 = 1$, which means that the temperature and the pressure are interdependent and one cannot be changed without affecting the other. There are two limits for the saturated mixtures, for example, saturated liquid and saturated vapor. Since, during phase change, the degree of freedom becomes one, we need only one thermodynamic property to define the state of a saturated liquid or saturated vapor as

$$h_{saturated\ liquid}(T, p) = h_f(T)$$

$$h_{saturated\ vapor}(T, p) = h_g(T)$$

For the mixtures between these two limits, we need to define the *quality*, which shows how far away is the current state from the saturated liquid state:

$$x = \frac{h - h_f}{h_{fg}}$$

where $h_{fg} = h_g - h_f$.

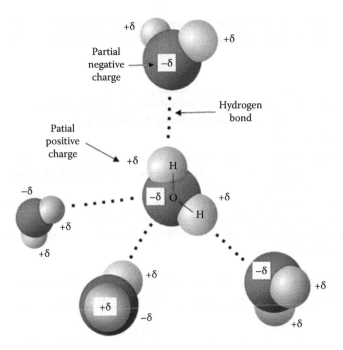

FIGURE 2.3 Schematic diagram of the hydrogen bonds of a water molecule.

The thermodynamic properties of water are dictated by its molecular structure. A water molecule is established by covalently bound two hydrogen and one oxygen atoms. Heat required to break 1 mol of molecules into their individual atoms is called bond energy. The hydrogen bonds are established between partially negatively charged oxygen of one water molecule and partially positively charged hydrogen atom of another water molecule. One water molecule may establish the maximum of four hydrogen bonds (Figure 2.3).

The strength of an O–H covalent bond in a water molecule is 456 kJ/bond, the O–H hydrogen has only about 3% of the strength of the covalent O–H bond, and therefore, boiling water at 100°C does not damage the molecular structure but break the hydrogen bonds only to evaporate it (Figure 2.4).

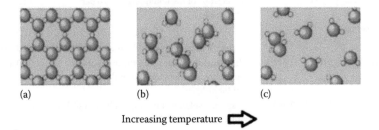

(a) (b) (c)

Increasing temperature ⇨

FIGURE 2.4 Schematic drawing of the orientation of the water molecules under different states. It is seen in the picture disorder increases with temperature. (a) Water molecules in solid state, (b) water molecules in liquid state, and (c) water molecules in vapor state.

We may use the MATLAB® m-function Xsteam to calculate the thermodynamic properties of steam according to the IAPWS IF-97 standard (Holmgren, 2007). Xsteam is a freeware.

Example 2.4: Determination of the Values of the Thermodynamic Properties of Water and Steam by Using the MATLAB m-Function Xsteam

Syntax, which may be used to obtain the numerical values of some of the thermodynamic properties of water and steam by using the MATLAB m-function Xsteam, is described in Example Table 2.4.1.

It should be recognized that entering one property, for example, temperature and pressure, are needed while working with the saturated water, but two parameters, for example, temperature and pressure, while working with superheated steam.

EXAMPLE TABLE 2.4.1 Numerical Values of Some of the Thermodynamic Properties of Water and Steam as Determined by Using the MATLAB m-Function Xsteam

Syntax	Returns What?	Numerical Value
Saturated water		
XSteam('tSat_p',1)	tSat: saturation temperature at 1 bar	99.6059°C
XSteam('pSat_t',90)	pSat: saturation pressure at 90°C	0.7018
XSteam('vL_p',70)	vL: specific volume of the liquid phase at 70 bar pressure	0.0014
XSteam('vV_p',70)	vV: specific volume of the vapor phase at 70 bar pressure	0.0274
XSteam('uL_t',95)	uV: internal energy of the liquid phase at 95°C	397.9306
XSteam('uV_t',95)	uV: internal energy of the vapor phase at 95°C	2.5000e+003
XSteam('hV_t',95)	hV: enthalpy of the vapor phase at 95°C	2.6676e+003
XSteam('hL_t',95)	hL: enthalpy of the liquid phase at 95°C	398.0185
XSteam('sV_p',70)	sV: entropy of the vapor phase at 70 bar pressure	5.8146
XSteam('sL_p',70)	sV: entropy of the liquid phase at 70 bar pressure	3.1220
Superheated steam		
XSteam('h_pt',1,20)	h: the enthalpy of water at 1 bar and 20°C	84.0118 kJ/kg
XSteam('rho_ph',1,3000)	rho: density of steam with the input of the pressure, p= 1 bar, and the enthalpy, 3000 kJ/kg	0.4056 kg/m³
XSteam('v_pt', 65,700)	v: specific volume of the vapor phase at 65 bar pressure and 700°C	0.0678
XSteam('v_pt', 700,65)	v: specific volume of the vapor phase at 700 bar pressure and 65°C	9.9110e−004

Example 2.5: Construction of the Saturated Water Table

(a) When we run MATLAB code E2.5.1, we obtain a saturated water table, where temperature varies from 5°C to 150°C; all the other thermodynamic properties corresponding to the given temperature are listed in the same row.

MATLAB CODE E2.5.1
Command Window

```
clear all
close all

% enter the temperature range
T = [5:5:150];

% tabulate the results:
fprintf('\n SATURATED WATER TABLE \n')
fprintf('\n  T  P_sat  v_sat_L  v_sat_V  u_sat_L  u_sat_V
  h_sat_L  h_sat_V  s_sat_L  s_sat_V') % column captions

fprintf('\n (oC)  (kPa)  (m3/kg)   (m3/kg) (kJ/kg) (kJ/kg)
  (kJ/kg) (kJ/kg)  (kJ/kg K)  (kJ/kg K) \n') % units

for i=1:length(T)
fprintf('%2.0f  %6.2f %6.4f  % 6.2f %5.1f       %7.1f  %5.1f
   %7.1f %6.2f   %7.2f \n',T(i), XSteam('pSat_t',T(i))*100,
   XSteam('vL_t',T(i)), XSteam('vV_t',T(i)),
   XSteam('uL_t',T(i)), XSteam('uV_t',T(i)),
   XSteam('hL_t',T(i)), XSteam('hV_t',T(i)),
   XSteam('sL_t',T(i)), XSteam('sV_t',T(i))); % P is in Bar
   in Xsteam, which is multiplied by 100 to convert in kPa
% you should notice that input of only one parameter is
   sufficient to print the others

end
```

When we run the code the followings will appear in the screen (columns are aligned manually in the text):

```
SATURATED WATER TABLE
```

T	P_sat	v_sat_L	v_sat_V	u_sat_L	u_ sat_V	h_ sat_L	h_ sat_V	s_ sat_L	s_ sat_V
(oC)	(kPa)	(m3/kg)	(m3/kg)	(kJ/kg)	(kJ/kg)	(kJ/kg)	(kJ/kg)	(kJ/ kg K)	(kJ/ kg K)
5	0.87	0.0010	147.02	21.0	2381.8	21.0	2510.1	0.08	9.02
10	1.23	0.0010	106.31	42.0	2388.7	42.0	2519.2	0.15	8.90

15	1.71	0.0010	77.88	63.0	2395.5	63.0	2528.4	0.22	8.78
20	2.34	0.0010	57.76	83.9	2402.4	83.9	2537.5	0.30	8.67
25	3.17	0.0010	43.34	104.8	2409.2	104.8	2546.5	0.37	8.56
30	4.25	0.0010	32.88	125.7	2415.9	125.7	2555.6	0.44	8.45
35	5.63	0.0010	25.21	146.6	2422.7	146.6	2564.6	0.51	8.35
40	7.38	0.0010	19.52	167.5	2429.4	167.5	2573.5	0.57	8.26
45	9.59	0.0010	15.25	188.4	2436.1	188.4	2582.5	0.64	8.16
50	12.35	0.0010	12.03	209.3	2442.8	209.3	2591.3	0.70	8.07
55	15.76	0.0010	9.56	230.2	2449.4	230.2	2600.1	0.77	7.99
60	19.95	0.0010	7.67	251.1	2455.9	251.2	2608.8	0.83	7.91
65	25.04	0.0010	6.19	272.1	2462.4	272.1	2617.5	0.89	7.83
70	31.20	0.0010	5.04	293.0	2468.9	293.0	2626.1	0.95	7.75
75	38.60	0.0010	4.13	313.9	2475.2	314.0	2634.6	1.02	7.68
80	47.41	0.0010	3.41	334.9	2481.6	334.9	2643.0	1.08	7.61
85	57.87	0.0010	2.83	355.9	2487.8	355.9	2651.3	1.13	7.54
90	70.18	0.0010	2.36	376.9	2494.0	377.0	2659.5	1.19	7.48
95	84.61	0.0010	1.98	397.9	2500.0	398.0	2667.6	1.25	7.42
100	101.42	0.0010	1.67	419.0	2506.0	419.1	2675.6	1.31	7.35
105	120.90	0.0010	1.42	440.1	2511.9	440.2	2683.4	1.36	7.30
110	143.38	0.0011	1.21	461.2	2517.7	461.4	2691.1	1.42	7.24
115	169.18	0.0011	1.04	482.4	2523.3	482.6	2698.6	1.47	7.18
120	198.67	0.0011	0.89	503.6	2528.9	503.8	2705.9	1.53	7.13
125	232.22	0.0011	0.77	524.8	2534.3	525.1	2713.1	1.58	7.08
130	270.26	0.0011	0.67	546.1	2539.5	546.4	2720.1	1.63	7.03
135	313.20	0.0011	0.58	567.4	2544.6	567.8	2726.9	1.69	6.98
140	361.50	0.0011	0.51	588.8	2549.6	589.2	2733.4	1.74	6.93
145	415.63	0.0011	0.45	610.2	2554.4	610.7	2739.8	1.79	6.88
150	476.10	0.0011	0.39	631.7	2559.0	632.3	2745.9	1.84	6.84

(b) When we run MATLAB code E2.5.2, we obtain a saturated water table, where pressure varies from 10 to 480 kPa; all the other thermodynamic properties corresponding to the given pressure are listed in the same row.

MATLAB CODE E2.5.2

Command Window

```
clear all
close all

% enter the pressure range (in Bar)
p = [0.1:0.2:4.8];

% tabulate the results:
fprintf('\n SATURATED WATER TABLE \n')
fprintf('\n p T_sat v_sat_L v_sat_V u_sat_L u_sat_V
  h_sat_L h_sat_V s_sat_L s_sat_V') % column captions
```

```
fprintf('\n  (kPa)  (oC)  (m3/kg)  (m3/kg)  (kJ/kg)  (kJ/kg)
(kJ/kg)  (kJ/kg)  (kJ/kg K)  (kJ/kg K) \n') % units

for i=1:length(p)
fprintf('%3.0f %2.0f %6.4f % 6.2f     %5.1f %7.1f %5.1f
%7.1f %6.2f %7.2f \n',p(i) *100, XSteam('tSat_p',p(i)),
XSteam('vL_p',p(i)), XSteam('vV_p',p(i)),
XSteam('uL_p',p(i)), XSteam('uV_p',p(i)),
XSteam('hL_p',p(i)),  XSteam('hV_p',p(i)),
XSteam('sL_p',p(i)), XSteam('sV_p',p(i))); % P is in Bar
in Xsteam, which is multiplied by 100 to convert in kPa
% you should notice that input of only one parameter is
sufficient to print the others
end
```

When we run the code the followings will appear in the
screen (columns are aligned manually in the text):

SATURATED WATER TABLE

p (kPa)	T_sat (oC)	v_sat_L (m3/kg)	v_sat_V (m3/kg)	u_sat_L (kJ/kg)	u_sat_V (kJ/kg)	h_sat_L (kJ/kg)	h_sat_V (kJ/kg)	s_sat_L (kJ/kg K)	s_sat_V (kJ/kg K)
10	46	0.0010	14.67	191.8	2437.2	191.8	2583.9	0.65	8.15
30	69	0.0010	5.23	289.2	2467.7	289.2	2624.6	0.94	7.77
50	81	0.0010	3.24	340.4	2483.2	340.5	2645.2	1.09	7.59
70	90	0.0010	2.36	376.6	2493.9	376.7	2659.4	1.19	7.48
90	97	0.0010	1.87	405.0	2502.1	405.1	2670.3	1.27	7.39
110	102	0.0010	1.55	428.7	2508.7	428.8	2679.2	1.33	7.33
130	107	0.0010	1.33	449.0	2514.3	449.1	2686.6	1.39	7.27
150	111	0.0011	1.16	466.9	2519.2	467.1	2693.1	1.43	7.22
170	115	0.0011	1.03	483.0	2523.5	483.2	2698.8	1.48	7.18
190	119	0.0011	0.93	497.6	2527.3	497.8	2703.9	1.51	7.14
210	122	0.0011	0.85	511.1	2530.8	511.3	2708.5	1.55	7.11
230	125	0.0011	0.78	523.5	2533.9	523.7	2712.7	1.58	7.08
250	127	0.0011	0.72	535.1	2536.8	535.4	2716.5	1.61	7.05
270	130	0.0011	0.67	546.0	2539.5	546.3	2720.0	1.63	7.03
290	132	0.0011	0.63	556.2	2542.0	556.5	2723.3	1.66	7.00
310	135	0.0011	0.59	565.9	2544.3	566.3	2726.4	1.68	6.98
330	137	0.0011	0.55	575.1	2546.5	575.5	2729.3	1.71	6.96
350	139	0.0011	0.52	583.9	2548.5	584.3	2732.0	1.73	6.94
370	141	0.0011	0.50	592.3	2550.4	592.7	2734.5	1.75	6.92
390	143	0.0011	0.47	600.4	2552.2	600.8	2736.9	1.77	6.90
410	145	0.0011	0.45	608.1	2553.9	608.6	2739.2	1.79	6.89
430	146	0.0011	0.43	615.6	2555.6	616.0	2741.3	1.80	6.87
450	148	0.0011	0.41	622.7	2557.1	623.2	2743.4	1.82	6.86
470	150	0.0011	0.40	629.7	2558.6	630.2	2745.3	1.84	6.84

Example 2.6: Use of the MATLAB m-Function Xsteam in Thermodynamic Analysis

An evacuated chamber (Example Figure 2.6.1) with perfectly insulated walls is connected to a superheated steam pipe at 178°C and 20 kPa. The valve is opened at $t = 0$ and steam flows until pressures in the chamber and the pipe are in equilibrium. The valve is turned off when $t = t_f$. Find the final temperature of the chamber.

Solution

Let us first write the mass balance as

$$m_{acc} = m_{in} - m_{out}$$

Since $m_{out} = 0$, the mass balance equation will become

$$m_{acc} = m_{in}$$

The energy balance for this system can be written in the integral form as

$$\left[m(u + e_p + e_k + pv)\right]_{in} - \left[m(u + e_p + e_k + pv)\right]_{out}$$
$$+ Q - W = \left[m(u + e_p + e_k)\right]_f - \left[m(u + e_p + e_k)\right]_0$$

This equation can be simplified, since $m_{out} = 0$, $Q = 0$, $W = 0$, $m_0 = 0$, and also e_p and e_k are negligible when compared to u and h:

$$[m(u + pv)]_{in} = [mu]_{acc}$$

Since $m_{in} = m_{acc}$ and $h = u + pv$, we may simplify this equation further as

$$h_{in} = u_{acc}$$

Superheated steam is coming from the steam pipe with its internal energy plus the pv work. The pv work will be added to the internal energy of the vapor in the

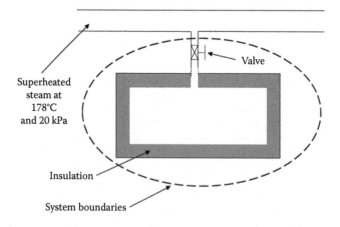

EXAMPLE FIGURE 2.6.1 The perfectly insulated steam chamber and the steam input line.

chamber. Since the pressure remains constant, temperature of the steam will increase, and it will be superheated. We will solve the rest of the problem by using the MATLAB code E2.6 and referring to XSteam:

MATLAB CODE E2.6

Command Window

```
clear all
close all

% enter the data

T = 178; % oC
p = 20/100; % kPa

h=XSteam('h_pt', p,T) % determine the enthalpy of the steam
  entering into the chamber (kJ/kg)

% determine the temperature of the superheated steam in the
  chamber (kJ/kg)
% you should notice that input of two parameter are needed
  to print the other

T=[170:1:350];
% tabulate u as a function of T at 20 kPa

fprintf('\n SUPER HEATED STEAM \n')

fprintf('\n   T (oC)   p (kPa) u(kJ/kg) \n ') % column captions

for i=1:1:length(T)
          error =abs(h- XSteam('u_pt', p,T(i)));
          if error < 1
          % P is in Bar in Xsteam, multiplied by 100 to convert
            in kPa

fprintf('%2.0f     %3.0f     %7.2f \n', T(i), p*100,
XSteam('u_pt', p,T(i)));
     end
end
```

When we run the code the followings will appear in the screen (columns are aligned manually):

```
h =
  2.8364e+003

 SUPER HEATED STEAM

 T (oC)      p (kPa)      u(kJ/kg)
   316        20          2836.81
```

The final temperature of the steam is calculated as 316°C.

Example 2.7: Estimation of the Energy of a Hydrogen Bond between Different Water Molecules

We will use the MATLAB m-function Xsteam to calculate the thermodynamic properties of the liquid water and steam. XSteam ('hL_t',100) and XSteam ('hv_t',100) returns $h_f = 419.1$ kJ/kg and $h_g = 2675.6$ kJ/kg.

$$\Delta h_{phase\ change} = h_g - h_f = 2256.5 \text{ kJ/kg}$$

$h_{phase\ change}$ is the amount of energy used to break the hydrogen bonds between the liquid water molecules during evaporation. $h_{phase\ change}$ is also called the latent heat of vaporization, since the temperature remains constant during the phase change. At 100°C one water molecule approximately makes 3.24 hydrogen bonds. Therefore, the energy required to break one bond is calculated as

$$\left(\text{Bond energy}\right) = \left(\frac{\Delta h_{phase\ change}}{\text{Number of bonds to be broken}}\right)$$

After substituting the numbers in the equation, we will have

$$\left(\text{Bond energy}\right) = \left(2256.5 \text{ kJ/kg}\right)\left(\frac{18\,g}{1\,mol}\right)\left(\frac{1\,kg}{1000\,g}\right)\left(\frac{mol}{3.24\,bonds}\right) = 12.54 \text{ kJ/bond}$$

Example 2.8: Boiling Point of a Solution

Calculate the boiling point of water containing 5% inert, a chemical that does not evaporate, at sea level.

One atmosphere is 0.987 bar. Vapor pressure of inert is negligible, when compared with that of water; therefore, the total vapor pressure on the 5% salt-containing solution will be $p = p_s(1-x)$. After rearranging this equation, we will obtain $p_s = p/(1-x) = 0.987/(1-0.05) = 1.039$ bar. We will calculate the boiling temperature under $p = 103.9$ kPa as 100.6789°C, after entering XSteam ('tSat _ p',1.039) to the command window of MATLAB.

Example 2.9: Calculation of the Internal Energy, Enthalpy, and Entropy with XSteam

MATLAB code E2.9 uses Xsteam to obtain the internal energy, enthalpy, and entropy at $p = 30$ kPa and $T = 625$°C, and $p = 30$ kPa and $T = 675$°C. The code also obtains the specific heat, c_p and c_v data from Xsteam; calculates u_{675}, h_{675}, and s_{675} from u_{625}, h_{625}, and s_{625} by using the following equations; and checks the agreement of the calculations with the values driven directly from Xsteam:

$$u_{675} = u_{625} + \int_{625}^{675} c_v(T)\,dT = u_{625} + \sum_{i=626}^{i=675} c_v(i)\left[T(i)-T(i-1)\right]$$

$$h_{675} = h_{625} + \int\limits_{625}^{675} c_p(T)\,dT = h_{625} + \sum_{i=626}^{i=675} c_p(i)\big[T(i) - T(i-1)\big]$$

$$s_{675} = s_{625} + \int\limits_{625}^{675} \frac{c_p(T)}{T}\,dT = s_{625} + \sum_{i=626}^{i=675} \frac{c_v(i)}{T(i)}\big[T(i) - T(i-1)\big]$$

MATLAB CODE E2.9

Command Window

```
clear all
close all
global cv

% print out values of u, h and s at 625 oC and 30 kPa

T = [625 675]; % oC
p= 30/100; % kPa, P is in Bar in Xsteam, divide in 100 to
  convert kPa into Bar

% tabulate u, h and s as a function of T at 30 kPa

fprintf('\n SUPER HEATED STEAM \n')

fprintf('\n T (oC) p (kPa) u(kJ/kg) h(kJ/kg) s(kJ/kg K) \n ')
  % column captions

for i=1:1:length(T)
  u(i)=XSteam('u_pt', p,T(i)); % determine the internal
    energy of the super-heated steam entering into the
    chamber (kJ/kg)

h(i)=XSteam('h_pt', p,T(i)); % determine the enthalpy of the
  super-heated steam entering into the chamber (kJ/kg)

s(i)=XSteam('s_pt', p,T(i)); % determine the enthalpy of the
  super-heated steam entering into the chamber (kJ/kg)

  % P is in Bar in Xsteam, multiplied by 100 to convert in kPa
  fprintf('%2.0f %3.0f %7.1f %7.1f %7.1f \n', T(i),
    p*100, XSteam('u_pt', p,T(i)), XSteam('h_pt',
    p,T(i)), XSteam('s_pt', p,T(i)))
end

% enter the lower limit of the integrals
u625=u(1); h625=h(1); s625=s(1);

% Calculate u, h and s
```

```
% enter the temperatures
T1 = [625:1: 675]; % oC

for i=2:1:length(T1)
  cv(i)=XSteam('Cv_pt', p,T1(i));
  cp(i)=XSteam('Cp_pt', p,T1(i));
  sv(i)=XSteam('s_pt', p,T1(i));
  delta_u(i)=cv(i)*(T1(i)-T1(i-1));
  delta_h(i)=cp(i)*(T1(i)-T1(i-1));
  delta_sv(i)=cp(i)*(T1(i)-T1(i-1))/(T1(i)+273);
end


u675=u625+sum(delta_u) % kJ/kg
h675=h625+sum(delta_h) % kJ/kg
s675=s625+sum(delta_sv) % kJ/kg K
```

When we run the code the followings will appear in the
screen (columns are aligned manually):

SUPER HEATED STEAM

T (oC)	p (kPa)	u(kJ/kg)	h(kJ/kg)	s(kJ/kg K)
625	30	3346.9	3761.4	9.7
675	30	3435.7	3873.2	9.8

```
u675 =
 3.4357e+003

h675 =
 3.8732e+003

s675 =
 9.8394
```

u675, h675, s675 are the same as the previously calculated
numbers

2.4 Tables of Experimentally Determined Thermodynamic Properties

Within the last decades, scientists performed elaborate experiments to determine the thermodynamic properties of various substances. Reference books such as NIST web-book and *Perry's Chemical Engineers' Handbook* (Perry et al., 1973) provide us a rich database of thermodynamic properties.

Table 2.3 lists thermodynamic properties of several chemical species under standard conditions, that is, 1 atm and 25°C. Under standard conditions, chemical species are considered to be pure, that is, $c_i = 1$.

TABLE 2.3 Enthalpy and Gibbs Free Energy of Formation and Absolute Entropy of Some Chemicals of Biological Importance under Standard Conditions at 1 atm and 25°C

Chemical Species	Formula	Δh_f^0 (kJ/mol)	s^0 (J/mol K)	Δg_f^0 (kJ/mol)
Acetic acid (l)	CH_3COOH	−484.2	159.8	−389.45
Ammonia (aq)	NH_3	−80.3	111.3	−26.8
Ammonium (aq)	NH_4^+	−132.8	112.8	−79.5
Bicarbonate (aq)	HCO_3^-	−691.1	95.0	−587.1
Carbon dioxide (g)	CO_2	−393.5	213.6	−300.4
Carbonate (aq)	$CO_3^=$	−676.3	−53.1	−528.1
Carbonic acid (aq)	HCO_3	−699.7	187.4	−623.2
Ethanol (l)	C_2H_5OH	−276.9	161.0	−174.2
Glucose (s)	$C_6H_{12}O_6$	−1274.5	212.1	−910.6
Acid hydrogen (aq)	H^+	0	0	0
Methane (g)	CH_4	−74.9	186.2	−50.8
Methanol (l)	CH_3OH	−238.7	126.8	−166.3
Oxygen (g)	O_2	0	205.0	0
Sucrose (s)	$C_{12}H_{22}O_{11}$	−2221.7	360.2	−1544.3
Water (v)	H_2O	−241.8	188.7	−174.1
Water (l)	H_2O	−285.8	69.9	−180.7

Source: Chang, R., *Chemistry*, McGraw-Hill, New York, 2005.

In biological systems, the environment usually acts as a buffer for H^+, so that the pH remains constant regardless of the number of H^+ consumed or generated during individual reactions. Similarly, ionic strength remains constant, too. In the Section 2.1, we have stated that for a single-phase mixture of *n*-species, thermodynamic state can be defined with $(f = 2-\pi + c)$ $n + 1$ properties. But for biochemical systems, where pH and ionic strength are strictly regulated and pressure remains constant, Alberty (2003) showed that Δh_f and Δg_f of the chemical species in the mixture can be calculated based on the temperature, pH, and ionic strength of the system. Table 2.4 lists an example of Alberty's calculations. Details of these calculations will be explained in Section 2.7.

One of the most important chemical reactions is combustion. During combustion, hydrocarbons react with oxygen and the end products of a complete combustion are CO_2 and H_2O. The enthalpy of reaction is referred to as the combustion enthalpy, Δh_c. Water produced as a result of the combustion process can be in vapor or liquid form. If the produced water is in vapor form, then Δh_c is referred to as the lower heating value (LHV). If water comes out of the combustion process in liquid phase, Δh_c increases due to the release of the enthalpy of vaporization of water, and then the enthalpy of this reaction is named as the higher heating value (HHV). Table 2.5 lists LHV and HHV of common fuels.

Tables 2.6 and 2.7 list chemical exergies of some chemicals and energy sources.

TABLE 2.4   Enthalpy of Formation and Gibbs Free Energy of Formation of Some
Metabolites at Standard Physiological Conditions

Chemical Species	$\Delta\, h_f^T$ (kJ/mol)	$\Delta\, g_f^T$ (kJ/mol)
Acetyl CoA	−1.30	−69.07
Adenosine diphosphate	−2644.19	−1376.62
Adenosine triphosphate	−3632.27	−2239.05
1,3-bisphophoglycerate	1.14	−2295.63
Citrate	−1524.00	−943.91
Iso-citrate	−3.97	−997.98
Coenzyme A	−1.82	−7.60
Carbon dioxide	−700.59	−541.25
Fructose 6-phosphate	−5.84	−1369.12
Fructose 1,6-biphosphate	−4.45	−2295.25
Fumarate	−779.05	−513.24
Glucose	−1267.40	−393.80
Glucose 6-phosphate	−2281.24	−1280.83
Glyceraldehyde phosphate	−3.23	−1131.90
Water$_{(l)}$	−286.70	−150.53
α-ketoglutarate	0.00	−659.09
Malate	−2.80	710.33
Nicotinamide adenine dinucleotide (oxidized)	−10.86	1100.24
Nicotinamide-adenine dinucleotide (reduced) (NADH)	−41.93	1165.09
Oxygen (aqueous)	−11.70	17.53
Oxaloacetate	0.87	−743.65
Phosphenolpyruvate	−0.37	−1237.25
2-phospho-D-glycerate	−1.23	−1395.64
3-phospho-D-glycerate	−1.23	−1401.80
Inorganic phosphate (P$_i$)	−1303.34	−1049.70
Pyruvate	−597.09	−341.02
Succinate	−915.46	−515.43
Succinyl-CoA	−3.56	−361.63
Lactate (LAC_c)	−688.38	−298.93

Source: Alberty, R.A., *Thermodynamics of Biochemical Reactions*, John Wiley & Sons, Hoboken, NJ, 2003.

Note: $T = 310.15$ K, pH = 7 and ionic strength $I = 0.18$ M.

TABLE 2.5 Lower Heating Value and Higher Heating
Value of Common Fuels

Chemical Species	HHV (MJ/kg)	LHV (MJ/kg)
Methane, $CH_{4(g)}$	55.50	50.0
Ethane	51.90	47.80
Propane	50.35	46.35
Butane	49.50	45.75
Diesel	44.80	43.40
Wood fuel	24.20	17.00

Source: Cengel, Y. and Boles, M., *Thermodynamics: An
Engineering Approach*, McGraw-Hill, Singapore, 2011.

TABLE 2.6 Standard Chemical Exergy of Some Biologically Important Chemicals at a T
of 298.15 K and a p of 101.325 kPa

Substance	ex^o_{chem} (MJ/mol)	Substance	ex^o_{chem} (MJ/mol)
$C_{(graphite)}$	410.26	$H_3PO_{4(s)}$	104.0
$CO_{2(g)}$	19.87	H_2S (g)	812.0
$C_6H_{12}O_{6\ (s)}$ (α-galactose)	2929.8	H_2SO_4 (g)	223.4
$C_{12}H_{22}O_{11\ (s)}$ (β-lactose)	5988.1	K	366.6
$C_{12}H_{22}O_{11\ (s)}$ (saccharose)	6007.8	Mg	633.8
$Ca_{(s)}$	712.4	$N_{2\ (g)}$	0.72
$CaCO_{3(s)}$ (aragonite)	1.0	$N_{2\ (atmospheric)}$	0.69
CaO	110.2	$NH_{3(g)}$	337.9
$Ca(OH)_{2(s)}$	53.7	$NH_4Cl_{(s)}$	331.3
$Ca_3(PO_4)_{2(s)}$	19.4	$NH_4NO_{3(s)}$	294.8
$H_{2(g)}$	236.1	$(NH_4)_2SO_{3(s)}$	660.6
H	331.3	$NaCl_{(s)}$	14.3
HCl(g)	84.5	$O_{2(g)}$	3.97
$HNO_{3(l)}$	43.5	$P_{(s),\ α,\ white}$	875.8
$H_2O_{(g)}$	9.5	$S_{(s),\ rhombic}$	609.6
$H_2O_{(l)}$	0.9	$SO_{2(g)}$	313.4

Source: Adapted from Szargut, J. et al., *Exergy Analysis of Thermal, Chemical, and Metallurgical
Processes*, Hemisphere Publishing, New York, 1988.

TABLE 2.7 Exergy of Biologically Important Energy Sources

Sunshine	0.52 kJ/m² s (Petela, 2008)
Electricity	1 MJ/MJ (Szargut, 1988)
Diesel oil	44.4 MJ/kg (Szargut, 1988)
Diesel oil	39.4 MJ/L (Hovelius and Hansson, 1999)
Natural gas	0.812 MJ/mol or 51.04 MJ/kg (Szargut, 1988)

Example 2.10: Calculation of the LHV and HHV of Methane

When we burn methane with oxygen, one of the products will be carbon dioxide and the other one will be water either in vapor or gas phase or a mixture of both. The reaction is as follows when water is in vapor phase:

(a) $CH_{4(g)} + 2O_{2,(g)} \rightarrow CO_{2(g)} + 2H_2O_{(v)}$

$$LHV = \Delta h_{f,CO_2\ (g)} + 2\Delta h_{f,H_2O\ (v)} - \left(\Delta h_{f,CH_4(v)} + 2\Delta h_{f,O_2(g)} \right)$$

Enthalpy of formation values of each species can be found in Table 2.3.

$$LHV = (-393.5) + 2(-241.8) - \left[(-74.9) + (0) \right] = -802.2\ kJ/mol$$

Molar weight of methane is 16.04 kg/kmol.

$$LHV = \frac{-802.2\ kJ/mol}{16.04\ g/mol} = 50.01\ MJ/kg$$

(b) When water leaves the system in liquid form, then

$$CH_{4,(g)} + 2O_{2,(g)} \rightarrow CO_{2,(g)} + 2H_2O_{(l)}$$

$$HHV = (-393.5) + 2(-285.8) - \left[(-74.9) + (0) \right] = -890.2\ kJ/mol$$

$$= \frac{-890.2\ kJ/mol}{16.04\ g\ mol} = 55.50\ MJ/kg$$

The calculations indicate that larger heat of combustion is released when liquid water, instead of water vapor leave the system. The calculated values are in good agreement with the experimentally determined data (Table 2.5).

Example 2.11: Enthalpy of Formation of Aqueous Glucose

The apparent reaction of the aerobic respiration of glucose is

$$C_6H_{12}O_{6(aq)} + 6O_{2(g)} \rightarrow 6CO_{2(g)} + 6H_2O_{(l)}$$

$$\Delta h_R = -2880\ kJ/mol\ glucose$$

In a system, where no work is transferred through the system boundaries, Δh_R is equal to the metabolic heat released as a result of respiration. Heat of formation of H_2O and CO_2 are given as

$$C_{graphite} + O_{2(g)} \rightarrow CO_{2(g)} \quad \Delta h_{f,CO_2} = -417\ kJ/mol$$

$$H_{2(g)} + \tfrac{1}{2}O_{2(g)} \rightarrow H_2O_{(l)} \quad \Delta h_{f,H_2O} = -286\ kJ/mol$$

We may estimate the enthalpy of formation of glucose from the metabolic heat released during respiration as

$$\Delta h_R = 6\Delta h_{f,CO_2} + 6\Delta h_{f,H_2O} - \Delta h_{f,glucose} - 6\Delta h_{f,O_2}$$

where

$\Delta h_{f,O_2} = 0$ by definition

$\Delta h_{f,glucose} = (2880) + (6)(-417) + (6)(-286) = -1338$ kJ/mol

This is the enthalpy of formation of crystalline glucose. When glucose dissolves, OH groups make hydrogen bond with water and each hydrogen bond will have 20 kJ/mol of energy. Therefore, the enthalpy of formation of dissolved glucose is

$$\Delta h_{f,glucose(aq)} = \Delta h_{f,glucose(s)} + (5)\Delta h_{f,hydrogen\ bond}$$

$$Dh_{f,glucose(aq)} = (-1338\ \text{kJ/mol}) + (5)(20\ \text{kJ/mol}) = -1238\ \text{kJ/mol}$$

$\Delta h_{f,glucose}$ calculated here is in very good agreement with $\Delta h_{f,glucose}$ given in the literature as –1262 KJ/mol.

The total bond energy of glucose, that is, heat needed to break glucose into its atoms, was calculated as 9230 kJ/mol. We may expect to utilize –9230 kJ of energy to synthesize glucose from its atoms, which is about seven times of its Δh_f. The difference is caused because of the difference in the starting materials, for example, the atoms C, H, and O versus the molecules CO_2 and H_2O in each process.

Example 2.12: Calculation of the Enthalpy of Formation and Gibbs Free Energy of Formation of *Saccharomyces cerevisiae*

Saccharomyces cerevisiae is a species of yeast. Being a microorganism, *S. cerevisiae* consists of many different chemical substances. It is nearly impossible to determine the exact concentration of each individual chemical species in a *S. cerevisiae*. Thus, we cannot define it as a mixture and calculate its Δh_f and Δg_f based on the thermodynamic properties of individual chemical species.

This example aims to demonstrate how the thermodynamic properties Δh_f, Δs_f, and Δg_f, which are essential for energy, entropy, and exergy analyses, can be obtained for microorganisms from the available data in the literature. For *S. cerevisiae*, we have limited data in the literature.

Battley et al. (1997) determined the unit carbon formula (UCF) for *S. cerevisiae* as

$$CH_{1.613}O_{0.557}N_{0.158}P_{0.012}S_{0.003}K_{0.022}Mg_{0.003}Ca_{0.001}$$

This formula is an empirical representation of the cellular structure of the cell, which means that within its structure for each carbon atom, there are 1.613 hydrogen, 0.557 oxygen, etc., atoms.

The UCF weight of S. *cerevisiae* cells is

$$UCF_{cell\ weight} = \sum_{i=1}^{n} \alpha_i aw_{component,i}$$

where

α_i is the number

$aw_{componet,i}$ is the atomic weight of the ith component within the cell; therefore,

$$UCF_{cell} = (1)(12) + (1.613)(1) + (0.557)(16) + (0.158)(14) + (0.012)(31)$$

$$+ (0.003)(32) + (0.022)(39) + (0.003)(24) + (0.001)(40)$$

$$= 26.2\ g/UCF.$$

Battley (1998) measured the enthalpy of combustion of S. *cerevisiae* in a calorimeter as −19.44 kJ/g. If we assume that in the calorimeter the microorganism is combusted completely with the stoichiometric amount of oxygen, then the following reaction can summarize the combustion:

$$CH_{1.613}O_{0.557}N_{0.158}P_{0.012}S_{0.003}K_{0.022}Mg_{0.003}Ca_{0.001(s)} + 1.51O_{2(g)}$$

$$\rightarrow CO_{2(g)} + 0.806H_2O_{(l)} + 0.079N_{2(g)} + 0.003P_4O_{10(cr)} + 0.003SO_{3(g)}$$

$$+ 0.011K_2O_{(cr)} + 0.003MgO_{(cr)} + 0.001\ CaO_{(cr)}$$

The reactants of this apparent reaction are the cell in solid state and oxygen (gas), and the products are carbon dioxide (gas), water (liquid), nitrogen and SO_3 (both of them are gas), and K_2O, MgO, and CaO (all in crystalline state). Based on the derived apparent combustion reaction, the enthalpy of combustion can be written in terms of the enthalpy of formation of the reactants and products:

$$\Delta h_{c,cell} = \Delta h_{f,CO_2(g)} + 0.806\Delta h_{f,H_2O(l)} + 0.079\Delta h_{f,N_2(g)}$$

$$+ 0.003\Delta h_{f,P_4O_{10}(cr)} + 0.003\Delta h_{f,SO_3(g)} + 0.011\Delta h_{f,K_2O(cr)}$$

$$+ 0.003\Delta h_{f,MgO(cr)} + 0.001\Delta h_{f,CaO(cr)} - [\Delta h_{f,cell(s)} + 1.510 h_{f,O2(g)}]$$

where

$$\Delta h_{c,cell} = (-19.44\ kJ/g)(26.2\ g/UCF) = -509.3\ kJ/UCF$$

The enthalpies of formation of all the chemicals involved in the combustion reaction, except the Δh_f of the cell, are available in Table 2.3. After substituting the numerical values, the enthalpy of formation is calculated as $\Delta h_{f,cell(s)} = -133.13\ kJ/UCF$.

Other data found in the literature are the heat capacity of lyophilized S. *cerevisiae* cells as a function of temperature. These data are employed to calculate the entropy of the cell as

$$s = \int_0^{T_f} \frac{c_p(T)dT}{T}$$

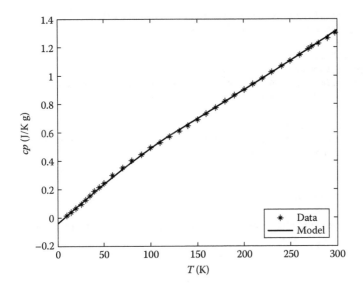

EXAMPLE FIGURE 2.12.1 Variation of c_p of lyophilized *Saccharomyces cerevisiae* cells with temperature. The MATLAB® code implies that $c_{p,cells} = 0.0064T^2 - 0.0529T^3$. (Data adapted from Battley, E.H. et al., *Thermochim. Acta*, 298, 37, 1997.)

The MATLAB code E2.12 employs the $c_p(T)$ values reported by Battley et al. (1997) to carry out integration; the initial condition $s = 0$ at $T = 0$ K is defined by the third law of thermodynamics (Example Figure 2.12.1).

MATLAB CODE E2.12
Command Window

```
clear all
close all
format compact

% enter the data
T=[10 15 20 25 30 35 40 45 50 60 70 80 90 100 110 120 130
   140 150 160 170 180 190 200 210 220 230 240 250 260 270
   273.15 280 290 298.15]; % temperatures (K)
cp=[0.014 0.035 0.063 0.093 0.124 0.155 0.186 0.215 0.244 0.299
    0.350 0.399 0.445 0.489 0.531 0.572 0.612 0.651 0.691 0.731
    0.774 0.819 0.863 0.904 0.944 0.985 1.026 1.066 1.107 1.149
    1.189 1.201 1.226 1.264 1.299]; % specific heats (J/K g)

% plot the data
plot (T,cp,'k*'); hold on;
xlabel ('T (K)');
ylabel ('cp (J/K g)')

%  fit a polynomial empirical model to the data
N=3; % determine N by trial and error to obtain good
   agreement between model & data
```

```
c = polyfit(T,cp,N) % the fitted polynomial is
   cpModel=c(1)+c(2)T+c(3)T^2+c(4)T^3

Tmodel=1:1:300;
cpModel=polyval(c,Tmodel);
plot (Tmodel,cpModel,'k-', 'LineWidth',2); hold on;
legend('data','model','Location','SouthEast')

for i=2:length(T)
   deltaS(i)=(cp(i)*(T(i)-T(i-1))/T(i)); % incremental
      entropy increase
end

S= sum(deltaS)
```

When we run the code the following lines and figure will
  appear in the screen:

```
c =
  0.0000   -0.0000 0.0064  -0.0529
S =
  1.3055
```

Entropy of lyophilized *S. cerevisiae* cells is calculated as s_{cell} = 1.305 J/kg at
T = 298.15 K with the MATLAB code, and it agrees well with the numerical
value reported as s_{cell} = 1.304 J/kg by Battley et al. (1997).

To calculate the Gibbs free energy of formation, let us assume that the
S. cerevisiae cells are synthesized by bringing its elements together:

$$C_{(s)} + \frac{1.613}{2}H_{2(g)} + \frac{0.557}{2}O_{2(g)} + \frac{0.158}{2}N_{2(g)} + \frac{0.012}{4}P_{4(s)} + 0.003S_{(s)} + 0.022K_{(s)}$$

$$+ 0.003Mg_{(s)} + 0.001Ca_{(s)} \rightarrow CH_{1.613}O_{0.557}N_{0.158}P_{0.012}S_{0.003}K_{0.022}Mg_{0.003}Ca_{0.001(s)}$$

Then the entropy of formation of the cell may be expressed as that of the synthesis reaction:

$$\Delta s_{f,cell} = \sum_{i=1}^{n} s_{cell} - \alpha_i \Delta s_{f,element,i}$$

where α_i is the number of the atoms of the *i*th element involved into the formation
reaction; the standard entropies at 298 K are given in Example Table 2.12.1.

EXAMPLE TABLE 2.12.1   Standard Entropies of the Atoms of *Saccharomyces cerevisiae*
at 298 K

Element	$C_{(s)}$	$H_{2(g)}$	$O_{2(g)}$	$N_{2(g)}$	$P_{4(s)}$	$S_{(s)}$	$K_{(s)}$	$Mg_{(s)}$	$Ca(s)$
Δs_f^0 (J/mol K)	5.73	130.5	199.6	191.0	164.4	31.9	64.2	32.7	41.6

Source: Data adapted from Mahan, B., *University Chemistry*, 2nd ed., Addison-Wesley Pub. Co.
Manila, 1969; GenChem Textbook, http://chemed.chem.wisc.edu/chempaths/GenChem-Textbook/
Standard-Molar-Entropies-984.html.

After substituting the numbers, we will have $\Delta s_{f,cell} = -149.7$ J/UCF K, which agrees well with the result calculated by Battley et al. (1997) as $\Delta s_{f,cell} = -151.4$ J/UCF K.

Gibbs free energy of formation can be calculated based on the enthalpy and entropy values as

$$\Delta g_{f,cell} = \Delta h_{f,cells} - T\Delta s_{f,cells}$$

After substituting $\Delta h_{f,cells} = -133.1$ kJ/UCF, $T = 298.15$ K, and $\Delta s_{f,cell} = -149.7$ kJ/UCF K, we will calculate that $\Delta g_{f,cells} = -88.5$ kJ/UCF.

Remark: Note that the calculated thermodynamic properties are based on the UCF of the dried *S. cerevisiae* cell. A living *S. cerevisiae* cell has nearly 70% water. Studies done with scanning electron microscope indicate that a *S. cerevisiae* cell has a spherical shape with a diameter of ca. 4.5 μm (Pringle et al., 1979). The average cell density is 1.1126 g/mL (Baldwin and Kubitschek, 1984). Therefore, the mass of a single *S. cerevisiae* cell is calculated as 4.0×10^{-14} kg. Since about 70% of the cell volume is water, and not synthesized by the cell, we will consider that only 30% of the cell mass contributes to enthalpy, entropy, and Gibbs free energy of formation:

$$\Delta h_{f,cell} = \left(-133.13 \text{ kJ/UCF}\right)\left(1000 \text{ J/kJ}\right)\left(\frac{\text{UCF}}{26.6 \text{ g}}\right)\left(4.0\times10^{-14} \text{ kg/cell}\right)\left(1000 \text{ g/kg}\right)\left(0.30\right)$$

$$= -6.0\times10^{-8} \text{ J/cell}$$

$$\Delta s_{f,cell} = \left(-151.4 \text{ J/UCF K}\right)\left(\frac{\text{UCF}}{26.6 \text{ g}}\right)\left(4.0\times10^{-14} \text{ kg/cell}\right)\left(1000 \text{ g/kg}\right)\left(0.30\right)$$

$$= -6.8\times10^{-11} \text{ J/cell K}$$

$$\Delta g_{f,cell} = \left(-88.5 \text{ kJ/UCF}\right)\left(\frac{\text{UCF}}{26.6 \text{ g}}\right)\left(4.0\times10^{-14} \text{ kg/cell}\right)\left(1000 \text{ g/kg}\right)\left(0.30\right)$$

$$= 4.0\times10^{-11} \text{ J/cell}$$

Example 2.13: Calculation of the Exergy of *Chlamydomonas reinhardtii* Lipid

Chemical exergy values of numerous chemical species are available in the literature. However, it is rather difficult to find thermodynamic properties for specific chemicals, especially biological hydrocarbons. This example aims to demonstrate an estimation method for the chemical exergy of a *Chlamydomonas reinhardtii* lipid. *C. reinhardtii* is a single-cell green alga, which is widely used for experimental studies.

Chemical exergy content of a substance can be estimated based on the known chemical exergy values of the products at the true dead state. Moran (1982) suggested the following formula to estimate the chemical exergy of liquid hydrocarbons of the type $C_zH_yO_x$:

$$ex_{Moran} = LHV\left(1.0401 + 0.01728\frac{y}{z} + 0.0432\frac{x}{z}\right)$$

where LHV stands for lower heating value. Szargut and Styrylska (1964) suggested an alternative formula:

$$ex_{Szargut\,and\,Strylska} = LHV\left(1.04224 + 0.011925\frac{y}{z} - \frac{0.042}{z}\right)$$

LHV for common fuels are listed in literature. If LHV data cannot be found from the open literature, then it can be estimated as

$$LHV = h_f^0 - \left((-393,520)z + (-241,820)\frac{y}{2}\right)$$

where h_f^0 is the standard enthalpy of formation. Since a precise measurement of h_f^0 for algal oil could not be found in the literature, h_f^0 is estimated based on the photosynthesis enthalpy. Enthalpy of photosynthesis Δh_r is 479.4 kJ/mol of C. Accordingly,

$$h_f^0 = 479.4 \times z$$

The *C. reinhardtii* lipid has an empirical formula of $C_{69}H_{98}O_6$ (Küçük et al., 2015). Accordingly, we can estimate its thermodynamic properties as

$$h_f^0 = 479.4 \times 69 = 33,078.6 \text{ kJ/mol}$$

$$LHV = h_f^0 - \left((-393,520)69 + (-241,820)\frac{98}{2}\right) = 39,035,139 \text{ kJ/kmol}$$

$$ex_{Moran} = 41,705,108 \text{ kJ/kmol} = 40,746 \text{ kJ/kg}$$

$$ex_{Szargut\,and\,Strylska} = 41,321,359 \text{ kJ/kmol} = 40,371 \text{ kJ/kg}$$

Results of the two estimation methods for the chemical exergy differ less than 1%. Example Table 2.13.1 lists $h_f^0 w$, LHV, and chemical exergy estimations for various biochemicals.

2.5 Group Contribution Method

Biological systems usually have very complex chemical structures; therefore, their thermodynamic properties, such as specific heat, enthalpy, and Gibbs free energy of formation, may not be found in the literature. In such cases, the detailed chemical formula of such chemicals may be used to list the groups contributing to their structure, and then the thermodynamic properties of these contributing methods are added up to estimate those of the entire structure. The molecular groups are not chosen randomly.

EXAMPLE TABLE 2.13.1 Standard Enthalpy of Formation (h_f^0), Lower Heating Value, and Chemical Exergy (ex_{ch}) Estimations for Various Biochemicals

						Chemical Exergy Calculated According to	
				h_f^0	LHV	Moran (1982)	Szargut and Styrylska
	C	H	O	(kJ/mol)	(kJ/kg)	(kJ/kg)	(1964) (kJ/kg)
Cell debris	4.17	8	1.25	1,999	26,162	28,417	27,602
Triacylglycerol	8	16	1	3,835	39,672	40,940	40,211
Lipid based on doconexent	69	98	6	3,3079	38,138	40,371	40,000
Lipid based on oleic acid	57	100	6	27,326	39,197	41,644	41,158
Lipid based on palmitic acid	51	98	6	24,449	39,566	42,111	41,563
Sugar	6	12	6	2,876.4	21,175	22,427	21,248
Algal biomass	1	1.83	0.48	479.4	26,309	26,890	25,157

For example, data of radical groups cannot be calculated separately, but taken as a whole. Benson (1965), Shieh and Fan (1982), Szargut et al. (1988), Domalski and Hearing (1993), Marrero and Gani (2001), and Gharagheizi et al. (2014) are among the researchers who made substantial contribution to the development of this method. Algae, yeast, bacteria, viruses, biopolymers, and bioactive chemicals such as vitamins are among the biomaterials for which the only information available is in the form of an empirical formula. For such case studies, Batteley and coworkers' studies (Battley, 1999; Battley and Stone, 2000) may be referred to estimate the entropy of formation of the biomass. Kopp's rule may be regarded as one of the pioneering studies of the group contribution methods for the estimation of the specific heat of solids and liquids. Kopp's rule says that specific heat of a solid compound is equal to the sum of the heat capacities of its constituting atoms as

$$c_p = \sum_{i=1}^{n} n_i c_{p,i}$$

In this equation, c_p of each molecule is determined experimentally and the number of each atom, n_i, in a molecule is set according to the molecular structure. The previously stated equation is written for large data sets that include as many molecules as possible. Later, the values of $c_{p,i}$ are determined via regression analysis to minimize the sum of the square error between the measured values of c_p and their numerical estimates (Hurst and Harrison, 1992). The values of the atoms of the original formulation of Kopp's rule are given in Table 2.8.

Example 2.14: Estimation of the Specific Heat of Liquid Aniline

Aniline is an organic compound with the formula C_6H_7N. It consists of a phenyl group attached to an amino group. It is a precursor to many industrial chemicals. We may estimate the specific heat of aniline with Kopp's rule as

$$c_p = \sum_{i=1}^{3} n_{atom} c_{p,atom} = (6)(12) + (7)(18) + (1)(33) = 231 \text{ kJ/mol}$$

TABLE 2.8   Kopp's Rule for Elemental Contribution Values of the Atoms in Solid and Liquid States

Atom	$c_{p,atom}$ Specific Heat of the Atom in Solid State (J/g atom K)	$c_{p,atom}$ Specific Heat of the Atom in Liquid State (J/g atom K)
H	9.6	18
B	11	20
C	7.5	12
O	17	25
F	21	29
Si	16	24
S	26	31
All the others	26	33

Source: Adapted from Hurst JE, Harrison BK, Estimation of liquid and solid heat capacities using a modified Kopp's rule. *Chemical Engineering Communications* 112, 21–30, 1992.

Kopp's rule provides only a rough estimate; c_p of the liquid aniline is given in the literature as 191.1 kJ/mol.

Hurst and Harrison (1992) suggested a set of values for $c_{p,atom}$ and claimed that the specific heats of the liquids and the solids may be estimated generally more accurately with their parameters. We will recalculate the specific heat of aniline after substituting $c_{p,C} = 13.08$ kJ/mol K, $c_{p,H} = 9.20$ kJ/mol K, and $c_{p,N} = 30.19$ kJ/mol K:

$$c_p = \sum_{i=1}^{3} n_{atom} c_{p,atom} = (6)(13.08) + (7)(9.20) + (1)(30.19) = 173.1 \text{ J/mol}$$

As it is seen here, Kopp's rule gave a closer estimate of the experimentally determined value of the $c_{p,aniline}$, but we may not claim a superiority of one method of estimation over the other one with such a small set of assessment.

Kopp's rule may be used to estimate the specific heats of numerous materials around the room temperature. The specific heat may also be related to temperature as $c_p = aT + bT$ constants of this empirical are available in the literature, as exemplified in Table 2.9.

Group contribution method proposed by Chueh and Swanson (1973a,b) gives accurate estimates of the specific heats of the organic liquids at 293.15 K:

$$c_p = \sum_{i=1}^{N_i} n_{atom} c_{p,atom} + 18.33 \times m$$

TABLE 2.9 Constants of the Empirical Equation $c_p = a + bT$ for Components of Wood Solids When c_p Is Expressed J/g K

Solid	a	b
α-Cellulose	−0.07146	4.14×10^{-3}
Lignin	−0.25660	3.22×10^{-3}

Source: Adapted from Domalski, E.S. et al., Thermodynamic data for biomass conversion and waste incineration, Solar Technical Information Program, U.S. Department of Energy, September 1986, available at http://www.nrel.gov/biomass/pdfs/2839.pdf, accessed August 16, 2014.

TABLE 2.10 Contributions of the Some of the Groups to the Specific Heat of the Biological Compounds

Group	$c_{p,group}$ (J/group K)	Group	$c_{p,group}$ (J/group K)
$-CH_3$	36.82	$-\overset{\mid}{\underset{\mid}{C}}-$	20.92
$-CH_2-$ (nonring element)	30.38	$-COOH$ (acid)	79.91
$-CH_2-$ (ring element)	29.94	$-NH_2$	58.58
$-OH$	44.77		

where m is the number of carbon groups requiring an additional contribution, which are those that are joined by a single bond to a carbon group, which in turn is connected to a third carbon group by a double or triple bond. If a carbon group meets this criterion in more than one ways, m should be increased by one for each of the ways. (Exceptions: CH_3 groups or carbon groups in a ring never require an additional contribution, and the *first* additional contribution for a $-CH_2-$ group is 10.46 J/mol K rather than 18.83 J/mol K. However, if the $-CH_2-$ group meets the criterion in a second way, the second additional contribution reverts to the 18.83 J/mol K value.) Contributions of some of the groups to the specific heat are given in Table 2.10.

Example 2.15: Estimation of the Thermodynamic Properties of Glucose

(a) Specific heat of glucose (Example Figure 2.15.1) may be estimated by using Kopp's rule by referring to the data presented in Example Table 2.15.1:

$$c_p = \sum_{i=1}^{N} n_{i,atom} c_{p,atom,i}$$

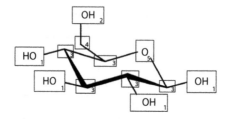

EXAMPLE FIGURE 2.15.1   The chemical structure of glucose.

EXAMPLE TABLE 2.15.1   Kopp's Rule for Elemental Contribution
Values of the Atoms of in Solid and Liquid States

Atom	Occurrence in ($C_6H_{12}O_6$)	$c_{p,atom,i}$ (J/mol K)
C	6	10.89
H	12	7.56
O	6	13.42
		$c_p = 236.6$ J/mol K

Source: Adapted from Green, D.W. and Perry, R.H., *Perry's Chemical Engineers' Handbook*, 7th ed., McGraw-Hill, New York, 1997.

Enthalpy of formation and the Gibbs free energy of formation of glucose are estimated with the group contribution method by using the data given in Example Table 2.15.2.

$$\Delta h_f^o = 68.29 - \sum_{i=1}^{n} n\Delta h_{f,i}^o = 68.29 - 1216.3 = -1148.0 \text{ kJ/mol}$$

EXAMPLE TABLE 2.15.2   Estimation of the Enthalpy of Formation and Gibbs Free Energy of Glucose

Group Number in the Figure	Group	Times of Occurrence in the Chemical Structure	Δh_f^o (kJ/mol)	Δg_f^o (kJ/mol)
1	–OH (ring)	4	−221.65	−197.37
2	–OH (alcohol)	1	−208.04	−189.20
3	CH	5	8.67	40.99
4	–CH$_2$–	1	−26.80	−3.68
5	–O–	1	−138.16	−98.22
		Total	$\sum_{i=1}^{n} n\Delta h_{f,i}^o = -1216.3$ kJ/mol	$\sum_{i=1}^{n} n\Delta g_{f,i}^o = -875.6$ kJ/mol

$$\Delta g_f^o = 53.88 - \sum_{i=1}^{n} n \Delta g_{f,i}^o = 53.88 - 875.6 = -821.7 \text{ kJ/mol}$$

The use of the constants 68.29 and 53.88 are suggested by Joback and Reid (1987).

(b) We will estimate the standard entropy of formation Δs_f^o of glucose from the following expression:

$$\Delta g_f^o = \Delta h_f^o - T \Delta s_f^o$$

$$\Delta s_f^o = \frac{\Delta h_f^o - \Delta g_f^o}{T} = \frac{-1148.0 - (-821.7)}{298} = -1.1 \text{ kJ/mol K}$$

The standard entropy of formation may be estimated with the following equation:

$$\Delta s_f^o = s_{compound}^o - \sum_{i=1}^{N} n_{element,i} s_{element,i}^o$$

where

$s_{compound}^o$ is the ideal gas absolute entropy of the compound at 298 K and 1 atm

$n_{element,i}$ and $s_{element,i}^o$ are the number of times of the occurrence and the absolute entropy of the ith element in its standard state at 298 K.

We may calculate $s_{compound}^o$ after substituting the numbers in the following equation by using the data given in Example Table 2.15.3:

$$s_{compound}^o = \Delta s_f^o + \sum_{i=1}^{N} n_{element,i} s_{element,i}^o = -1.1 + 10.6 = 9.5 \text{ kJ mol K}$$

EXAMPLE TABLE 2.15.3 Estimation of the Absolute Entropy of the Elements of Glucose

Molecule	$s_{element,i}^o$ (Green and Perry, 1997)	$n_{element,i}$	$(s_{element,i}^o)(n_{element,i})$
C	5.74	6	63.1
H_2	130.57	6	1240.4
O_2	205.04	3	922.7
			$\sum_{i=1}^{N} (n_{element,i})(s_{element,i}^o) = 10,589 \text{ J/mol K}$
			$= 10.6 \text{ kJ/mol K}$

EXAMPLE TABLE 2.15.4   Standard Chemical Exergies of the Constituent Elements of Glucose

Molecule	$ex^o_{element,i}$ (kJ/mol) (Exergoecology Portal, 2013)	$n_{element,i}$	$(ex^o_{element,i})(n_{element,i})$
C	410.25	6	2461.5
H_2	236.10	6	1416.6
O_2	3.97	3	10.1
			$\sum_{i=1}^{N}(n_i)(ex^o_{ch,i})=3888.2$ kJ/mol

(c) The standard chemical exergy of the glucose may be calculated by means of the exergy balance of a reversible formation reaction by following the same procedure by Szargut et al. (2005):

$$ex^o_{ch} = \Delta g^o_f + \sum_{i=1}^{N}(n_i)(ex^o_{element,i})$$

where

ex^o_{ch} is the standard chemical exergy of the compound
Δg^o_f is the standard Gibbs energy of formation
n_i is the number of the atoms of the ith element in the chemical formula
N is the number of the different atoms
$ex^o_{ch,i}$ is the standard chemical exergy of the ith element as listed in Example Table 2.15.4.

After substituting the data in the formula, we will calculate the standard exergy of glucose as

$$ex^o_{ch} = \Delta g^o_f + \sum_{i=1}^{N}(n_i)(ex^o_{ch,i}) = (-821.7)+(3888.2)=3066.5 \text{ kJ/mol}$$

Example 2.16: Estimation of the Specific Heat Capacity, Absolute Entropy, and Standard Chemical Exergy of *Neisseria meningitidis* Serogroup C Antigen (Subunit *n*-Acetylneuraminic Acid)

Neisseria meningitidis produces serogroup C antigen, a polymer of 285 repeating units of *n*-acetylneuraminic acid (Example Figure 2.16.1, $C_{11}H_{19}NO_9$). The chemical structure of *n*-acetylneuraminic acid is given in Figure 2.16.1. Since it has a sophisticated chemical structure, we will calculate its thermodynamic properties by employing the group contribution method (Değerli et al., 2015).

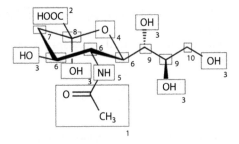

EXAMPLE FIGURE 2.16.1 The chemical structure of *n*-acetylneuraminic acid.

(a) Estimation of the Specific Heat of *n*-Acetylneuraminic Acid and Serogroup C Antigen

The contributing groups of the specific heat are its atoms. We will use Kopp's rule with the specific heat of the atoms as suggested by Hurst and Herrison (Example Table 2.16.1) to calculate the specific heat of a repeating unit ($C_{11}H_{19}NO_9$):

$$c_p = \sum_{i=1}^{N} n_{i,atom} c_{p,atom,i}$$

The specific heat of the antigen, for example, the total of the specific heat of the 285 repeating units is

$$c_{p,antigen} = \sum_{i=1}^{N} n_{repeating\ units} c_{p,\ repeating\ unit}$$

$$= (285)(403)$$

$$= 114{,}855 \text{ J/mol} \cdot \text{K}$$

$$= 115 \text{ kJ/mol} \cdot \text{K}$$

EXAMPLE TABLE 2.16.1 Atomic Specific Heats as Suggested by Hurst and Herrison

Atom	Occurrence in ($C_{11}H_{19}NO_9$)	$c_{p,atom,i}$ (J/mol K)
C	11	10.89
H	19	7.56
O	9	13.42
N	1	18.74
		$c_p = 403$ J/mol K

Source: Green, D.W. and Perry, R.H., *Perry's Chemical Engineers' Handbook*, 7th ed., McGraw-Hill, New York, 1997.

(b) Estimation of the Absolute Entropy $s^o_{compound}$ of n-Acetylneuraminic Acid and Serogroup C Antigen

Similar to the previous example, the standard entropy of formation is calculated as

$$\Delta s^o_f = s^o_{compound} - \sum_{i=1}^{N} n_{element,i} s^o_{element,i}$$

And the standard Gibbs free energy of formation is

$$\Delta g^o_f = \Delta h^o_f - T\Delta s^o_f$$

Contributions of the first-order groups for the calculation of Δh^o_f and Δg^o_f of n-acetylneuraminic acid are adapted from Marrero and Gani (2001) as described in Example Table 2.16.2.

We will calculate Δs^o_f from the following expression:

$$\Delta s^o_f = \frac{\Delta h^o_f - \Delta g^o_f}{T} = \frac{-1638.9 - (1194.3)}{298} = -1.5 \text{ kJ/mol K}$$

The absolute entropy of the elements is given in Example Table 2.16.3.

We may calculate $s^o_{compound}$ for n-acetylneuraminic acid after substituting the numbers in the following equation:

$$s^o_{compound} = \Delta s^o_f + \sum_{i=1}^{N} n_{element,i} s^o_{element,i} = -1.5 + 2.3 = 0.8 \text{ kJ/mol K}$$

EXAMPLE TABLE 2.16.2   Contribution of the First-Order Groups for the Calculation of Δh^o_f and Δg^o_f as Suggested by Marrero and Gani (2001)

Group Number in the Figure	Group	Times of Occurrence in the Chemical Structure	Δh^o_f (kJ/mol)	Δg^o_f (kJ/mol)
1	CH_3CO	1	−180.604	−120.667
2	COOH	1	−389.931	−337.090
3	OH	5	−178.360	−144.051
4	−O−	1	−137.353	−114.062
5	NH (cyclic)	1	23.138	72.540
6	−CH (cyclic)	3	−12.464	6.107
7	−CH$_2$ (cyclic)	1	−18.575	13.287
8	C (cyclic)	1	−2.098	−0.193
9	−CH	2	−7.122	8.254
10	−CH$_2$	1	−20.829	8.064
			$\Delta h^o_f = -1638.9$ kJ/mol	$\Delta g^o_f = -1194.3$ kJ/mol

EXAMPLE TABLE 2.16.3 Absolute Entropy of the Elements of *n*-Acetylneuraminic Acid and Serogroup C Antigen

Molecule	$s^o_{element,i}$ (Green and Perry, 1997)	$n_{element,i}$	$(s^o_{element,i})(n_{element,i})$
C	5.74	11	63.1
H_2	130.57	19/2	1240.4
O_2	205.04	9/2	922.7
N_2	191.50	1/2	95.8
		$\sum_{i=1}^{N}(n_{element,i})(s^o_{element,i}) = 2.3$ kJ/mol K	

Then, we may estimate $s^o_{antigen}$ as

$$s^o_{antigen} = \left(n_{repeating\ units}\right)\left(\Delta s^o_{f,repeating\ unit}\right) = (285)(0.8) = 228 \text{ kJ mol K}$$

(c) Estimation of the Chemical Exergy of *n*-Acetylneuraminic Acid and Serogroup C Antigen

The standard chemical exergy of the *n*-acetylneuraminic acid may be calculated by means of the exergy balance of a reversible formation reaction by following the same procedure by Szargut et al. (2005):

$$ex^o_{ch} = \Delta g^o_f + \sum_{i=1}^{N}\left(n_i\right)\left(ex^o_{element,i}\right)$$

where

ex^o_{ch} is the standard chemical exergy of the compound
Δg^o_f is the standard Gibbs free energy of formation
n_i is the number of the atoms of the *i*th element in the chemical formula
N is the number of the different atoms
$ex^o_{ch,i}$ is the standard chemical exergy of the *i*th element as listed in Example Table 2.16.4.

EXAMPLE TABLE 2.16.4 Chemical Exergy of the Elements of *n*-Acetylneuraminic Acid and Serogroup C Antigen

Molecule	$ex^o_{element,i}$ (kJ/mol) (Exergoecology Portal, 2013)	$n_{element,i}$	$(ex^o_{element,i})(n_{element,i})$
C	410.25	11	4513
H_2	236.10	19/2	2243
O_2	3.97	9/2	18
N_2	0.72	1/2	0.4
		$\sum_{i=1}^{N}(n_i)(ex^o_{ch,i}) = 6774.0$ kJ/mol	

After substituting the data in the formula, we will calculate the standard exergy of *n*-acetylneuraminic acid as

$$ex^o_{ch} = \Delta g^o_f + \sum_{i=1}^{N} (n_i)(ex^o_{ch,i}) = (-1194.3) + (6774.0) = 5579.7 \text{ kJ/mol}$$

All the data are multiplied by the number of 285 repeating units to find the thermodynamic data:

$$ex^o_{ch,\,antigen} = \sum_{i=1}^{N} n_{repeating\ units} ex^o_{ch,repeating\ unit} = (285)(5,579.7) = 1,590,215 \text{ kJ/mol}$$

Example 2.17: Calculation of the Enthalpy, Gibbs Energy, and Entropy of Formation of Octopine at 298 Kelvin and 100 kPa (1 Atmosphere) with the Group Contribution Method

Octopine ($C_9H_{18}N_4O_4z$) functions in the muscle tissue of invertebrates, such as octopus, in a way analogous to that of lactic acid in the vertebrate tissues (Hockachka et al., 1977). Its molecular structure is sketched in Example Figure 2.17.1.

(a) Calculation of the Enthalpy and the Gibbs Free Energy of Formation of Octopine at 298 K and 100 kPa (1 atm)

The atomic group contributions in the structure of octopine are listed in Example Table 2.17.1. (Data adapted from Perry et al., 1997.)

Enthalpy of formation and Gibbs free energy of formation at standard conditions (T = 298 K, and p = 100 kPa e.g., 1 atm) are calculated as (Green and Perry, 1997)

$$\Delta g^o_f = 53.88 + \sum_{i=1}^{n} n_i \Delta g_i$$

$$\Delta h^o_f = 68.29 + \sum_{i=1}^{n} n_i \Delta h_i$$

Therefore, $\Delta g^o_{f,298}$ is found as −218.97 kJ/mol and $\Delta h^o_{f,298}$ is −601.13 kJ/mol.

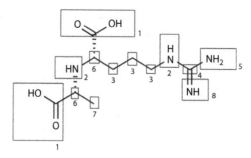

EXAMPLE FIGURE 2.17.1   Molecular structure of octopine.

EXAMPLE TABLE 2.17.1 Atomic Group Contributions in the Structure of Octopine

Group Number in the Figure	Group	Times of Occurrence in the Chemical Structure	$\Delta h^\circ_{f, group}$ (kJ/mol)	$\Delta g^\circ_{f, group}$ (kJ/mol)
1	— COOH	2	−426.72	−387.87
2	$\mid$ — NH	2	53.47	80.39
3	— CH_2 —	3	−20.64	8.42
4	$=C—$ $\mid$	1	83.99	92.36
5	— NH_2	1	−22.02	14.07
6	$\mid$ — CH $\mid$	2	29.89	58.36
7	— CH_3	1	−76.45	−43.96
8	$=$ NH	1	93.70	119.66

$$\Delta h^\circ_f = \sum_{i=1}^{N} n_{occurance} \Delta h^\circ_{group} \qquad \Delta g^\circ_f = \sum_{i=1}^{N} n_{occurance} \Delta g^\circ_{group}$$

$$= -669.4 \text{ kJ/mol} \qquad = 272.9 \text{ kJ/mol}$$

(b) Calculation of the Entropy of Formation of Octopine at 298 K and 100 kPa (1 Atmosphere)

We will calculate Δs°_f from the following expression:

$$\Delta s^\circ_f = \frac{\Delta h^\circ_f - \Delta g^\circ_f}{T} = \frac{(-601.13)-(-218.97)}{298} = -1.28 \text{ kJ/mol K}$$

The entropy of the compound can be calculated from the following equation:

$$\Delta s^\circ_f = s^\circ_{compound} - \sum_{i=1}^{N} n_{element, i} s^\circ_{element, i}$$

The calculation of the sum of the absolute entropy of the elements is given in Example Table 2.17.2.

After substituting the numbers in the equation, we will obtain

$$s^\circ_{compound} = \Delta s^\circ_f + \sum_{i=1}^{N} n_{element, i} s^\circ_{element, i} = -1.28 + 2.02 = 0.74 \text{ kJ/mol K}$$

EXAMPLE TABLE 2.17.2 Sum of the Absolute Entropy of the Elements in the Structure of Octopine

Molecule	$s^o_{element,i}$ (Green and Perry, 1997)	$n_{element,i}$	$(s^o_{element,i})(n_{element,i})$
C	5.74	9	51.7
H_2	130.57	9	1175.1
O_2	205.04	2	410.8
N_2	191.50	2	383.0
		$\sum_{i=1}^{N}(n_{element,i})(s^o_{element,i}) = 2021$ J/mol K	
		$= 2.02$ kJ/mol K	

(c) Estimation of the Chemical Exergy of Octopine

$$ex^o_{ch} = \Delta g^o_f + \sum_{i=1}^{N}(n_i)\left(ex^o_{ch,i}\right)$$

where

$ex^o_{ch,i}$ is the chemical exergy of the compound at $T = 298.15$ K and $p = 100$ kPa (kJ/mol)

$ex^o_{ch,i}$ is the absolute exergy of element i at $T = 298.15$ K and $p = 100$ kPa (kJ/mol)

N is the number of the different elements in the compound

n_i is the number of moles of element i in 1 mol of a compound

Contribution of each group to the exergy of the structure is given in Example Table 2.17.3.

EXAMPLE TABLE 2.17.3 Contribution of Each Group to the Exergy of the Structure of Octopine

Molecule	$ex^o_{ch,i}$ (kJ/mol) (Exergoecology Portal, 2013)	$n_{element,i}$	$(ex^o_{element,i})(n_{element,i})$
C	410.3	9	3692.7
H_2	236.1	9	2124.9
O_2	3.97	2	7.9
N_2	0.72	2	1.4
		$\sum_{i=1}^{N}(n_i)(ex^o_{chem,i}) = 5827$ kJ/mol	

Source: Data adapted from Szargut, J. et al., Towards an international reference environment of chemical exergy, Elsevier Science, 2005, http://www.exergoecology.com/papers/towards_int_re.pdf.

After substituting the numbers, we will have

$$ex^o_{ch} = \Delta g^o_f + \sum_{i=1}^{N} (n_i)(ex^o_{ch,i}) = -218.97 + 5827 = 5608 \text{ kJ/mol}$$

Example 2.18: Estimation of the Thermodynamic Properties of Chitin

Chitin $(C_8H_{13}O_5N)_n$ is a long-chain polymer found in the cell walls of fungi and of the exoskeletons of crabs, lobsters, and shrimps and insects. Its structure is depicted in Example Figure 2.18.1, and the enthalpy and the Gibbs free energy of formation of chitin are detailed in Example Table 2.18.1:

$$\Delta h^o_{298} = 68.29 + \sum_{i=1}^{n} n_i \Delta h_i$$

$$\Delta h^o_{298} = 68.29 + (-843.56) = -775.27 \text{ kJ/mol}$$

$$\Delta g^o_{298} = 53.88 + \sum_{i=1}^{n} n_i \Delta g_i$$

$$\Delta g^o_{298} = 53.88 + (-451.49) = -397.61 \text{ kJ/mol}$$

In order to find out the standard molar entropy, entropy of formation of chitin is needed (Example Table 2.18.2):

$$\Delta s^o_f = \frac{\Delta h^o_f - \Delta g^o_f}{T} = \frac{(-775.27) - (-397.61)}{298} = -1.27 \text{ kJ/mol K}$$

The standard entropy of formation may be estimated with the following equation:

$$s^o_{compound} = \Delta s^o_f + \sum_{i=1}^{N} n_{element,i} s^o_{element,i} = -1.27 + 1.50 = 0.23 \text{ kJ mol K}$$

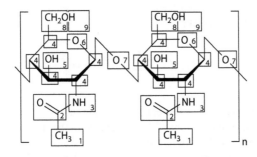

EXAMPLE FIGURE 2.18.1 Structure of chitin and its contributing groups.

EXAMPLE TABLE 2.18.1   Calculation of the Enthalpy and the Gibbs Free Energy of Chitin

Groups Number in the Figure 2	Group	Times of Occurrence in Chemical Structure	$\Delta h^o_{f,\ group}$ (kJ/mol)	$\Delta g^o_{f,\ group}$ (kJ/mol)
1	—CH$_3$ (nonring)	1	−76.45	−43.96
2	—C=O (nonring)	1	−133.22	−120.50
3	—NH— (nonring)	1	53.47	89.39
4	— CH (ring)	5	8.67	40.99
5	—OH (ring)	1	−221.65	−197.37
6	—O— (ring)	1	−138.16	−98.22
7	—O— (nonring)	1	−133.22	−105.00
8	—CH$_2$— (nonring)	1	−20.64	8.42
9	—OH (nonring)	1	−208.04	−189.20
			$\sum (\Delta h^o_f) = -834.56$ kJ/mol	$\sum (\Delta g^o_f) = -451.49$ kJ/mol

EXAMPLE TABLE 2.18.2   Calculation of the Absolute Entropy of the Chitin

Molecule	$s^o_{element,i}$ (Green and Perry, 1997)	$n_{element,i}$	$\left(s^o_{element,i}\right)\left(n_{element,i}\right)$
C	5.74	8	45.92
H$_2$	130.57	13/2	848.705
O$_2$	205.04	5/2	512.6
N$_2$	191.50	1/2	95.75
			$\sum \left[\left(s^o_{element,i}\right)\left(n_{element,i}\right)\right]$
			= 1503 J/mol K

The standard chemical exergy of the chitin may be calculated by means of the exergy balance of a reversible formation reaction by following the same procedure by Szargut et al. (2005):

$$ex^o_{ch} = \Delta g^o_f + \sum_{i=1}^{N} (n_i)\left(ex^o_{element,i}\right)$$

where

Δg^o_f is the standard Gibbs free energy of formation

n_i is the number of the atoms of the ith element in the chemical formula

N is the number of the different atoms

$ex^o_{ch,i}$ is the standard chemical exergy of the ith element as listed in Example Table 2.18.3

EXAMPLE TABLE 2.18.3   Calculation of the Standard Exergy of the Chitin

Molecule	$ex^o_{element,i}$ (kJ/mol) (Exergoecology Portal, 2013)	$n_{element,i}$	$\left(ex^o_{element,i}\right)\left(n_{element,i}\right)$
C	410.25	8	3282.00
H_2	236.10	13/2	1534.65
O_2	3.97	5/2	9.93
N_2	0.72	1/2	0.36
			$\sum_{i=1}^{N}\left(n_i\right)\left(ex^o_{chem,i}\right) = 4826.94$ kJ/mol

After substituting the data in the formula, we will calculate the standard exergy of chitin as

$$ex^o_{ch} = \Delta g^o_f + \sum_{i=1}^{N}\left(n_i\right)\left(ex^o_{ch,i}\right) = \left(-397.61\right) + \left(4826.94\right) = 4429.33 \text{ kJ/mol}$$

2.6 Thermodynamic Properties of Mixtures

Almost all actual systems are mixtures. An ideal mixture is defined as a mixture with zero enthalpy and zero volume of mixing. The closer these properties to zero, the more *ideal* the behavior of the mixture becomes. The ideal mixture assumption is valid if the chemical structures of the species in the mixture are similar. With similar chemical structures, attractive and repulsive forces between the molecules of the substances in the mixture would be comparable. Thus $dh_{mix} = 0$. Mixtures of species with different chemical structures are called real mixtures with $dh_{mix} \neq 0$ and $dv_{mix} \neq 0$.

Example 2.19: Total Volume after Mixing Balls of the Same and Different Diameters

If we mix balls with the same diameter, then the final volume would be the sum of the original volumes (Example Figure 2.19.1).

Analogically, if we mix gases with similar chemical properties, then the extensive properties of the mixtures can be calculated as the sum of the extensive properties of the species. Therefore, any solution property, M, may be related with those of the partial properties, $\overline{M}_i$, as $M = \sum_{i=1}^{n} \overline{M}_i$ where the volume, v; internal energy, u; enthalpy, h; entropy, s; or Gibbs free energy, G, may be substituted for M. Such mixtures are ideal mixtures.

In Figure 2.19.2 on the right-hand side, we fill the volume V_1 with larger balls, in a way that we cannot fit any more balls to the circle. Then, we add smaller balls of volume V_2, since the smaller balls will fit into the empty spaces between the larger balls, in such a way that $V_1 + V_2 = V_1$.

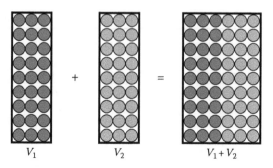

EXAMPLE FIGURE 2.19.1   Mixing of the balls with the same diameter.

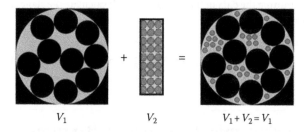

EXAMPLE FIGURE 2.19.2   Mixing of the balls with larger diameter with the balls with smaller diameter.

When we mix two fluids with fundamentally different chemical structures, we observe that their thermodynamic properties will change in a way that their final property will actually be different than the sum of the pure chemical properties. Such mixtures are nonideal or real mixtures.

In a similar way, we may also visualize the increase of the entropy upon mixing two pure compounds:

Example 2.20: Entropy of Mixing

When we mix pure molecules of X located in 5×6 sites and pure molecules of 0 located in 2×6 sites, we will obtain 7×6 sites to locate the molecules as described in Example Figure 2.20.1.

Upon mixing entropy increases, we may calculate it with Boltzmann's equation. In Chapter 1, Boltzmann's equation was given as

$$s = -k_B \sum_i p_i \ln p_i$$

For an ideal gas, this equation can be simplified as

$$s = -k_B \sum \ln \Omega$$

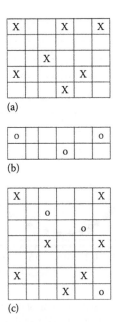

(a)

(b)

(c)

EXAMPLE FIGURE 2.20.1 Schematic description of mixing of chemicals o and X. (a) Solution of the chemicals X, (b) solution of the chemicals o, and (c) Mixture of the solutions of the chemicals o and X.

where Ω is the thermodynamic probability. Then for this system,

$$\Delta s = k_B \left[\ln(\Omega_x) + \ln(\Omega_O)\right]_{final} - k_B \left[\ln(\Omega_x) + \ln(\Omega_O)\right]_{initial}$$

$$= \left(1.38 \times 10^{-23} \text{ kJ/K}\right)\left\{\left[\ln(42^7) + \ln(42^3)\right] - \left[\ln(30^7) + \ln(12^3)\right]\right\}$$

$$= \left(1.38 \times 10^{-23} \text{ kJ/K}\right)\left\{\left[7\ln(42) + 3\ln(42)\right] - \left[7\ln(30) + 3\ln(12)\right]\right\}$$

$$= 8.45 \times 10^{-23} \text{ kJ/K}$$

where

$k_B = 1.38 \times 10^{-23}$ J/K is Boltzmann's constant

Ω_O (number of microstates available)[number of Os] is the number of the microstates available in a macrostate for arranging the Os (a similar expression is also written for Xs)

Example 2.21: Calculations with Ethanol–Water Mixtures

Volume change upon mixing of ethanol and water is depicted in Example Figure 2.21.1. Density of water and ethanol at 25°C are 997.1 and 785.2 kg/m³, respectively.

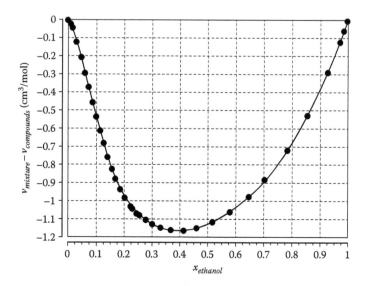

EXAMPLE FIGURE 2.21.1 Specific volume of ethanol water mixtures is always less than the sum of the pure component. (Data taken from Dortmund data bank; $T = 25°C$; $x_{ethanol}$ is the mole fraction of ethanol in the mixture.)

(a) Calculate the amounts of ethanol and water to obtain 2 L of 30% ethanol and 70% water mixture (mole percentages) at 25°C.

$$V_{water} = \frac{1}{\rho_{water}} = \left(\frac{1}{997.1 \text{ kg/m}^3}\right)\left(\frac{0.018 \text{ kg}}{1 \text{ mol}}\right) = 1.8\times10^{-5} \text{ m}^3/\text{mol}$$

$$V_{ethanol} = \frac{1}{\rho_{ethanol}} = \left(\frac{1}{785.2 \text{ kg/m}^3}\right)\left(\frac{0.046 \text{ kg}}{1 \text{ mol}}\right) = 5.9\times10^{-5} \text{ m}^3/\text{mol}$$

$$V_{compounds} = (x_{water}V_{water} + x_{ethanol}V_{ethanol})$$

$$= \left[(0.70)\left(1.8\times10^{-5} \text{ m}^3/\text{mol}\right)+(0.30)\left(5.9\times10^{-5} \text{ m}^3/\text{mol}\right)\right]$$

$$= 3.0\times10^{-5} \text{ m}^3/\text{mol}$$

This is the total volume of the compounds, without the effect of the volume change upon mixing. Data given in Example Figure 2.21.1 state that the percent volume decrease in the mixture is

$$\frac{V_{mixture} - V_{compounds}}{V_{compounds}} = \frac{(-1.13 \text{ cm}^3/\text{mol})}{(3.0\times10^{-5} \text{ m}^3/\text{mol})}\left(\frac{\text{m}^3}{10^6 \text{ cm}^3}\right)(100) = -3.8\%$$

Therefore, we should prepare $(2 \text{ L})(1 + 0.038) = 2.08 \text{ L}$ of mixture to compensate for the decrease of the volume. The total number of the moles of the mixture therefore should be

$$n_{total} = \left(\frac{2.08 \text{ L}}{3.0 \times 10^{-5} \text{ m}^3/\text{mol}} \right) \left(\frac{1 \text{ m}^3}{1000 \text{ L}} \right) = 69.3 \text{ mol}$$

$$n_{water} = (0.70)(n_{total}) = (0.70)(69.3) = 48.5 \text{ mol}$$

$$m_{water} = (48.5 \text{ mol})(18 \text{ g/mol}) = 873 \text{ g}$$

$$n_{ethanol} = (0.30)(n_{total}) = (0.30)(69.3) = 20.8 \text{ mol}$$

$$m_{ethanol} = (20.8 \ mol)(46 \text{ g/1 mol}) = 956.8 \text{ g}$$

$$V_{ethanol} = (20.8 \text{ mol})(5.9 \times 10^{-5} \text{ m}^3/\text{mol})(100 \text{ cm/1 m})^3 = 1227 \text{ cm}^3$$

$$V_{water} = (48.5 \text{ mol})(1.8 \times 10^{-5} \text{ m}^3/\text{mol})(100 \text{ cm/1 m})^3 = 873 \text{ cm}^3$$

We should add 1227 cm³ ethanol to 873 cm³ water to obtain the required solution.

(b) Alcohol content of wine is expressed in volume percentage. Calculate the mole fraction of ethanol in a wine containing 12% ethanol.

If we neglect the contribution of soluble solids, then wine is produced by adding 12 cm³ of ethanol to 88 cm³ of water.

$$\rho_{water} = 997.1 \text{ kg/m}^3 = (997.1 \text{ kg/m}^3)(1 \text{ mol/0.018 kg}) = 55,388 \text{ mol water/m}^3$$

$$\rho_{ethanol} = 785.2 \text{ kg/m}^3 = (785.2 \text{ kg/m}^3) \left(\frac{1 \text{ mol}}{0.046 \text{ kg}} \right) = 17,070 \text{ mol ethanol/m}^3$$

The volume fraction of ethanol in the mixture is (12 cm³ ethanol)/(12 cm³ ethanol + 88 cm³ water); we may calculate its molar fraction as

$$\frac{(12 \text{ cm}^3)(17,070 \text{ mol ethanol/m}^3)(\text{m}^3/10^6 \text{cm}^3)}{(12 \text{ cm}^3)(17,070 \text{ mol ethanol/m}^3)(\text{m}^3/10^6 \text{ cm}^3) + (88 \text{ cm}^3)(55,388 \text{ mol water/m}^3)(\text{m}^3/10^6 \text{ cm}^3)} =$$

$$\frac{0.21 \text{ mol ethanol}}{0.21 \text{ mol ethanol} + 4.87 \text{ mol water}} = 4 \text{ mol}\%$$

implying that $x_{ethanol} = 0.04$.

(c) If 1000 L of wine with $x_{ethanol,1} = 0.02$ is mixed with 2000 L of wine with $x_{ethanol,2} = 0.05$, what is the total volume of the mixture? We still assume that wine is a mixture of water and ethanol only.

(i) With the wine that has $x_{ethanol,1} = 0.02 v_{compounds,1}$, then

$$V_{compounds} = x_{water,1} V_{water,1} + x_{ethanol,1} V_{ethanol,1}$$

$$= (0.98)(1.8 \times 10^{-5}) + (0.02)(5.9 \times 10^{-5})$$

$$= 1.9 \times 10^{-5} \ m^3/mol = 19.0 \ cm^3/mol$$

and

$$v_{mixture} - v_{compounds} = -0.1 \ cm^3/mol$$

Therefore, after mixing the compounds, we will have

$$v_{mixture} = 19.0 - 0.1 = 18.9 \ cm^3/mol$$

In 1000 L of wine, we will have

$$(1,000 \ L)\left(\frac{1,000 \ cm^3}{1 \ L}\right)\left(\frac{mol}{18.9 \ cm^3}\right) = 52,910 \ mol$$

Then $n_{water} = (52,910)(0.98) = 51,852$ mol of water and $n_{ethanol} = (52,910) - (51,852) = 1,058$ mol of ethanol.

(ii) With the wine that has $x_{ethanol,2} = 0.05 v_{compounds,2}$, then

$$V_{compounds} = x_{water,2} V_{water,2} + x_{ethanol,2} V_{ethanol,2}$$

$$= (0.95)(1.8 \times 10^{-5}) + (0.05)(5.9 \times 10^{-5})$$

$$= 2.0 \times 10^{-5} \ m^3/mol = 20.0 \ cm^3/mol$$

and $v_{mixture} - v_{compounds} = -0.13 \ cm^3/mol$; therefore, after mixing the compounds, we will have $v_{mixture} = 20.0 - 0.13 = 19.9 \ cm^3/mol$.

In 2,000 L of wine, we will have

$$(2,000 \ L)\left(\frac{1,000 \ cm^3}{1 \ L}\right)\left(\frac{mol}{19.9 \ cm^3}\right) = 100,502 \ mol$$

Then $n_{water} = (100,502)(0.95) = 95,477$ mol of water and $n_{ethanol} = (100,502) - (95,477) = 5,025$ mol of ethanol.

iii. We will add up the number of the moles of water coming from both mixtures to find out the number of moles of water and ethanol in the final mixture:

$$n_{water} = 51,825 \ (\text{from 2\% solution}) + 95.477 \ (\text{from 5\% solution})$$

$$= 147,302 \ mol$$

$$n_{ethanol} = 1058 \text{ (from 2\% solution)} + 5025 \text{ (from 5\% solution)}$$

$$= 6083 \text{ mol}$$

then we calculate mole fraction of ethanol in the mixture as

$$x_{ethanol} = \frac{n_{ethanol}}{n_{water} + n_{ethanol}} = \frac{(6,083)}{(147,302)+(6,083)} = 0.04$$

With 4.0% ethanol, $v_{compounds}$ is

$$v_{compounds} = x_{water} v_{water} + x_{ethanol} v_{ethanol}$$

$$= (0.96)(1.8 \times 10^{-5}) + (0.04)(5.9 \times 10^{-5})$$

$$= 1.96 \times 10^{-5} \text{ m}^3/\text{mol} = 19.6 \text{ cm}^3/\text{mol}$$

$$v_{mixture} - v_{compounds} = -0.19 \text{ cm}^3/\text{mol}$$

therefore, after mixing the compounds, we will have

$$v_{mixture} = 19.6 - 0.19 = 19.4 \text{ cm}^3/\text{mol}$$

and the total volume of the mixture will be

$$V_{mixture} = v_{mixture} \, n_{mixture}$$

$$= (19.4 \text{ cm}^3/\text{mol})\left[(147,302)+(6,083)\right]$$

$$= 2,975,669 \text{ cm}^3 = 2,976 \text{ L}$$

We mixed together 1000 + 2000 = 3000 L of wine and obtained 2976 L mixture.

(iv) When we mix two substances, rearrangement of the molecules in the mixture may lead to generation or consumption of heat, which is referred to as the heat of mixing. If the mixture absorbs heat from the environment, the process is called endothermic mixing, and if heat is released upon mixing, the process is referred to as exothermic mixing. Calculate the heat that would be released while preparing 3,000 L of wine by mixing 6,083 mol of pure ethanol with 147,302 mol of pure water by using the data given in Example Figure E.2.21.2.

It is already calculated that when we mix 147,302 mol of water and 6,083 mol of ethanol, we will have 2,976 L of mixture, and we read from Example Figure 2.21.2 that $\Delta h_{mixing} = -350$ J/mol. Then, we will calculate the total heat released as

$$(-350 \text{ J/mol}) \, (6,083 \text{ mol} + 147,302 \text{ mol}) \left(\frac{3,000}{2,976}\right)\left(\frac{1 \text{ MJ}}{10^6 \text{ J}}\right) = -53.7 \text{ MJ}$$

implying that 53.7 MJ of heat is released upon mixing water and ethanol.

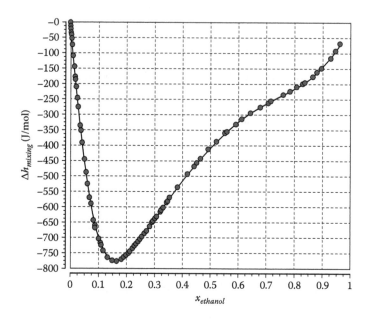

EXAMPLE FIGURE 2.21.2 Mixing enthalpy of water and ethanol at 25°C. (Data taken from Dortmund data bank; $T = 25°C$; $x_{ethanol}$ is the mole fraction of ethanol in the mixture.)

(v) Calculate the entropy of mixing of 147,302 mol of water ($T_{water} = 6°C$) and 6,083 mol of ethanol ($T_{ethanol} = 8°C$).

There are three things going on here. First, there is an entropy change associated with equilibrating the temperatures of the two solutions. Second, there is an entropy change associated with the heat released upon mixing of water and mixing, and third, there is entropy generation upon mixing of the two liquids. First, we will calculate the equilibrium temperature of the combined system:

$$\Delta q = 0 = n_{water}\, c_{water}\left(T_{mixture} - T_{water}\right) + n_{ethanol}\, c_{ethanol}\left(T_{mixture} - T_{ethanol}\right)$$

and then we may calculate $T_{mixture}$ as

$$T_{mixture} = \frac{n_{water}\, c_{water}\, T_{water} + n_{ethanol}\, c_{ethanol}\, T_{wethanol}}{n_{water}\, c_{water} + n_{ethanol}\, c_{ethanol}}$$

We may estimate the specific heats of liquid water and ethanol with Kopp's rule as

$$c_{water} = n_{Hydrogen}\, c_{p,\,Hydrogen} + n_{Oxygen}\, c_{p,\,Oxygen} = (2)(17.97) + (1)(25.08) = 61.02\ \text{J/mol K}$$

$$c_{ethanol} = n_{Carbon}\, c_{p,Carbon} + n_{Hydrogen}\, c_{p,Hydrogen} + n_{Oxygen}\, c_{p,Oxygen}$$

$$= (2)(11.70) + (6)(17.97) + (1)(25.08)$$

$$= 156.3\ \text{J/mol K}$$

After substituting the numbers in equation we will have

$$T_{mixture} = \frac{(147{,}302\ \text{mol})(61.02\ \text{J/mol K})(279\ \text{K}) + (6{,}083\ \text{mol})(156.3\ \text{J/mol K})(281\ \text{K})}{(147{,}302\ \text{mol})(61.02\ \text{J/mol K}) + (6{,}083\ \text{mol})(156.3\ \text{J/mol K})}$$

$$T_{mixture} = 279.1\ \text{K}$$

Entropy generation associated with temperature change of each chemical is $\Delta s = n\displaystyle\int_{T_{initial}}^{T_{final}} c\, dT/T$, and the total of water and ethanol upon mixing is

$$\Delta s_{T-change} = \Delta s_{T-change-water} + \Delta s_{T-change-ethanol}$$

$$= n_{water} c_{water} \ln\left(\frac{T_{mixture}}{T_{water}}\right) + n_{ethanol} c_{ethanol} \ln\left(\frac{T_{mixture}}{T_{ethanol}}\right)$$

and after substituting the numbers, we will have

$$\Delta s_{T-change} = (147{,}303\ \text{mol})(61.02\ \text{J/mol K})\ln\left(\frac{279.1}{279}\right)$$

$$+ (6{,}083\ \text{mol})(156.3\ \text{J/mol K})\ln\left(\frac{279.1}{281}\right)$$

$$\Delta s_{T-change} = 3231 - 6450 = -4119\ \text{J/K}$$

$$\Delta s_{T-change} = \frac{4{,}119\ \text{J/K}}{(147{,}303 + 6{,}083)\ \text{mol}} = -0.027\ \text{J/mol K}$$

Entropy generation due to the heat generation upon mixing for $x_{ethanol} = 0.293$ (read from Example Figure 2.21.2) is

$$\Delta s_{heat\ generation} = \frac{\Delta h_{mixing}}{T_{mixing}} = \left|\frac{(-53.7\ \text{MJ/mol})}{279.1\ \text{K}}\right| = 0.19\ \text{MJ/mol K}$$

It was explained in Section 1.7 that the entropy of an ideal gas may be calculated with Boltzmann's formula as

$$s = k_B \ln(\Omega)$$

where $k_B = 13 \times 10^{-23}$ J/K. Boltzmann's equation may be adapted to calculate the entropy of mixing of two ideal liquids as

$$\Delta s_{mixing} = -k_B \left(n_{water} + n_{ethanol} \right) \left[x_{water} \ln \left(x_{water} \right) + x_{ethanol} \ln \left(x_{ethanol} \right) \right]$$

where

$$x_{water} = \frac{n_{water}}{n_{water} + n_{ethanol}}$$

$$x_{ethanol} = \frac{n_{ethanol}}{n_{water} + n_{ethanol}}$$

Upon mixing of 147,303 mol of water with 6,083 mol of ethanol, we will have $x_{water} = 0.04$ and $x_{ethanol} = 0.96$, and after substituting the numbers in the equation, we will have

$$\Delta s_{mixing} -13 \times 10^{-23} \left(147,303 + 6,083 \right) \left[\left(0.04 \right) \ln \left(0.04 \right) + \left(0.96 \right) \ln \left(0.96 \right) \right]$$

$$= -3.3 \times 10^{-18} \text{ J/K}$$

The total entropy change in this process is

$$\Delta s_{total} = \Delta s_{T-change} + \Delta s_{heat\ generation} + \Delta s_{mixing}$$

$$\Delta s_{total} = \left(-4179 \text{ J/mol K} \right) + \left(0.19 \text{ MJ/mol K} \right) + \left(-3.3 \times 10^{-18} \text{ J/mol K} \right) = 0.19 \text{ MJ/mol K}$$

2.7 Thermodynamic Properties of Biochemical Mixtures under Physiological Conditions

Metabolic processes occur in cells, which involve mixtures of numerous metabolites. We can consider metabolic fluids (such as blood, extracellular fluid) as mixtures and calculate the thermodynamic properties of mixtures based on the properties of individual chemicals that make up this mixture. The enthalpy, entropy, and Gibbs free energy of the individual chemicals have to be evaluated under the *physiological conditions*. Tables 2.3 and 2.4 list Δh_f and Δg_f of various chemicals under standard conditions and under physiological conditions, respectively. In many engineering processes, *standard conditions* are taken as the reference. Under standard conditions, $p = 1$ atm, $T = 25°C$. In biochemical processes, in addition to temperature and pressure, pH and ionic strength (I) also play important roles. By changing pH and/or I, rates of the biochemical reactions may change dramatically, and metabolism may undergo a different pathway. Therefore, in biochemical systems, in the calculation of Δh_f and Δg_f, the effect of pH and I must be taken into account.

Alberty (2004a,b) defined the thermodynamic properties of metabolites undergoing biochemical processes under physiological conditions as *transformed* properties and denoted those with the superscript T to distinguish these from the properties at the standard conditions. The transformed properties are calculated based on the properties under standard conditions using the Van't Hoff equation and the extended Debye–Hückel theory. The transformed enthalpy of formation of the biochemical species i is defined as

$$\Delta h_{f,i}^T = \Delta h_{f,i}^0 + \frac{RT^2\beta(T)I^{1/2}}{1+BI^{1/2}}\left(z_i^2 - n_{H,i}\right)$$

where

$\beta(T)$ is a function of temperature
z is the valence
n_H is the number of the H^+ atoms
B is an empirical constant

The transformed Gibbs free energy of formation of the biochemical species i is defined as

$$\Delta g_{f,i}^T = \frac{T}{T_0}\Delta g_{f,i}^0 + \left(1-\frac{T}{T_0}\right)\Delta h_{f,i}^0 - \frac{RT\alpha(T)I^{1/2}}{1+BI^{1/2}}\left(z_i^2 - n_{H,i}\right) - n_{H,i}\Delta g_{f,H^+}^0 + 2.303 n_{H,i}RTpH$$

where $\alpha(T)$ is a function of temperature. Note that in the biochemical reactions, H^+ is not balanced, because the cellular fluid acts as a buffer for H^+. In the cell, at constant pH, different valences of some chemical species coexist as a mixture. Alberty uses the term *pseudoisomers* to define the various forms with different valences of the same species. For example, at pH = 7.0, ATP has three pseudoisomers, ATP^{4-}, ATP^{3-}, and ATP^{2-}, which coexist in equilibrium. Therefore, the standard Gibbs free energies of these species are calculated as (Alberty, 2004a,b)

$$\Delta g_{f,i}^0 = -RT\ln\sum_{j=1}^{N_{isomer}}\exp\left(-\frac{\Delta g_{f,j}^0}{RT}\right)$$

where

j is the subscript representing the pseudoisomers
N_{isomer} is the total number of the pseudoisomers

Some of the glycolytic enzymes, such as hexokinase and phosphofructokinase, are activated by binding ions, and then the enzyme–substrate complex is formed. Therefore, concentration of the ions like Mg^{2+}, K^+, Na^+, and Ca^{2+} changes the thermodynamic state; hence, the enthalpy and Gibbs free energy values of the metabolite change. The effect of the Mg^{2+} ion on standard transformed Gibbs energy of formation is calculated as

$$\Delta g_{f,i,Mg}^T = -n_{i,Mg}\left[\Delta g_{f,Mg2+}^0 + 2.303\,RT\log\left(c_{Mg2+}\right)\right]$$

Similar equations are established also for K^+, Na^+, and Ca^{2+}. The general formula to calculate standard transformed Gibbs energy of formation is

$$\Delta g_{f,i}^T = \frac{T}{T_0}\Delta g_{f,i}^0 + \left(1-\frac{T}{T_0}\right)\Delta h_{f,i}^0 - \frac{RT\alpha(T)I^{1/2}}{1+BI^{1/2}}\left(z_i^2 - n_{H,i}\right)$$

$$-\sum_{j=1}^{j=N}n_{M,i}\Delta g_{f,M,j^+}^0 + 2.303\sum_{j=1}^{j=N}n_{M,i}RTpM_j$$

where $M_j (1 \leq j \leq N)$ represents all the ions present in the system. At 298.15 K, values of the standard Gibbs free energy of formation are $\Delta g_f^0 (H^+) = 0$ kJ/mol, $\Delta g_f^0 (Mg^{2+}) = -455.11$ kJ/mol, $\Delta g_f^0 (K^+) = -283.45$ kJ/mol, $\Delta g_f^0 (Na^+) = -262.06$ kJ/mol, and $\Delta g_f^0 (Ca^{2+}) = -553.91$ kJ/mol (Weast and Astle, 1985).

The chemical exergy of the metabolite i under the physiological conditions can be calculated as a sum of its transformed Gibbs free energy of formation and the standard chemical exergy of the elements found in the structure of this metabolite:

$$ex_{ch,i} = \Delta g_{f,i}^T + \sum_{k=1}^{N_{elements}} \nu_k ex_{ch,k}^0$$

where

$ex_{ch,k}^0$ is the standard chemical exergy of the kth element
$N_{elements}$ is the total number of elements that make up the metabolite

Since mixing is an irreversible process, exergy of a metabolite i decreases as its concentration decreases. The decrease in a metabolites exergy due to mixing can be calculated as

$$ex_i = RT \ln \left(\frac{c_i}{c_t} \right)$$

The total concentration of the cellular components is $c_t = 1$.

Thus, the total exergy of the metabolite i is

$$ex_i (T, I, pH, c_{i,system}) = \underbrace{\frac{T}{T_0} \Delta g_{f,i}^0 + \left(1 - \frac{T}{T_0} \right) \Delta h_{f,i}^0}_{\text{Effect of temperature}}$$

$$\underbrace{- \frac{RT\alpha(T) I^{1/2}}{1 + BI^{1/2}} \left(z_i^2 - n_i \right)}_{\text{Effect of ionic strength}}$$

$$\underbrace{- n_{H,i} \Delta g_{f,H^+}^0 + 2.303 n_{H,i} RTpH}_{\text{Effect of pH}}$$

$$\underbrace{+ \sum_{k=1}^{N_{elements}} \nu_k ex_{ch,k}^0}_{\text{Chemical exergy of elements}}$$

$$\underbrace{+ RT \ln(c_{i,system})}_{\text{Effect of dilution}}$$

Example 2.22: Calculation of the Standard Gibbs Free Energy of ATP in a Dilute Aqueous Solution

Under the physiological range of pH, ATP appears in a cell as ATP^{4-}, ATP^{3-}, and ATP^{2-}. Thus, the concentration of ATP is

$$c_{ATP} = c_{ATP^{4-}} + c_{HATP^{3-}} + c_{H_2ATP^{2-}}$$

The standard Gibbs free energy of formation is calculated in the following expression (Table 2.23.1):

$$\Delta g^0_{f,ATP} = -RT \ln \left(\exp\left(-\frac{\Delta g^0_{ATP^{4-}}}{RT} \right) + \exp\left(-\frac{\Delta g^0_{ATP^{3-}}}{RT} \right) + \exp\left(-\frac{\Delta g^0_{ATP^{2-}}}{RT} \right) \right)$$

$$= -2838.19 \text{ kJ/mol}$$

Example 2.23: Transformation of the Gibbs Free Energy of Formation of Glucose from the Standard Conditions to the Physiological Conditions Prevailing in a Cell

Under the standard conditions (temperature = T_0, concentration = 1 M, $p = 1$ atm), the Gibbs free energy of formation Δg^0_f is given as

$$\Delta g^0_{f,glucose}(298.15 \text{ K}, 1 \text{ atm}) = -915.90 \text{ kJ/mol}$$

The physiological conditions prevailing in a cell are $T = 310.15$ K, pH = 7, and ionic strength $I = 0.18$ M. Genç et al. (2013a) calculated the difference between the Gibbs free energies of formation of glucose at the standard physiological condition (Δg^T) and the standard conditions by employing the following expression:

$$\Delta g^T = \frac{T}{T_0} \Delta g^0_{f,glucose} + \left(1 - \frac{T}{T_0}\right) \Delta h^0_{f,glucose}$$

$$- \frac{RT\alpha(T)I^{1/2}}{1 + BI^{1/2}} \left(z_i^2 - n_{H,i}\right) - n_{H,i}\Delta g^0_{f,H^+} + 2.303 n_H RTpH$$

EXAMPLE TABLE 2.23.1 Data for Dilute Aqueous Solutions at 298.15 K and Zero Ionic Strength

	Δg^0_f (kJ/mol)	Δh^0_f (kJ/mol)	z_i	$n_{H,i}$
ATP^{4-}	−2768.1	−3619.21	−4	12
ATP^{3-}	−2811.48	−3612.91	−3	13
ATP^{2-}	−2838.18	−3627.91	−2	14

Source: Adapted from Alberty, R.A., *Thermodynamics of Biochemical Reactions*, John Wiley & Sons, Hoboken, NJ, 2003.

The effect of temperature to Gibbs free energy is employed. The following expression shows thermodynamic values of glucose under the standard conditions (temperature $(T_0) = 298.15$ K in a dilute aqueous solution at zero ionic strength, $p = 1$ atm), the Gibbs free energy of formation of glucose $\left(\Delta g^0_{f,\,glucose} = -915.90 \text{ kJ/mol}\right)$, and enthalpy of formation $\left(\Delta h^0_{f,\,glucose} = -1262.19 \text{ kJ/mol}\right)$ taken from Alberty (2003):

$$= \underbrace{\frac{T}{T_0}\Delta g^0_{f,\,glucose} + \left(1 - \frac{T}{T_0}\right)\Delta h^0_{f,\,glucose}}_{\text{effect of temperature}}$$

$$= \frac{298.15}{310.15}(-915.90) + \left(1 - \frac{298.15}{310.15}\right)(-1262.19)$$

$$= -901.96$$

The coefficient $RT\alpha(T)$ (kJ mol$^{-3/2}$ kg$^{1/2}$) in the equation for the transformed Gibbs free energy of formation of a species is estimated:

$$RT\alpha(T) = 9.2043 \times 10^{-3}T - 1.28467 \times 10^{-5}T^2 + 4.95199 \times 10^{-8}T^3$$

The effect of ionic strength ($I = 0.18$) is employed. Where $z_i w$ is the valence of glucose and n_H is the number of moles of the H$^+$ atoms, respectively, B is an empirical constant (1.6 L$^{1/2}$ mol$^{-1/2}$) and $\Delta g^0_{f,\,H^+}$ is the Gibbs free energy of formation of the H$^+$ atom:

$$= \underbrace{\frac{RT\alpha(T)I^{1/2}}{1 + BI^{1/2}}\left(z^2_{glucose} - n_{H,\,glucose}\right)}_{\text{effect of ionic strength}}$$

$$= \frac{3.09651(0.18)^{1/2}}{1 + B(0.18)^{1/2}}\left(0^2 - 12\right)$$

$$= -9.39$$

The effect of pH (pH = 7) is employed where $\Delta g^0_{f,\,H^+}$ is the Gibbs free energy of formation of the H$^+$ atom, n_H is the number of moles of the H$^+$ atoms, and R is the gas constant:

$$= \underbrace{-n_{H,\,glucose}\Delta g^0_{f,\,H^+} + 2.303 n_{H,\,glucose}RTpH}_{\text{effect of pH}}$$

$$= \underbrace{-12(0) + (2.303)(12)\,(8.31451/1000)(310.15)(\log(10)^{-7})}_{\text{effect of pH}}$$

$$= \underset{\text{effect of pH}}{-498.774}$$

After substituting the numbers in this expression, they obtained

$$\Delta g^T_{f,\,glucose} = -393.80 \text{ kJ/mol}$$

Example 2.24: Calculation of the Exergy of Glucose under the Physiological Conditions Prevailing in a Cell

The standard chemical exergy of glucose $\left(ex_{ch,glucose}^0\right)$ is calculated from its elements as

$$6C + 6H_2 + 3O_2 \rightarrow C_6H_{12}O_6$$

$$ex_{ch,glucose}^0 = 6ex_{ch,C}^0 + 6ex_{ch,H_2}^0 + 3ex_{ch,O_2}^0$$

$$ex_{ch,glucose}^0 = (6)(410.26) + 0 + (3)(3.97)$$

$$ex_{ch,glucose}^0 = 2473.47 \text{ kJ/mol}$$

When concentration of glucose in a cell is 1.2 mM and that of the solution under standard conditions is 1 M, the concentration effect will be

$$ex_{conc} = RT \ln\left(\frac{c_{glucose,system}}{c_{standard}}\right)$$

$$ex_{conc} = (8.31451\frac{J}{mol\ K}1000)(310.25\ K)\ln\left(\frac{1.2}{1}\right)$$

$$ex_{conc} = -17.34 \text{ kJ/mol}$$

We will add up all the terms to calculate the exergy of glucose in the cell as

$$ex_{glucose} = \Delta g_{f,glucose}^T + ex_{ch}^0 + ex_{conc}$$

$$ex_{glucose} = -393.80 + 2473.47 + -17.34 = 2062.33 \text{ kJ/mol}$$

Example 2.25: Transformation of the Enthalpy of Formation of Glucose from the Standard Conditions to the Physiological Conditions Prevailing in a Cell

Under the standard conditions (temperature = T_0, concentration = 1 M, $p = 1$ atm), the enthalpy of formation Δh_f^0 of glucose is given as

$$\Delta h_{f,glucose}^0(298.15\ K, 1\ atm) = -1262.19 \text{ kJ/mol}$$

The physiological conditions prevailing in a cell are $T = 310.15$ K, pH = 7, and ionic strength $I = 0.18$ M. The transformed enthalpy of formation under the physiological conditions for glucose is defined as

$$\Delta h_{f,glucose}^T = \Delta h_{f,glucose}^0 + \frac{RT^2\beta(T)I^{1/2}}{1+BI^{1/2}}\left(z_{glucose}^2 - n_{H,glucose}\right)$$

where

$$RT^2\beta(T) = -1.28466T^2 10^{-5} + 9.90399T^3 10^{-8}$$

$$= -1.28466(310.15)^2 10^{-5} + 9.90399(310.15)^3 10^{-8}$$

$$= 1.71903$$

In this equation, z_i is the valence of glucose, n_H is the number of moles of the H^+ atoms, B is an empirical constant (1.6 $L^{1/2}$ mol$^{-1/2}$), and $\Delta h^0_{f,H^+}$ is the standard enthalpy of formation of the H^+ atom:

$$= \underbrace{\frac{RT^2(\frac{\partial \alpha}{\partial T})_P I^{1/2}}{1 + BI^{1/2}}}_{\text{effect of ionic strength}} \left(z^2_{glucose} - n_{H, glucose} \right)$$

$$= \frac{1.71903(0.18)^{1/2}}{1 + 1.6(0.18)^{1/2}} \left(0^2 - 12 \right)$$

$$= -5.2131$$

The transformed enthalpy of formation is

$$\Delta h^T_{f, glucose} = -1262.19 - 5.21$$

$$= -1267.40 \text{ kJ/mol}$$

Raoult's law and Henry's law and a correlation (Antoine's equation) are used frequently in conjunction with the phase equilibria in mixtures. Schematic description of the vapor phase, which is in equilibrium with its vapor, is given in Figure 2.5.

At equilibrium, total vapor pressure of the vapor mixture is calculated as

$$p_{Total} = \sum_{i=1}^{m} x_i p^{sat}_i$$

where p^{sat}_i and x_i are the vapor pressure of the pure component i and its mole fraction in the liquid mixture, respectively.

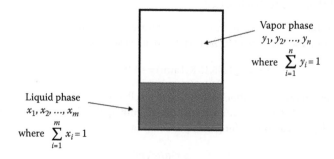

FIGURE 2.5   Schematic description of vapor and liquid phases at equilibrium.

It should be noticed that p_i^{sat} changes with temperature and must be calculated at the same temperature as p_{Total}. We may use the Antoine equation to predict the saturated vapor pressure of pure components as

$$\ln\left(p^{sat}\right) = A - \frac{B}{T+C}$$

where

A, B, and C are constants (Table 2.11)

T is temperature in °C

p^{sat} is the saturation vapor pressure in kPa

When the nth component of the system is gas, then Henry's law is used instead of Raoult's law:

$$p_{Total} = \sum_{i=1}^{n-1} x_i p^{sat}{}_i + Hx_n$$

where, H is the Henry's constant (Table 2.12). At room temperature, a gas will always be in the gaseous state, however vapor may co-exist in both liquid and vapor phases.

TABLE 2.11 Constants of the Antoine Equation for Some Biologically Important Chemicals $\ln(p^{sat}) = A-(B/T + C)$ (temperature T is in °C)

	A	B	C	Temperature Range (°C)
Acetic acid	15.0717	3580.80	224.650	24–142
Ethanol	16.8958	3795.17	230.918	3–96
Methanol	16.5785	3638.27	239.500	−11–83
Water	16.3872	3885.70	230.170	0–200

Source: Smith, J.M. et al., *Introduction to Chemical Engineering Thermodynamics*, 7th ed., McGraw Hill, New York, 2005.

TABLE 2.12 Henry's Law Constants for Solubility of Gases in Water at Low to Moderate Pressures H (kPa)

	17°C	27°C	37°C	47°C	57°C	67°C
Air	62×10^5	74×10^5	84×10^5	92×10^5	99×10^5	104×10^5
CO_2	12.8×10^4	17.1×10^4	21.7×10^5	27.2×10^5	32.2×10^5	
O_2	38×10^5	45×10^5	52×10^5	57×10^5	61×10^5	65×10^5
N_2	76×10^5	89×10^5	101×10^5	110×10^5	118×10^5	124×10^5

Source: Adapted from Çengel, Y.A. and Boles, M.A., *Thermodynamics: An Engineering Approach*, 7th ed., McGraw-Hill, Boston, MA, 2007.

Note: Henry's law is expressed as $P = Hx$, where partial pressure P is in Pa and x is in mole fractions.

Example 2.26: Calculation of the Composition of the Vapor Phase in Equilibrium with the Liquid Phase of Known Composition

An alcoholic drink contains 12 mol % ethanol in the liquid phase, and the rest may be assumed as water (Example Figure 2.26.1). Calculate the mole fractions of ethanol and water in the vapor phase at 8°C.

Ethanol–water mixture obeys Raoult's law. We may use the Antoine equation to predict the saturated vapor pressure of ethanol and water as

$$\ln\left(p^{sat}\right) = A - \frac{B}{T+C}$$

where

A, B, and C are constants

T is temperature in °C

p^{sat} is the saturation vapor pressure in kPa

We will read from Table 2.11 that $A = 16.3872$, $B = 3885.70$, and $C = 230.170$ for water. After substituting the constants in the Antoine equation, we will calculate $p_{water}^{sat} = 1.075$ kPa. Antoine equation constants for ethanol are $A = 16.8958$, $B = 3795.17$, and $C = 230.918$, and then we will calculate $p_{ethanol}^{sat} = 2.748$ kPa. Partial pressures of water vapor and ethanol in the vapor phase are $p_{water} = x_{water} p_{water}^{sat} = (0.88)(1.075) = 0.94$ kPa and $p_{ethanol} = x_{ethanol} p_{ethanol}^{sat} = (0.12)(2.748) = 0.33$ kPa, and then we may calculate fractions of water and ethanol in the vapor phase as

$$y_{water} = \frac{p_{water}}{p_{Total}} = \frac{(0.94\ \text{kPa})}{(0.94+0.33)\ \text{kPa}} = 0.74$$

and

$$y_{ethanol} = \frac{p_{ethanol}}{p_{Total}} = \frac{(0.33\ \text{kPa})}{(0.94+0.33)\ \text{kPa}} = 0.26$$

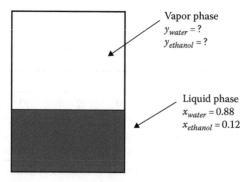

EXAMPLE FIGURE 2.26.1   Schematic description of the ethanol and water mixture vapor and liquid phases at equilibrium.

(b) Solve this problem once more by assuming that the liquid phase contains 14% inert solids, for example, sugars and chemicals, providing aroma and color.

$$x_{water} = (0.88)\left(\frac{100-14}{100}\right) = 0.76$$

and

$$x_{ethanol} = (0.12)\left(\frac{100-14}{100}\right) = 0.10$$

and then we may calculate the saturation pressure of each component as

$$p_{water} = x_{water} \ p_{water}^{sat} = (0.76)(1.075) = 0.82 \text{ kPa}$$

$$p_{ethanol} = x_{ethanol} \ p_{ethanol}^{sat} = (0.10)(2.748) = 0.28 \text{ kPa}$$

and then we may calculate fractions of water and ethanol in the vapor phase as

$$y_{water} = \frac{p_{water}}{p_{Total}} = \frac{(0.82 \text{ kPa})}{(0.82+0.28) \text{ kPa}} = 0.75$$

and

$$y_{ethanol} = \frac{p_{ethanol}}{p_{Total}} = \frac{(0.28 \text{ kPa})}{(0.82+0.28) \text{ kPa}} = 0.25$$

Example 2.27: Calculation of the Composition of the Phases in a Carbonated Beverage with Henry's Law

(a) Carbonated beverage contains CO_2(H = 9.9 10^7 Pa) dissolved in water. Determine composition of the vapor and liquid phases in a sealed container at 10°C.

We will make trial-and-error solution, after assuming that $x_{CO_2} = 0.01$. In the liquid phase, we have $x_{water} + x_{CO_2} = 1$; a similar equation is valid in the vapor phase as $y_{water} + y_{CO_2} = 1$. We may use the Antoine equation to predict the saturated vapor pressure of water as

$$\ln\left(p^{sat}\right) = A - \frac{B}{T+C}$$

where
A, B, and C are constants
T is temperature in °C
p^{sat} is the saturation vapor pressure in kPa

We will read from Table 2.11 that $A = 16.3872$, $B = 3885.70$, and $C = 230.170$. After substituting the constants in the Antoine equation, we will calculate $p_{water}^{sat} = 1.23$ kPa.

In the vapor phase, we have

$$p_{Total} = x_{water} p_{water}^{sat} + x_{CO_2} H = (0.99)(1.23) + (0.01)(990 \times 10^2) = 991.2 \text{ kPa}.$$

Since we have $y_{water} p_{Total} = x_{water} p_{water}^{sat}$, we may calculate y_{water} as

$$y_{water} = \frac{x_{water} p_{water}^{sat}}{p_{Total}} = \frac{(0.99)(1.23 \text{ kPa})}{(991.2 \text{ kPa})} = 0.0012$$

We also have $y_{CO_2} p_{Total} = p_{CO_2} = x_{CO_2} H_{CO_2}$, and then we may calculate y_{CO_2} as

$$y_{CO_2} = \frac{x_{CO_2} H_{CO_2}}{p_{Total}} = \frac{(0.01)(990 \times 10^2 \text{ kPa})}{(991.2 \text{ kPa})} = 0.998$$

We will check to see $y_{water} + y_{CO_2} = 1$. After substituting $y_{water} = 0.0012$ and $y_{CO_2} = 0.9988$ into this equation, we will get $0.0012 + 0.9988 = 1$, which confirms that the initial assumption $x_{CO_2} = 0.01$ is correct.

(b) Solve this problem once more by assuming that the liquid phase contains 10% sugar.

Total mass fraction of the species in the solution is $x_{H_2O} + x_{CO_2} + x_{sugar} = 1$, where $x_{sugar} = 0.10$, $x_{H_2O} = (0.90)(0.99) = 0.891$, and $x_{CO_2} = (0.90)(0.01) = 0.009$. We also have $y_{H_2O} + y_{CO_2} = 1$. Henry's law requires $p_{Total} = x_{water} p_{water}^{sat} + x_{CO_2} H = 892.1$ kPa. Therefore, we may calculate

$$y_{H_2O} = \frac{x_{H_2O} p_{H_2O}}{p_{Total}} = \frac{(0.89)(990 \times 1.23 \text{ kPa})}{(892.1 \text{ kPa})} = 0.0012$$

and

$$y_{CO_2} = 1 - 0.0012 = 0.9988.$$

Biological liquids like cellular plasma, blood, and milk can be considered as liquid mixtures. In order to determine the thermodynamic properties of biological liquids, we need to determine the composition, and then mixture properties can be calculated based on the properties of the species.

Blood is comprised of plasma and cells, which are platelets, red blood cells, and white blood cells. Forty-five percent of blood is cells and 50% is plasma. Plasma is the liquid component of the blood. Ninety percent of plasma is water and 10% is dissolved minerals that are nutrients (proteins, salts, and glucose), wastes (e.g., urea, creatinine), hormones, and enzymes.

Milk is actually a dilute aqueous solution. For example, raw cow milk contains 88.4% water. The rest of the components are 3.3% protein, 3.6% fat, and 4.7% carbohydrates. These percentages may vary for milk from different species (cow, sheep, human, etc.) and even the same species living under different environmental conditions. However, the main ingredient remains water, and therefore, milk can be treated as a dilute aqueous solution, which is ideal.

Numerous different types of proteins, fats, and carbohydrates coexist in milk. For simplicity (and because of lack of information), let us assume that all proteins are polymers of alanine, all fats are based on oleic acid, and all milk carbohydrates are beta-lactose. Now, our job is reduced to determine the thermodynamic properties of alanine, oleic acid fat, and beta-lactose. If data are available in the literature, then this is usually the most reliable source. For example, the enthalpy of formation and the Gibbs free energy of beta-lactose are given as 15.06 MJ/kg and 17.49 MJ/kg, respectively (Szargut et al., 1988).

2.8 Questions for Discussion

Q2.1 By referring to the bond energies, estimate the enthalpy of formation of aqueous fructose (Question Figure 2.1.1) with the reaction

$$6CO_2 + 6H_2O \rightarrow C_6H_{12}O_6 + 6O_2$$

Q2.2 Estimate the total bond energy of aqueous adenine (Question Figure 2.2.1). What is the difference between the bond energy you have calculated with the enthalpy of formation of aqueous adenine?
Additional data: Bond energy of $C=N$ is 615 kJ/mol and bond energy of $C-N$ is 305 kJ/mol.

Q2.3 Phase diagram of carbon dioxide is given in Question Figure 2.3.1.
(a) Determine the T and P range, where CO_2 is a supercritical fluid.
(b) Determine the critical and the triple points.
(c) At 250 K, at what pressure solid and liquid CO_2 coexist?
(d) A fermentation process is carried on at $T = 25°C$ and $p = 10 \times 10^5$ Pa. How many cubic meters of CO_2 will be released when 1000 mol of O_2/day is supplied, when respiratory quotient is 0.90?
Data: $R = 8.32$ kPa m^3/kmol K, 1×10^5 Pa = 1 bar = 1 atm, respiratory quotient = (moles of CO_2 released)/(moles of O_2 provided).

QUESTION FIGURE 2.1.1   Chemical structure of fructose.

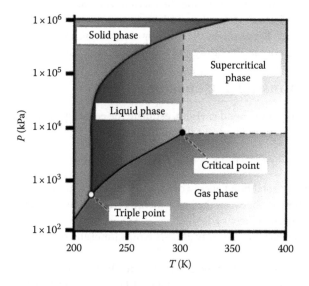

QUESTION FIGURE 2.2.1   Chemical structure of adenine.

QUESTION FIGURE 2.3.1   Phase diagram of carbon dioxide.

(e) Fermentation broth consists of 20% nonvolatile solids and the rest is water. What will be the vapor pressure of water on the fermentation broth?

(f) If the exhaust stream consists of 5% O_2, and the rest is water vapor and CO_2($T = 25°C$ and $p = 3 \times 10^5$ Pa), calculate the fraction of the carbon dioxide in the fermentation broth. (You may obtain the data from the previous examples.)

Q2.4. (a) By referring to Table 2.12, calculate the amounts of dissolved nitrogen, oxygen, and carbon dioxide in a lake (assume pure water) at 27°C under atmospheric pressure in a location where air is made of 79% nitrogen, 20.5% oxygen, and 0.5% carbon dioxide.

(b) What is the O_2-to-N_2 ratio of liquid in the lake at 17°C?

(c) What is the ratio of the dissolved O_2 in liquid water at 17°C and 27°C?

(d) Would the fish survive better in the lake at 17°C or 27°C? Explain by referring to the results of your calculations.

Q2.5 *Escherichia coli* cell is cylindrical in shape with a diameter of 1.0 μm and a length of 2.6 μm (Hahn et al., 2002). Its UCF is $CH_{1.59}O_{0.374}N_{0.263}P_{0.023}S_{0.006}$ (Battley, 1993), and its density is 1.105 g/mL (Martinez-Salas et al., 1981). The enthalpy of combustion and the entropy of formation are reported to be $\Delta h_{c,cell} = -5.3 \times 10^{-11}$ J/cell (Cordier et al., 1987) and $\Delta s_{f,cell} = -4.0 \times 10^{-13}$ J/cell K (Battley, 1993), respectively. Estimate its entropy and Gibbs free energy of formation.

Q2.6 Methanol vapor is a real gas. Calculate the increase of the specific volume of methanol, under 20 atm pressure when we increase temperature of methanol from 210°C to 290°C.

Q2.7 When the peptide $C_{4.793}H_{7.669}O_{1.396}N_{1.370}S_{0.046}$ is burned in a calorimeter in accordance with the reaction

$$C_{4.793}H_{7.669}O_{1.396}N_{1.370}S_{0.046} + 7.805\ O_2 \rightarrow 4.793\ CO_2$$

$$+\ 3.1265\ H_2O\ +\ 3.1270\ NO_3^-$$

$$+\ 0.046\ SO_4^{2-}\ +\ 1.416\ H^+$$

Q2.8 (a) An average human weighing 70 kg has a blood volume of 5.00 L and body temperature of 37°C. After assuming that the blood has the same properties as water, calculate the number of moles of N_2 that may dissolve in the bloodstream.
Data: $x_{N_2} = 0.78$ and $x_{O_2} = 0.22$ in the air; the person is at the sea level and under the atmospheric pressure of 1 atm (1×10^5 Pa).
(b) The maximum extent of the depth with air-based scuba equipment is about 150 m. Calculate the number of moles of N_2 dissolved in the bloodstream under these conditions. Pressure at 150 m depth may be calculated as $P = \rho h g$
Data: $\rho = 1.05$ kg/cm³, $h = 150$ m, $g = 981$ cm/s²
Conversion factor: 1 Pa = 1 kg/m s².
(c) If the diver goes from 150 m of depth water to the surface in a very short time, what would be the volume of the nitrogen (real gas) released?
(d) If all the nitrogen will be released as gas at 150 m under the sea level, what would be the volume of the nitrogen (real gas) released?

3

Energy Conversion Systems

3.1 Nonbiological Fundamentals

Thermodynamics has emerged as a science in an attempt to understand energy conversion principles in mechanical devices. In today's modern society, mechanical energy conversion systems are omnipresent. Turbomachinery such as turbines, pumps, and compressors enable us to transfer energy between mechanical systems and fluids. Generators convert mechanical work into electricity. Electricity is consumed in mechanical or electrical devices. Mechanical energy conversion systems can be categorized as work-producing and work-consuming systems (Figure 3.1).

In the following, we will discuss energy cycles, which are used frequently for engineering purposes. Work-producing systems make use of renewable or nonrenewable energy reservoirs to produce useful work. In conventional work-producing systems, such as thermal power plants, the first step is to burn a fuel (coal, diesel oil, hydrogen, etc.) to generate heat. The enthalpy of combustion, Δh_c, is fed into a *heat engine*, which converts part of the inflowing heat into useful work. According to the energy balance, if the heat engine operates at steady state, then the enthalpy of combustion has to be equal to the sum of the heat released and work produced in the cycle. In conventional power plants, work production occurs via rotating the shaft of a turbine.

The operation principle of a work-producing system is very similar to the work production in biological systems. In the cellular energy metabolism, the source of high energetic chemical is nutrients instead of fuels. Nutrients are burned, and the enthalpy of this reaction is equal to the sum of the metabolic heat released and the biological work performed. Work production in biological systems can occur through different mechanisms, which will be discussed and exemplified in Chapter 4. Biothermodynamics makes use of the theories that are developed for mechanical devices to understand biological systems (Figure 3.2).

3.1.1 Heat Engines

A heat engine is defined as a cyclic device that produces useful work from heat. Heat is a dissipative form of energy; therefore, all of the inflowing heat cannot be converted into work. Part of the inflowing heat is rejected to the environment as waste heat, as shown in Figure 3.3.

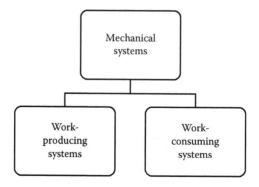

FIGURE 3.1   Classification of mechanical systems based on their thermodynamic aspects.

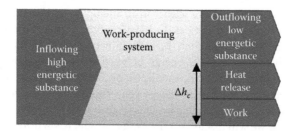

FIGURE 3.2   Energy flow in a work-producing system.

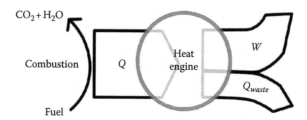

FIGURE 3.3   Energy flow in a heat engine.

A heat engine undergoes four processes:

1. Heat addition
2. Expansion
3. Heat removal
4. Compression

This sequence is called a cycle. A thermodynamic cycle is completed when the state of the working fluid is the same as the initial state. A mechanical cycle is completed when the mechanism returns to its original position. A cycle is referred to as a closed cycle, if the working fluid remains inside the engine throughout its cycle. A cycle can

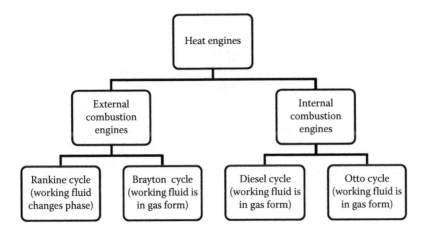

FIGURE 3.4   Classification of heat engines.

also be open, if there is a mass transfer through the system boundaries. For example, in car engines, air is taken into cylinders at the beginning of each cycle, and at the end of a cycle, air is thrown out through the exhaust ducts. And the thermodynamic and mechanic cycle begins again with fresh air.

A heat engine has to involve devices that make heat addition, expansion, heat removal, and compression possible. For example, in a car engine, all of the four processes occur inside piston-cylinder devices. Heat addition is actually the combustion of the fuel inside the cylinders. Expansion and compression processes occur due to the motion of the piston, and heat removal step is actually throwing the air and burned fuel mixture out of the cylinder. In a power generation facility, a boiler may be used to add heat to the heat engine, and usually turbines are employed to expand the working fluid. Heat exchangers are used to cool down the fluid, and pumps are employed to compress the fluid back into the boiler.

Heat engines are categorized based on the location of the combustion and the type of the working fluid (Figure 3.4).

A thorough discussion of heat engines can be found in mechanical engineering thermodynamics textbooks, such as Cengel and Boles (2011) and Bejan (2006). Furthermore, there is a vast number of books specialized on different types of heat engines. Here, we will limit our discussion on the fundamentals of heat engines, which will be visualized by analyzing Rankine cycles in detail. Important thermodynamic concepts like the Carnot cycle and first and second law efficiencies will be defined. The last section of this chapter is devoted to analogies between the mechanical and biological systems.

3.1.2 Rankine Cycle

A schematic drawing of a simple Rankine cycle is given in Figure 3.5.

A simple Rankine cycle includes four mechanical devices:

1. *Pump*: The working fluid, which is in liquid form and at a low pressure (thermodynamic state 1), is pressurized and fed into the boiler. In an ideal Rankine cycle, pump operates isentropically.

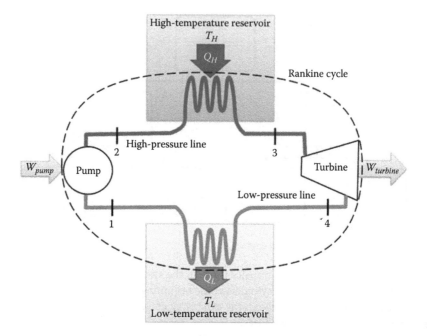

FIGURE 3.5 Schematic drawing of a Rankine cycle. Since Rankine cycles operate at steady state, mass balance becomes $\dot{m}_1 = \dot{m}_2 = \dot{m}_3 = \dot{m}_4 = \dot{m}$.

2. *Boiler*: As the pressurized working fluid, which is in liquid form (thermodynamic state 2), flows through the boiler, it is vaporized. At the outlet of the boiler, the working fluid is in superheated vapor form (thermodynamic state 3). The heat added to the fluid in the boiler is referred to as Q_H. In an ideal Rankine cycle, there is no friction loss inside the boiler, that is, the boiler operates isobarically.

3. *Turbine*: The working fluid at a high pressure and a high temperature flows into the turbine. In the turbine, part of the inflowing energy of the fluid is transformed into shaft work. In an ideal turbine, fluid elements follow the shape of the turbine blades perfectly. There are no frictional losses, no turbulence, no noise, and no vibration. An ideal turbine operates isentropically.

4. *Condenser*: The working fluid flows from the turbine into the condenser in vapor form (thermodynamic state 4). Condenser changes the phase of the fluid to liquid by extracting heat. The heat rejected from the fluid in the condenser is referred to as Q_L. An ideal condenser operates isobarically.

The *high-temperature heat reservoir* that transfers Q_H to the boiler and the *low-temperature heat reservoir* that extracts Q_L from the condenser are not taken into the system boundaries that define the Rankine cycle. Thermal power plants operating as Rankine cycles employ large combustion chambers, where the fuel (coal, natural gas, etc.) is fed at a steady rate.

Geometrical design of the combustion chamber may change from application to application but usually, the design is made to assure that the flame remains at the center of the chamber to protect the chamber from high thermal stress. A pipeline surrounds the flame. Heat is transferred from the flame to the water inside the pipeline. Under normal operating conditions, combustion occurs steadily and the flame temperature remains constant.

In an effort to decrease the dependence on fossil energy resources, renewable energy sources began to replace the combustion process. Geothermal resources, waste heat of cement plants, and oils heated via solar energy serve as renewable high-temperature heat reservoirs. Examples of low-temperature heat reservoirs are atmospheric air and seawater. A medium can be defined as heat reservoir, only if its temperature remains constant, independent of the amount of the heat transferred.

Energy, entropy, and exergy balances for steady flow devices are

$$\sum_{in} \dot{m}\left(h + \frac{v^2}{2} + g_a h_l\right) - \sum_{out} \dot{m}\left(h + \frac{v^2}{2} + g_a h_l\right) + \dot{Q} - \dot{W} = 0$$

$$\sum_{in} \dot{m}s - \sum_{out} \dot{m}s + \sum_{j} \frac{\dot{Q}_j}{T_{b,j}} + \dot{S}_{gen} = 0$$

$$\sum_{in} \dot{m}ex - \sum_{out} \dot{m}ex + \sum_{j}\left(1 - \frac{T_0}{T_{b,j}}\right)\dot{Q}_j + \dot{Ex}_{destr} = 0$$

For the devices in the Rankine cycle, this equation can be simplified since kinetic and potential energy of the fluid are negligibly low compared to its enthalpy and there is only one inlet and one outlet:

$$h_{in} - h_{out} + q - w = 0$$

$$s_{in} - s_{out} + \frac{q}{T} + s_{gen} = 0$$

In the pump, fluid is incompressible, and there is no heat transfer. Thus,

$$w_{pump} = h_2 - h_1 = v_1(p_2 - p_1)$$

$$s_{gen,pump} = s_2 - s_1$$

For the boiler,

$$q_H = h_3 - h_2$$

$$s_{gen,boiler} = s_3 - s_2 - \frac{q_H}{T_H}$$

Note that here q_H is transferred at the temperature T_H. In other words, the system boundary is in the thermal reservoir, and boundary temperature remains at the constant temperature T_H.

For the turbine,

$$w_{turbine} = h_4 - h_3$$

$$s_{gen,turbine} = s_4 - s_3$$

For the condenser,

$$q_L = h_4 - h_1$$

$$s_{gen,condenser} = s_1 - s_4 + \frac{q_L}{T_L}$$

Note that for saturated liquid at temperature T and pressure p, the thermodynamic properties can be approximated as only a function of temperature. So, at state one, thermodynamic properties are

$$h_1 = h_{f@T_1}$$

$$s_1 = s_{f@T_1}$$

$$v_1 = v_{f@T_1}$$

Note that the thermodynamic properties at the turbine inlet are determined from the superheated water vapor table for specified temperature and pressure values of turbine inlet:

$$h_3 = h_{@T_3, p_3}$$

$$s_3 = s_{@T_3, p_3}$$

The net work generated through a cycle is

$$w_{net} = w_{turbine} - w_{pump}$$

And the net heat absorbed is

$$q_{net} = q_H - q_L$$

The thermal efficiency of the cycle can be calculated as

$$\eta_{th} = \frac{w_{net}}{q_{in}}$$

This efficiency is also referred to as the first law efficiency or the energy efficiency.

These equations show that the heat q_H is the driving force for the cycle and absorbed by the working fluid as shown in the energy flow diagram in Figure 3.6. In this energy flow diagram, the blue lines show how the enthalpy of the working fluid changes as it flows through the devices in the Rankine cycle. Boiler increases the enthalpy of the fluid and turbine extracts work from the inflowing enthalpy. The final two processes (condensing and pumping) are performed to return the fluid's thermodynamic state back to the state at the inlet of the boiler.

One of the frequently asked questions is why do we need the condenser. Theoretically, a condenser can be taken out of the cycle, then fresh fluid has to be fed into the pump, and the vapor (which is at a relatively high temperature, as will be shown in the following examples) at the outlet of the turbine is released to the environment. However, the energetic and environmental cost of such a modification is so high that it justifies the investment and operation costs of the addition of a condenser into the cycle.

As the energy flow diagram shows, the working fluid is the *energy carrier* in this cycle. The choice of the working fluid depends on the operating conditions of the Rankine cycle. In a conventional thermal power plant, where q_H is generated via

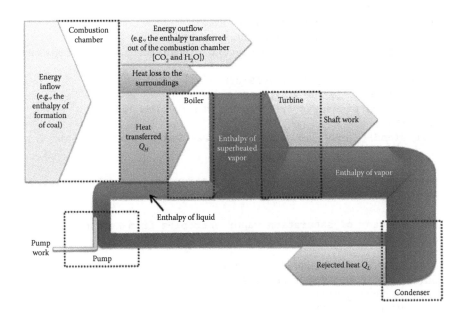

FIGURE 3.6 Detailed energy flow diagram.

combustion of a fossil fuel, the temperature in the combustion chamber can be as high as 1000°C. With this T_H, water can be superheated at pressures as high as a few MPa's. But if the temperature of the thermal reservoir is low, for example, if a geothermal reservoir at $T_H = 180°C$ is used to extract q_H, then water cannot be superheated at high pressures. Under such conditions, organic fluids such as chlorofluorocarbons (refrigerants) are employed, because these have lower critical temperature and pressure. In order to visualize the effect of the thermodynamic properties of the working fluid, T–s (temperature vs. entropy) diagrams can be studied. The temperature versus entropy diagram (Figure 3.3) of water is plotted by using the MATLAB m-function XSteam (Holmgren, 2007; Kitchin, 2011) in MATLAB code 3.1:

MATLAB CODE 3.1

```
Command Line

clear all
close all

% enter the data
T = linspace(0,500,100); % temperatures
PinBar= [0.01 0.05 0.1 0.5 1 4 10 40 100 250 500 1000]; % Pressure
  in bar

hold all

for P = PinBar
% get a vector of entropies from Xsteam (in bar)
  S = arrayfun(@(t) XSteam('s_PT',P,t),T);
  % adjust figure position to prevent misplacement of the pressure
  labels
  set(gca,'Position',[0.13 0.09 0.8 0.7])
  plot(S,T,'k-','LineWidth',1.85)
% convert the pressures from bar to kPa for the label
  PSI=P*100;
  text(S(end),T(end),sprintf('%2.1e kPa',PSI),'rotation',90)
end

% get a vector of saturation liquid entropies from Xsteam
sliq = arrayfun(@(t) XSteam('sL_T',t),T);

% get a vector of saturation vapor entropies from Xsteam
svap = arrayfun(@(t) XSteam('sV_T',t),T);

plot(svap,T,'r-', 'LineWidth',1.85)
plot(sliq,T,'b-','LineWidth',1.85)
xlabel('S (kJ/kg K)')
ylabel('T (^\circC)')
```

When we run the code, Figure 3.7 will appear in the screen:

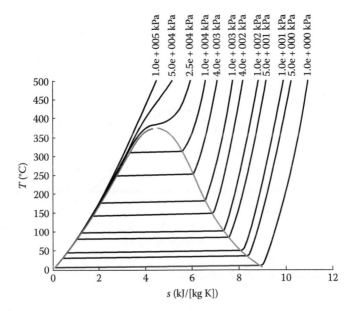

FIGURE 3.7   Temperature versus the entropy plot for steam with constant pressure curves. Note that the constant pressure curves in the subcooled liquid region are so close to each other, so that they seem to overlap. This means that in the subcooled liquid region, a pressure increase hardly causes an increase in the temperature or entropy.

Example 3.1: Efficiency of a Thermal Power Plant

This example aims to illustrate the thermodynamic analysis of a Rankine cycle. Consider that the Rankine cycle is ideal, that is, both the turbine and the pump operate isentropically, and there is no pressure drop in the pipelines, the boiler, and the condenser. The boiler produces steam at 560°C and 3 MPa. At the outlet of the condenser, water is subcooled to 30.01°C at 100 kPa. Calculate q_H, q_L, $w_{turbine}$, w_{pump} and the thermodynamic efficiency of the cycle.

In analyses of heat engines, it is useful to sketch a *T–s* diagram and indicate the thermodynamic states of the streams on the *T–s* diagram as shown in Example Figure 3.1.1.

At the outlet of the boiler, T_3 and p_3 are given; thus, the rest of the thermodynamic properties can be found from steam tables or Xsteam. For example, the *NIST Chemistry WebBook* lists (The National Institute of Standards and Technology, 2014):

T (°C)	p (MPa)	v (m³/kg)	u (kJ/kg)	h (kJ/kg)	s (J/g K)	c_v (J/g K)	c_p (J/g K)	Phase
560	3	0.12599	3214.3	3592.3	7.4041	1.7457	2.2576	Vapor

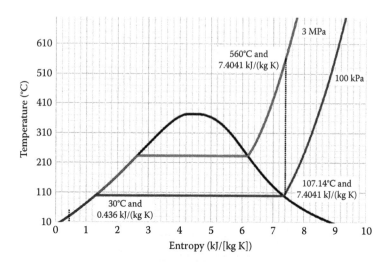

EXAMPLE FIGURE 3.1.1   *T–s* diagram.

Since the turbine operates isentropically, $s_4 = s_3$ and pressure is given as $p = 0.1$ MPa. Thus, at point 4:

T (°C)	p (MPa)	v (m³/kg)	u (kJ/kg)	h (kJ/kg)	s (J/g K)	c_v (J/g K)	c_p (J/g K)	Phase
107.88	0.1	1.7344	2518.6	2692.0	7.4041	1.5335	2.0476	Vapor

Through the condenser, pressure remains constant but temperature drops to 30°C:

T (°C)	p (MPa)	v (m/kg)	u (kJ/kg)	h (kJ/kg)	s (J/g K)	c_v (J/g K)	c_p (J/g K)	Phase
30.01	0.1	0.0010044	125.76	125.86	0.43686	4.1172	4.1798	Liquid

Through the pump, pressure is increased isentropically. The thermodynamic properties are listed for $s_1 = s_2 = 0.43686$ and $p_2 = 3$ MPa as:

T (°C)	p (MPa)	v (m³/kg)	u (kJ/kg)	h (kJ/kg)	s (J/g K)	c_v (J/g K)	c_p (J/g K)	Phase
30.07	3	0.0010031	125.76	128.77	0.43686	4.1077	4.1719	Liquid

Once the thermodynamic properties of each stream are obtained, we can calculate the heat and work transfer for the devices:

$$q_H = 3592.3 - 128.77 = 3463.53 \text{ kJ/kg}$$

$$W_{turbine} = 3592.3 - 2692.0 = 900.3 \text{ kJ/kg}$$

$$q_L = 2692.0 - 125.86 = 2566.14 \text{ kJ/kg}$$

$$w_{pump} = 128.77 - 125.86 = 2.91 \text{ kJ/kg}$$

These results show that out of the 3463.53 kJ/kg heat extracted from the high-temperature reservoir, 2566.14 kJ/kg is wasted as q_L. The net work obtained from this cycle is $w_{turbine} - w_{pump}$ = 900.3–2.91 = 897.39 kJ/kg. The thermal efficiency of this cycle is

$$\eta = \frac{897.39}{3463.53} = 26\%$$

Remark: Note that the values listed earlier are experimental data taken from steam tables. In Chapter 2, we stated that subcooled liquids are incompressible. Based on this assumption, we can write the enthalpy change in the pump as

$$h_2 - h_1 = (u_2 + p_2 v_2) - (u_1 + p_1 v_1)$$

$$= (u_2 - u_1) + (p_2 v_2 - p_1 v_1)$$

For incompressible substances,

$$v = \text{constant} = c(T_2 - T_1) + v(p_2 - p_1)$$

The temperature change in turbomachinery for incompressible substances is usually negligibly small. Experimental data support the negligible ΔT assumption. Here, the experimental data show that there is only a temperature increase of 0.06°C between the inlet and outlet of the pump. Thus, for incompressible water, we can write the enthalpy change through the pump as

$$h_2 - h_1 = v(p_2 - p_1)$$

The enthalpy change calculated based on this assumption is

$$h_2 - h_1 = 0.0010044(3000 - 100) = 2.913 \text{ kJ/kg}$$

The enthalpy change calculated based on the experimental data was calculated as 2.91 kJ/kg. Thus, the incompressible assumption for subcooled liquids is a valid assumption.

The energy balances demonstrate that the enthalpy difference between the inlet and outlet of each device either represent the heat transfer or the work transfer in the corresponding device. Entropy equations demonstrate that the entropy difference between the inlet and outlet of each device represent the sum of the entropy transferred and the entropy generated. The enthalpy versus entropy diagram, as described in Figure 3.8, is a very useful plot, since it visualizes the results of both the first and second laws of thermodynamics simultaneously. Figure 3.8 represents the *h–s* diagram of the Rankine cycle analyzed in the previous example. Note that since the analyzed Rankine cycle is ideal, there is no entropy generation. The difference in the entropy values of the fluid between different states arises only due to the heat absorbed (in the boiler) or the heat released (in the condenser), which are equal to each other.

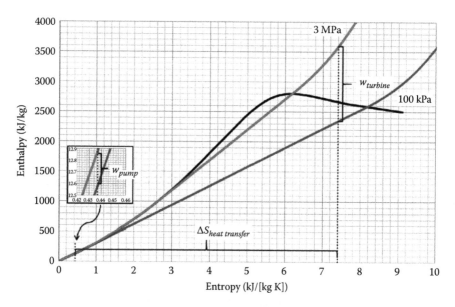

FIGURE 3.8   The *h–s* diagram.

Example 3.2: Improving the Efficiency of a Thermal Power Plant

The answer to the question *How we may improve the efficiency of the thermal power plant?* is presented in this example.

To increase the thermal efficiency, we have to increase the amount of work produced per kJ of extracted heat. There are several modifications to do that, such as increasing the boiler pressure and adding reheaters and feed water heaters. A thorough discussion on the possible improvement methods can be found in basic mechanical engineering literature. Here, we are going to limit our discussion on the effect of increasing the turbine inlet temperature and decreasing the condenser pressure.

(a) Increasing T_3

One possible modification is to superheat the steam to higher temperatures. If the steam would be heated to 980°C at 3 MPa, then its enthalpy would increase to 4584.3 kJ/kg. This modification would require an increase in the heat addition as $q_H = 4584.3-128.77 = 4455.53$ kJ/kg, and result in an increase in the turbine work as $w_{turbine} = 4584.3-3162.3 = 1422.0$ kJ/kg. The thermal efficiency rises to 31.85%. Example Figure 3.2.1 shows the *T–s* diagram of the modified Rankine cycle and the increase in the net work.

In practice, large turbine inlet temperatures are omitted, since turbine blades cannot withstand high temperatures. Modern technology allows high temperatures up to 620°C. In some high-tech applications, special materials or blade cooling technologies are employed to enable turbines to endure higher temperatures.

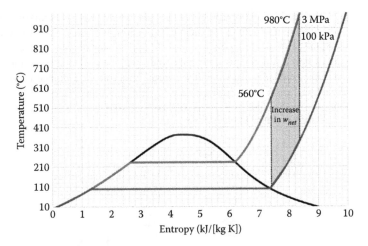

EXAMPLE FIGURE 3.2.1 *T–s* diagram describing that an increase in the turbine inlet temperature leads to increases in q_H, q_L, and $w_{turbine}$.

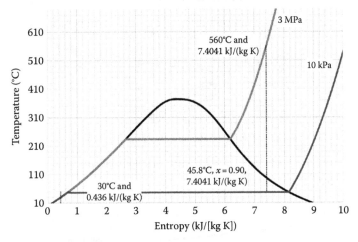

EXAMPLE FIGURE 3.2.2 *T–s* diagram describing that a decrease in p_4 causes an increase in thermal efficiency.

(b) Decreasing p_4

Another modification is to decrease the condenser pressure (Example Figure 3.2.2). By reducing the condenser pressure to 10 kPa, water at the outlet of the turbine becomes a saturated mixture with a quality of $x = 0.90$. Thermodynamic properties at point 4 can be calculated as $h_4 = 2344.69$ kJ/kg, $T_4 = 45.806°C$. Via this modification, q_H remains the same, but $w_{turbine}$ increases to 1247.61. The heat rejected (q_L) decreases, since the condenser inlet enthalpy is decreased. The pump work increases, but since the fluid is incompressible, increase in the pump work is very small. The new pump work is $w_{pump} = 0.001(3000-10) = 2.99$ kJ/kg instead of 2.91 kJ/kg. As a result, the thermal efficiency rises to 36%.

One disadvantage of this modification is to have a saturated mixture at the outlet of the turbine. Consider a small control volume of vapor in the turbine as shown in Example Figure 3.2.3. If this small volume of vapor flows into a low-temperature zone and condenses, then the volume it occupies suddenly decreases. The abrupt change in the occupied volume

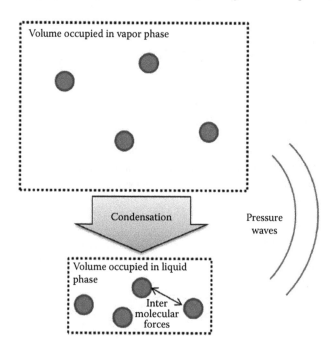

EXAMPLE FIGURE 3.2.3 The schematic description of the condensation of a small vapor bubble.

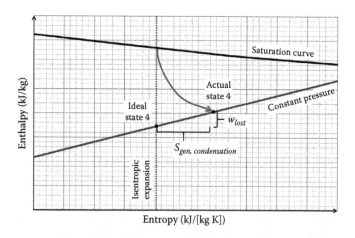

EXAMPLE FIGURE 3.2.4 Condensation in the metastable region.

initiates pressure waves. Fluctuations in pressure can apply large forces on the solid surfaces inside the turbine.

Condensation of very small vapor bubbles in restricted spaces occurs in a so-called metastable thermodynamic process, where entropy increases as shown in Example Figure 3.2.4.

In order to minimize the condensation losses, large moisture contents in the turbine are avoided. But slightly wet steam at the turbine outlet is preferred since it provides larger cycle efficiency and some flexibility in the operational range.

In a typical fossil fuel–fired power plant, only one-third of the energy present in the fuel is turned into useful energy, and two-thirds of the energy is wasted. One of the major sources of the waste is Q_L, which is dumped into cold water sinks, for example, ocean, river, or seas. Sophisticated Rankine cycle plants, which operate very cold ocean water, reach efficiencies up to 42%. With the engineering improvements within the last decades, technological advancement of the conventional power plants reached a maturity. Today, intense research and development studies are performed to obtain even very small increases in the cycle's thermodynamic efficiency. To increase the *overall* efficiency of a plant, however, can be easier, if the waste heat is used. The first power plant, which was built by Edison in the 1880s, used the waste heat Q_L to warm nearby factories and houses. Most of the modern power plants are established far from the cities; therefore, it is not feasible to use their waste heat in the houses (Casten and Schewe, 2009). One of the solutions found to this problem is building small (3–10,000 J/s) power plants in the facilities, which uses the electric power. These small power plants may be connected with the local grid. Although such small power plants make it possible to use Q_L with other purposes, they are rather expensive to build.

The *T–s* diagram of an actual Rankine cycle is shown in Figure 3.9. Note that this is a schematic drawing and not to scale. In this figure, the distance between the isobars in

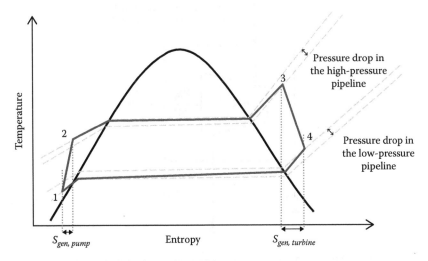

FIGURE 3.9 *T–s* diagram of an actual Rankine cycle (not to scale).

the compressed liquid region and the pressure losses are sketched much larger than the actual values to be able to demonstrate the losses.

Example 3.3: Efficiency of an Actual Rankine Cycle

In an actual Rankine cycle, steam enters into the turbine at 700°C and 8 MPa. At the outlet of the turbine, the pressure is measured as 20 kPa and the fluid is in the form of saturated vapor. Calculate the specific work of this turbine, and the mass flow rate of the steam required to perform a total work of 300 MJ.

The turbine can be considered as an open system as sketched in Example Figure 3.3.1:

Thermodynamic properties of steam at 700°C and 8 MPa are determined with MATLAB code E3.3.a:

MATLAB CODE E3.3.a

```
Command Window

clear all
close all

% print out values of u, h and s at 625 oC and 30 kPa

T = 700; % oC
p= 8000/100; % kPa, P is in Bar in XSteam, divide in 100 to
    convert  kPa into Bar
```

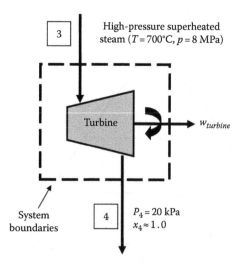

EXAMPLE FIGURE 3.3.1   Schematic description of the mass input and output and work production in the turbine.

```
% tabulate u, h and s as a function of T at 30 kPa

fprintf('\n SUPER HEATED STEAM \n')

fprintf('\n T (oC)  p (kPa)  u(kJ/kg)  h(kJ/kg)  s(kJ/kg K)
  \n ') % column captions

  u=XSteam('u_pt', p,T); % determine the internal energy of
  the super-heated steam entering into the chamber (kJ/kg)

h=XSteam('h_pt', p,T); % determine the enthalpy of the
  super-heated steam entering into the chamber (kJ/kg)
s=XSteam('s_pt', p,T); % determine the enthalpy of the
  super-heated steam entering into the chamber (kJ/kg)

  % P is in Bar in Xsteam, multiplied by 100 to convert in kPa
  fprintf('%2.0f  %3.0f  %7.1f  %7.1f  %7.1f \n', T,
    p*100, XSteam('u_pt', p,T), XSteam('h_pt',
    p,T), XSteam('s_pt', p,T))
```

When we run the code the followings will appear in the
  screen (columns are aligned manually):

```
SUPER HEATED STEAM


T (oC)    p (kPa)    u(kJ/kg)    h(kJ/kg)    s(kJ/kg K)
 700       8000       3443.8      3882.4        7.3
```

We will refer to XSteam to determine the thermodynamic properties of the
saturated water under $p = 20$ kPa of pressure with the MATLAB code E3.3.b:

MATLAB CODE E3.3.b

Command Window

```
clear all
close all

% enter the pressure range (in Bar)
P = [0.1:0.05:0.3];

% tabulate the results:
fprintf('\n SATURATED WATER TABLE \n')
fprintf('\n  P   T_sat v_sat_L v_sat_V u_sat_L u_sat_V
  h_sat_L h_sat_V s_sat_L s_sat_V') % column captions

fprintf('\n (kPa)  (oC)  (m3/kg)  (m3/kg)  (kJ/kg)  (kJ/kg)
  (kJ/kg)  (kJ/kg)  (kJ/kg K)  (kJ/kg K) \n') % units

for i=1:length(P)
```

```
fprintf('%3.0f %2.0f %6.4f % 6.2f %5.1f  %7.1f  %5.1f %7.1f
  %6.2f %7.2f \n',P(i) *100, XSteam('tSat_p',P(i)),
  XSteam('vL_p',P(i)), XSteam('vV_p',P(i)),
  XSteam('uL_p',P(i)), XSteam('uV_p',P(i)),
  XSteam('hL_p',P(i)),     XSteam('hV_p',P(i)),
  XSteam('sL_p',P(i)), XSteam('sV_p',P(i))); % P is in Bar in
  XSteam, which is multiplied by 100 to convert in kPa

end
```

When we run the code, the following will appear in the screen (columns are aligned manually in the printout):

SATURATED WATER TABLE

P (kPa)	T_sat (oC)	v_ sat_L (m3/kg)	v_ sat_V (m3/kg)	u_ sat_L (kJ/kg)	u_ sat_V (kJ/kg)	h_ sat_L (kJ/kg)	h_ sat_V (kJ/kg)	s_ sat_L (kJ/kg K)	s_ sat_V (kJ/kg K)
10	46	0.0010	14.67	191.8	2437.2	191.8	2583.9	0.65	8.15
15	54	0.0010	10.02	225.9	2448.0	225.9	2598.3	0.75	8.01
20	60	0.0010	7.65	251.4	2456.0	251.4	2608.9	0.83	7.91
25	65	0.0010	6.20	271.9	2462.4	271.9	2617.4	0.89	7.83
30	69	0.0010	5.23	289.2	2467.7	289.2	2624.6	0.94	7.77

At the exit of the turbine, the working fluid is a mixture of both saturated water and vapor. The enthalpy and entropy of the saturated mixture are calculated as

$$h_4 = 2608.9 \text{ kJ/kg}$$

$$s_4 = 7.91 \text{ kJ/(kg K)}$$

The specific work produced in this turbine is (Example Figure 3.3.2)

$$w = h_3 - h_4 = 3882.4 - 2608.9 = 1273.5 \text{ kJ/kg}$$

In order to produce a total work of 300 MJ, the mass flow rate of the steam should be

$$\dot{m} = \frac{W}{w} = 235 \text{ kg/s}$$

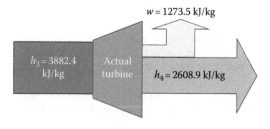

EXAMPLE FIGURE 3.3.2 Schematic description of the energy input, output, and work production in the actual turbine.

Note that the entropy generation in this turbine is

$$s_{gen} = s_4 - s_3 = 7.91 - 7.3 = 0.61 \, kJ/(kg \, K)$$

If this turbine would generate no entropy, that is, function without losses, then the thermodynamic state at the outlet would be different. For an isentropic turbine $s_{gen} = 0$, which leads to

$$s_{4s} = s_3 = 7.3 \, kJ/(kg \, K)$$

Since s_g at 20 kPa is larger than the calculated s_{4s}, the outlet stream of the isentropic turbine should be a saturated mixture with a quality of

$$x_{4s} = \left(\frac{s_{4s} - s_{g@20 \, kPa}}{s_{g@20 \, kPa} - s_{f@20 \, kPa}} \right) = 0.91$$

Enthalpy at the point 4s is calculated as

$$h_{4s} = x_{4s} h_{g@20 \, kPa} + (1 - x_{4s}) h_{f@20 \, kPa}$$
$$= 2391.73 \, kJ/kg$$

Then the maximum work that an ideal turbine would produce under these conditions is calculated as

$$w_{max} = h_3 - h_{4s} = 3882.4 - 2391.73 = 1490.67 \, kJ/kg$$

These results may also be shown schematically as follows (Example Figure 3.3.3):

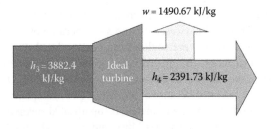

$w = 1490.67$ kJ/kg

$h_3 = 3882.4$ kJ/kg

Ideal turbine

$h_4 = 2391.73$ kJ/kg

EXAMPLE FIGURE 3.3.3 Schematic description of the energy input, output, and work production in the ideal turbine.

The second law or isentropic efficiency of a turbine can be defined as

$$\eta_{II,\,turbine} = \frac{w_{actual}}{w_{max}}$$

The second law efficiency of the investigated turbine is

$$\eta_{II,\,turbine} = \frac{w_{actual}}{w_{max}} = \frac{1273.5}{1490.67} = 85\%$$

An ideal turbine would decrease the need for the steam flow rate to 201.25 kg/s.

Note that both the ideal and actual turbines are considered as adiabatic turbines. The lost work in the actual turbine is not released as heat to the surroundings but used to increase the entropy content of the outflowing stream ($s_4 > s_{4s}$).

3.1.3 Sources of the Irreversibilities

Thermodynamic analyses show that the efficiency of an ideal Rankine cycle can be increased by varying the operating conditions. The next question that arises is: What is the limit for the maximum efficiency? What is the maximum work that we can produce from a thermal reservoir at T_H by extracting 1 kJ of Q_H?

In order to answer these questions, let us investigate the energy and entropy flow in the Rankine cycle in detail. To achieve the maximum efficiency, loss of work potential into dissipative forms of energy, that is, irreversibilities, should be avoided. Irreversibilities can occur both inside and outside the system boundaries. Internal losses occur due to turbulence, friction, condensation shock, etc. External losses occur because of the heat transfer across a finite temperature difference. Irreversibilities are measured with entropy generation. For the two processes, which involve work transfer, that is, pump and turbine, entropy balances are given as

$$s_{gen,\,pump} = s_1 - s_2$$

$$s_{gen,\,turbine} = s_4 - s_3$$

To avoid entropy generation in these devices, processes should occur isentropically, that is, the fluid's thermodynamic state should change so that its entropy remains constant ($s_1 = s_2$ and $s_3 = s_4$). In actual devices, $s_1 < s_2$ and $s_3 < s_4$, because of the internal irreversibilities. Some of the internal irreversibilities can be prohibited via careful design. For example, if the inner surface of the pipelines through which the working fluid is flowing is perfectly smooth, then the viscous dissipation can be minimized. On the other hand, some of the losses are tolerated because of feasibility. For example, entropy generation due to condensation shock can be eliminated, if superheated steam leaves the turbine. However, this will limit operational flexibility and reduce the total work produced. Sometimes economic, technologic, and environmental benefits justify losses in thermodynamic efficiency.

For the two heat transfer processes (boiler and condenser), entropy balances are

$$S_{gen, boiler} = s_3 - s_2 - \frac{q_H}{T_H}$$

$$S_{gen, condenser} = s_1 - s_4 + \frac{q_L}{T_L}$$

Heat transfer processes can occur in an internally reversible manner, if dissipative forms of energy transfer (such as fluid friction with the pipe surface, turbulence, noise, etc.) can be eliminated. This would mean that the boiler increases fluid's entropy (s_3-s_2) by the same amount as the condenser decreases fluid's entropy (s_1-s_4).

Note that both T_L and T_H are the temperatures of the thermal reservoirs, not the temperature of the working fluid. The system boundaries of the boiler and the condenser are placed on the outer surfaces of these devices. In order to illustrate the heat transfer across the finite temperature difference between T_H and T_{fluid} and the entropy generation related to it, Figure 3.10 shows a schematic drawing of the boiler. For this simplified analysis boiler is considered as an axisymmetric pipeline, where the temperature of the fluid changes only in the direction of the flow, that is, any temperature gradient in the direction normal to the flow is neglected. The temperature of the outer surface of the pipe wall is constant at T_H. The temperature of the inner surface changes in the flow direction. At the inlet, it is nearly at T_2 (pump outlet temperature). As heat transfer occurs, temperature of the liquid water increases up to the saturation temperature. Then, boiling begins and the temperature remains constant until the fluid's state achieves saturated vapor. After this point temperature increases again until exits the boiler at T_3. The amount of the entropy transfer through the inner surface of the wall

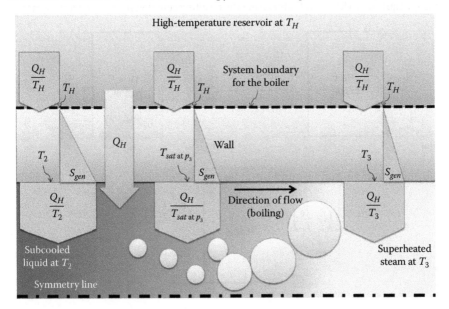

FIGURE 3.10   Entropy generation due to heat transfer across a finite temperature difference.

changes in the flow direction. As a result, entropy generation in the wall is a function of the position. The total entropy generated in the pipe wall remains inside the system boundaries.

This analysis shows that a reversible heat transfer can only be achieved isothermally. In order to avoid s_{gen}, T_H and T_{fluid} have to be approximately the same. Heat transfer across an infinitely small temperature difference would eliminate the entropy generation, but it would also take unfeasibly long time periods.

In summary, the total entropy generation of a Rankine cycle is

$$s_{gen} = s_{gen, pump} + s_{gen, turbine} + s_{gen, boiler} + s_{gen, condenser}$$

If the Rankine cycle is ideal, that is, there are no internal irreversibilities, then the entropy generation in the cycle is

$$s_{gen} = \frac{Q_L}{T_L} - \frac{Q_H}{T_H}$$

This result indicates that the total entropy generated in an ideal Rankine cycle is the difference between the amounts of the entropy transfers due to heat transfer. Even though this result is achieved from the analysis of a Rankine cycle, the result is valid for any heat engine. In the heat addition process, an entropy of Q_H/T_H is transferred into the cycle. In steady-state operation, this inflowing entropy should flow out of the system boundaries, which occurs via the heat rejection process (condenser). If the entropy rejected in the condenser Q_L/T_L is the same amount of Q_H/T_H, then system operates reversibly. But if entropy is generated within the system boundaries, then Q_L/T_L becomes larger than Q_H/T_H.

Figure 3.11 visualizes energy, entropy, and exergy flow in heat engines. The equations for the exergy balance are not written here explicitly, but these can be derived by multiplying the entropy equations with T_0 and subtracting the resulting equation from the energy equation.

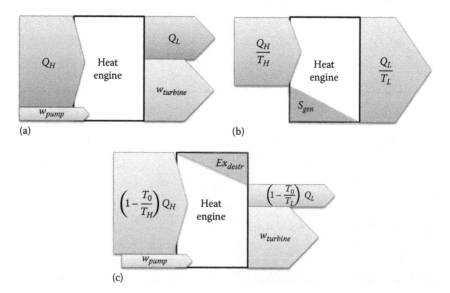

FIGURE 3.11 (a) Energy, (b) entropy, and (c) exergy flow in an internally reversible Rankine cycle.

3.1.4 Carnot Engine

Carnot engine is a theoretical, reversible heat engine that eliminates all the irreversibilities mentioned earlier. It was proposed by the French engineer Sadi Carnot in 1824. The principles of Carnot engine are later used by Clausius and Kelvin to formalize the second law of thermodynamics and define the absolute temperature scale.

Carnot engine operates with the following four processes:

1. Isothermal heat addition
2. Isentropic expansion
3. Isothermal heat rejection
4. Isentropic compression

For an internally reversible heat engine, we already showed that the entropy generation is

$$s_{gen} = \frac{Q_L}{T_L} - \frac{Q_H}{T_H}$$

If the heat transfer processes can be performed isothermally, then the cycle becomes completely reversible since

$$s_{gen} = 0, \quad \text{if } \frac{Q_L}{T_L} = \frac{Q_H}{T_H}$$

This means that the working fluid is nearly at T_H during the heat addition process and nearly at T_L during the heat rejection process. Whether an isothermal heat transfer process is technologically feasibile or not is beyond our focus. Carnot heat engine is a conceptual cycle. Carnot cycle may not be applicable but it is one of the most important concepts of thermodynamics, because of the following reasons:

1. It helps us to assess the feasibility of potential energy resources.
2. It gives us a reference to compare existing systems and pinpoint the problematic areas.
3. It serves as a model to understand relatively little known phenomena like natural or biological systems.

A Carnot engine is schematically sketched as in Figure 3.12.

Since there are no internal or external irreversibilities in the Carnot engine, it produces work at the maximum efficiency. The energy, entropy, and exergy flow diagrams for a Carnot engine are given in Figure 3.13.

3.1.5 Absolute Temperature

Lord Kelvin made use of the principles of the Carnot engine to define an absolute temperature scale. Carnot engine shows that to obtain a reversible heat transfer,

$$\frac{Q_L}{T_L} = \frac{Q_H}{T_H}$$

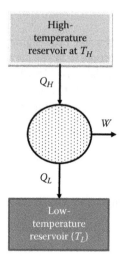

FIGURE 3.12 Schematic drawing of the Carnot engine.

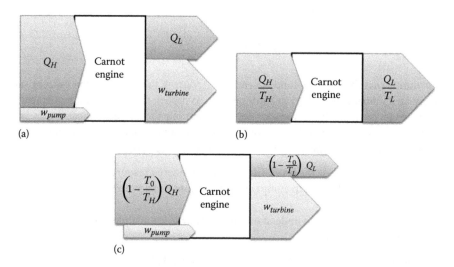

FIGURE 3.13 (a) Energy, (b) entropy, and (c) exergy flow in Carnot cycle, which is both internally and externally reversible.

Kelvin showed mathematically that a temperature scale, which satisfies

$$\frac{T_H}{T_L} = \frac{Q_H}{Q_L}$$

can be defined. This temperature scale is called *absolute*, and its unit is Kelvin (K). Figure 3.14 demonstrates the linear relationship between the heat transfer and the absolute temperature scale. With extrapolation, the position of the absolute zero temperature is found as −273.15°C.

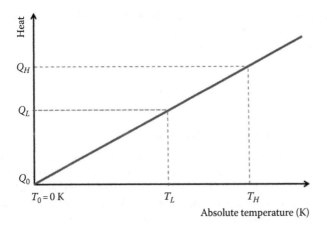

FIGURE 3.14 Definition of the absolute temperature scale.

Note that Kelvin scale is defined only based on thermodynamic principles and does not depend on any material property.

3.1.6 Second Law Efficiency

The thermodynamic efficiency is defined as

$$\eta_{th} = \frac{w_{net}}{q_{in}}$$

$$\eta_{th} = \frac{q_H - q_L}{q_H} = 1 - \frac{q_L}{q_H}$$

By using the definition of the absolute temperature, we can write the efficiency for a Carnot cycle as

$$\eta_{Carnot} = 1 - \frac{T_L}{T_H}$$

Carnot efficiency depends only on the absolute temperatures of the heat reservoirs. Properties of the heat engine like the type of the working fluid, maximum or minimum pressure, material of the turbine blades, etc., are irrelevant for the Carnot efficiency. Carnot efficiency shows us the thermodynamically possible limit. For example, if we have a geothermal reservoir at a temperature of $T_H = 150°C$ in a coastal region where seawater is available at $T_L = 5°C$, then the best work-producing cycle that we can build would have an efficiency of $\eta_{Carnot} = 1-((5 + 273.15)/(150 + 273.15)) = 34\%$. A fossil fuel–based power plant with $T_H = 800°C$ in the same location would have a maximum efficiency of $\eta_{Carnot} = 1-((5 + 273.15)/(800 + 273.15)) = 74\%$.

Note that only for a Carnot engine (i.e., an internally and externally reversible heat engine as shown in Figure 3.12) $T_L/T_H = Q_L/Q_H$. For actual heat engines, $T_L/T_H \neq Q_L/Q_H$. The second law efficiency of a heat engine is defined as

$$\eta_{II} = \frac{\eta_{th}}{\eta_{Carnot}}$$

Example 3.4: Efficiency of an Actual Thermal Power Plant

A power plant produces 750 MW of net work, by burning coal. The high-temperature reservoir is at 650°C. Seawater is used as the low-temperature reservoir, which is at 8°C. The second law efficiency of the power plant is 80%. How much heat is discarded to the sea? Repeat the analysis, if T_H is increased to 750°C and the second law efficiency remained constant at 80%.

(a) When $T_H = 650°C$,

$$\eta_{Carnot} = 1 - \frac{T_L}{T_H} = 1 - \frac{8 + 273}{650 + 273} = 0.70$$

$$\eta_{th} = 0.80\eta_{Carnot} = (0.80)(0.70) = 0.56$$

Since $\eta_{th} = w/|q_H|$, we calculate

$$Q_H = \frac{W}{\eta_{th}} = \frac{(750)}{(0.56)} = 1339 \text{ MW}$$

We calculate the amount of the heat discarded to the sea after applying the first law of thermodynamics to the cycle as

$$Q_L = Q_H - W = 589 \text{ MW}$$

Note that $Q_L/Q_H = 0.44$, implying that 44% of the energy extracted from the coal is dumped into the sea.

(b) When $T_H = 750°C$,

$$\eta_{Carnot} = 1 - \frac{T_L}{T_H} = 1 - \frac{8 + 273}{750 + 273} = 0.73$$

$$\eta_{th} = 0.80\eta_{Carnot} = (0.80)(0.73) = 0.58$$

$$Q_H = \frac{W}{\eta_{th}} = \frac{(750)}{(0.58)} = 1293 \text{ MW}$$

and

$$Q_L = Q_H - W = 543 \text{ MW}$$

These results show that Q_L decreases and η_{Carnot} increases with increasing T_H.

Example 3.5: Losses Related to Imperfect Fuels

Natural energy resources may contain humidity and impurities. The mass fraction and the chemical structure of the impurities are very important, since these can cause serious damage in the combustion chamber. Furthermore, dangerous exhaust gases can be formed. Humidity in fossil fuels can also cause problems and decrease in the power plants' thermodynamic efficiency. In this example, we will analyze the environmental impact of a thermal power plant, which uses moist lignite as the energy resource.

For this analysis, we will use the thermodynamic data calculated for the power plant in the previous example, which absorbs a heat of 1293 MW in the boiler.

Assume that the lignite includes a legible amount of impurities. The following data are reported for a power plant running on lignite (Kolovos et al., 2002):

Moisture content on dry basis = (water/dry lignite) × 100 = 50.9%

Ash content on dry basis = (ash/dry lignite) × 100 = 33.1%

CO_2 emission rate = (weight of CO_2 emission/weight of dry lignite) × 100 = 8.6%

The amount of water in 1 ton of moist lignite is calculated as

$$\frac{m_{water}}{m_{dry\ lignite}} = 0.509; \quad \text{therefore,} \quad \frac{m_{water}}{0.509} = m_{dry\ lignite}$$

$$m_{water} + m_{dry\ lignite} = 1 \text{ ton}$$

$$m_{water} + \frac{m_{water}}{0.509} = 1 \text{ ton; we may calculate from this equation that}$$

$m_{water} = 0.337$ ton and $m_{dry\ lignite} = 1 - 0.337 = 0.663$ ton, implying that 1 ton of moist lignite contains 0.337 ton of water and 0.663 ton of dry matter.

When moist lignite is fed into the combustion chamber, a portion of the combustion heat is used up to evaporate and superheat the moist inside the lignite. Assuming that the moist lignite is fed into the chamber at 5°C and 1 atm and the exhaust gases are discarded at T_H and 1 atm, then the heat used to evaporate and superheat the water inside the lignite is calculated as

$$\dot{Q}_{water} = \dot{m}_{water}\left(h_{@1\ atm\&T_H} - h_{f@5°C}\right)$$

We need to first determine the $h_{f@5°C}$ and $h_{@1\ atm\&T_H}$ with Xsteam:

```
XSteam('hL_t',5) returns 21.0 kJ/kg
XSteam('h_pt',10,750) returns 4039.3 kJ/kg
```

$$q_{water} = 337 \, \frac{\text{kg water}}{\text{ton moist lignite}} (4039.1 - 21.0) \, \frac{\text{kJ}}{\text{kg water}}$$

$$= 1354.17 \, \frac{\text{kJ}}{\text{ton moist lignite}}$$

The required total amount of the heat released by combustion is then

$$\dot{m}_{moist\,lignite}\, y\Delta h_{c,\,lignite} = (1-y)\dot{m}_{moist\,lignite}(h_{@1\,atm\&T_H} - h_{f@5°C}) + \dot{Q}_H$$

Heating value of dry lignite is 15 MJ/kg. The mass flow rate of the moist lignite required for this power plant is calculated as

$$\dot{m}_{moist\,lignite} = \frac{\dot{Q}_H}{y\Delta h_{c,\,lignite} - (1-y)(h_{@1\,atm\&T_H} - h_{f@5°C})}$$

$$= \frac{1293 \text{ MW}}{0.663 \cdot 15 \text{ MJ/kg} - 0.337 \cdot 1.354 \text{ MJ/kg}}$$

$$= 136.27 \text{ kg/s}$$

The amount of ash, which would be produced in the combustion chamber, is calculated as

$$\left(\frac{\dot{m}_{ash}}{\dot{m}_{dry\,lignite}}\right)\left(\frac{\dot{m}_{dry\,lignite}}{\dot{m}_{moist\,lignite}}\right)\dot{m}_{moist\,lignite} = (0.331)(0.663)136.27$$

$$= 29.90 \text{ kg/s}$$

Remark: The heat used to evaporate the water content of the lignite can be saved, if lignite is dried before it is fed into the combustion chamber. If waste heat can be employed for this purpose, then the overall efficiency of the power plant can be enhanced, and also CO_2 and ash emission would be decreased. If 85% of the lignite water content can be removed, then the mass flow rate of the

$$\dot{m}_{moist\,lignite} = \frac{\dot{Q}_H}{y\Delta h_{c,\,lignite} - (1-y)0.15(h_{@1\,atm\&T_H} - h_{g@100°C})}$$

$$= \frac{1293 \text{ MW}}{0.663 \cdot 15 \text{ MJ/kg} - 0.337 \cdot 0.15 \cdot ((4039.3 - 2674.9)/1000) \text{ MJ/kg}}$$

$$= 130.92 \text{ kg/s}$$

The amount of ash, which would be produced in the combustion chamber, is decreased to

$$\left(\frac{\dot{m}_{ash}}{\dot{m}_{dry\,lignite}}\right)\left(\frac{\dot{m}_{dry\,lignite}}{\dot{m}_{moist\,lignite}}\right)\dot{m}_{moist\,lignite} = (0.331)(0.663)130.92$$

$$= 28.73 \text{ kg/s}$$

3.1.7 Heat Pumps

Heat pumps are thermodynamic cycles that consume work to transfer heat from a low-temperature reservoir to a high-temperature reservoir. Air-conditioning devices and refrigerators operate as heat pump cycles. We may consider the heat engine and the heat pump as *opposites* for practical purposes (Figures 3.15 and 3.16).

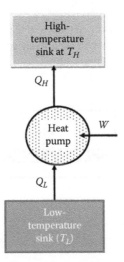

FIGURE 3.15 Heat engine and the heat pump are reverse processes. Heat engine converts heat into work, when heat flows from the high-temperature sink at T_H to the low-temperature sink at T_L, while the heat pump consumes work to pump heat from low-temperature sink at T_c to the high-temperature sink at T_h.

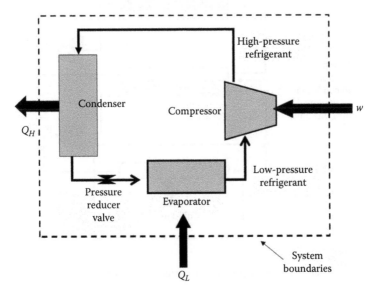

FIGURE 3.16 Schematic drawing of a heat pump.

The *coefficient of performance* (COP) is defined as the ratio of the heat removed from the cold reservoir to the work input as

$$\mathrm{COP} = \frac{Q_L}{W}$$

When the temperature T_H of the hot reservoir and temperature T_L of the cold reservoir are too far away, then we may employ a cascade refrigeration system, where the cooling is achieved through smaller steps (Figure 3.17).

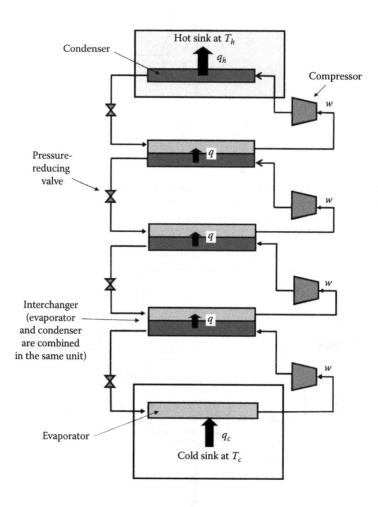

FIGURE 3.17   Schematic drawing of a cascade refrigeration system.

Example 3.6: Coefficient of Performance of Ammonia–Carbon Dioxide Cascade Refrigeration System

Getu and Bansal (2008) suggested a carbon dioxide (low-temperature circuit) and ammonia (high-temperature circuit) cascade refrigeration system for supermarkets (Example Figure 3.6.1).

The condenser of the high-temperature circuit running with ammonia is placed on the roof. Evaporator of the low-temperature circuit running with carbon dioxide is located in the cold room. The temperature versus entropy and the pressure versus enthalpy plots for the system are given in Example Figure 3.6.2.

For the ammonia circuit $T_L = -5°C$ and $T_H = 40°C$, for the carbon dioxide circuit $T_H = 0°C$ and $T_L = -30°C$. Assume that the condensers and the evaporators are isobaric, and the compressors are isentropic units. The data that are presented in Example Table 3.6.1 are extracted from Example Figure 3.6.2.

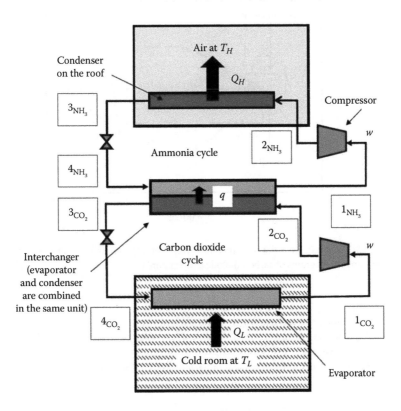

EXAMPLE FIGURE 3.6.1 Schematic drawing of the ammonia and carbon dioxide cascade refrigeration system.

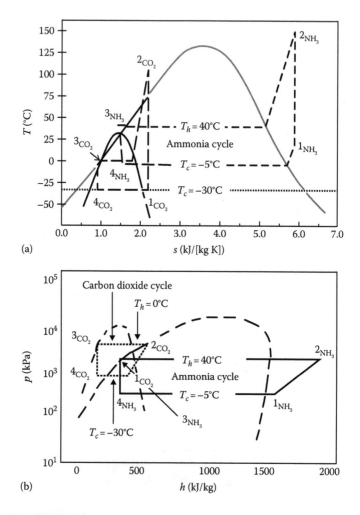

EXAMPLE FIGURE 3.6.2 Temperature versus entropy (a) and pressure versus enthalpy (b) diagrams of carbon dioxide and ammonia.

We should have $(q_{evaporator})_{ammonia} = (q_{condenser})_{carbon\ dioxide}$.

Therefore, $(m_{NH_3})(h_{4NH_3} - h_{1NH_3}) = (m_{CO_2})(h_{4CO_2} - h_{1CO_2})$, and if $m_{CO_2} = 1\ kg/s$, we must have $m_{NH_3} = 0.33\ kg/s$.

We may calculate the coefficient of performance of the system as

$$COP = \frac{(q_{evaporator})_{CO_2}\, m_{CO_2}}{(w_{compressor})_{NH_3}\, m_{NH_3} + (w_{compressor})_{CO_2}\, m_{CO_2}} = \frac{(350)(1)}{(250)(0.33) + (125)(1)} = 1.7$$

EXAMPLE TABLE 3.6.1 Thermodynamic Data Extracted from Example Figure 3.6.2 and Calculation of the Compressor Works and the Heat Loads of Each Heat Exchanger

Circuit Point	p (kPa)	T (°C)	s (kJ/kg K)	h (kJ/kg)	w or q (kJ/kg)
1_{CO_2}	1000	−30	2.2	475	$w_{compressor} = h_{2CO_2} - h_{1CO_2} = 125$
2_{CO_2}	7000	110	2.2	600	$q_{condenser} = h_{3CO_2} - h_{2CO_2} = -475$
3_{CO_2}	7000	0	0.9	125	$w_{walf} = h_{4CO_2} - h_{3CO_2} = 0$
4_{CO_2}	1000	−30	0.9	125	$q_{evaporator} = h_{1CO_2} - h_{4CO_2} = 350$
1_{NH_3}	400	10	6.0	1650	$w_{compressor} = h_{2NH_3} - h_{1NH_3} = 250$
2_{NH_3}	4000	160	6.0	1900	$q_{condenser} = h_{3NH_3} - h_{2NH_3} = -1550$
3_{NH_3}	4000	40	0.9	350	$w_{walf} = h_{4NH_3} - h_{3NH_3} = 0$
4_{NH_3}	500	−5	1.6	350	$q_{evaporator} = h_{1NH_3} - h_{4NH_3} = 1300$

Example 3.7: Drying Heat Pump

Drying is an energy-intensive process. Heat pump dryers are among the equipment employed for the drying of biological, for example, agricultural, food, wood, and microbial commodities (Cardona et al., 2002; Hepbasli et al., 2010; Minea, 2013a,b). A simple heat pump dryer is pictured in Example Figure 3.7.1.

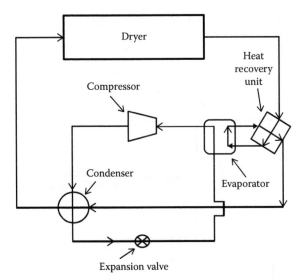

EXAMPLE FIGURE 3.7.1 A simple heat pump dryer is pictured in the following.

EXAMPLE TABLE 3.7.1   Exergy of the Electricity and Exergy Destruction in Each Unit of the Dryer

Unit	Exergy of the Fuel Used for Running the Unit (kJ/s)	Exergy Destroyed in the Unit (kJ/s)
Dryer	5.983	0.274
Heat recovery unit	1.193	0.701
Evaporator	0.528	0.070
Compressor	6.513	3.101
Condenser	1.734	0.232
Expansion valve	4.577	1.221
Total	20.528	5.599

Source: Data adapted from Hepbasli, A. et al., *Dry. Technol.*, 28, 1385, 2010.

Exergy of the product leaving each unit and the exergy of the fuel, for example, electricity for running each unit, and exergy destroyed in each unit were reported in an experimental heat pump dryer as given in Example Table 3.7.1.

We may calculate the exergy efficiency of the process as

$$\eta = 1 - \frac{ex_{destroyed}}{ex_{fuel}} = 1 - \frac{5.599}{20.528} = 0.73$$

3.2 Carnot Engine Analogies in Natural and Biological Phenomena

Even though Carnot cycle attracted little attention when it was proposed by Nicolas Léonard Sadi Carnot in 1824 (Carnot, 1824), it became one of the fundamental theories of thermodynamics in the forthcoming years. There are numerous studies in the

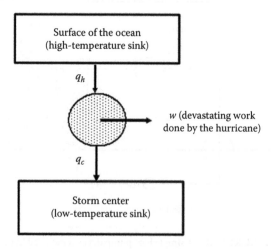

FIGURE 3.18   Schematic drawing of the energy uptake from the sun to maintain life on earth in analogy with the Carnot model.

literature, where relatively less known phenomena are modeled by establishing analogy with the Carnot cycle. Carnot cycle has been a model to describe natural phenomena, too. For example, Patzek (2004) describes the energy uptake from the sun to maintain life on earth in analogy with the Carnot model as described in Figure 3.18:

Another example is the Carnot analogy of the hurricane work production (Emanuel, 1991), as described in Example 3.8.

Example 3.8: Estimation of the Devastating Power of a Hurricane

Hurricane is a specific name given to the storms in the Atlantic and eastern Pacific. There are similar storms in different parts of the earth caused by similar reasons. A mature hurricane may be regarded as a Carnot cycle driven by the disequilibrium between the tropical oceans and atmosphere (Emanuel, 1991). We may describe it schematically as given in Example Figure 3.8.1.

This model suggests that air gets heat and collects entropy on the surface of the ocean, at T_h, then ascends adiabatically to the higher levels of the atmosphere, where entropy is lost with electromagnetic radiation to the universe at constant temperature T_c. Heat received by the air at the surface of the ocean may be expressed approximately as $q_h = T_h(s_h - s_c)$, and heat released to the universe at the higher levels of the atmosphere may be approximated as $q_c = T_c(s_c - s_h)$. In order to complete the cycle, air goes back to the surface of the earth; meanwhile, the vortex is generated, which performs the devastating work, for example, destroys the buildings. When the temperature on the surface of the ocean is 30°C, and the temperature of the higher level of the atmosphere is $T_c = -60$°C. The maximum work production of the hurricane will be

$$\eta_{carnot\ cycle} = 1 - \frac{T_c}{T_h} = 1 - \frac{-60 + 273}{30 + 273} = 0.30$$

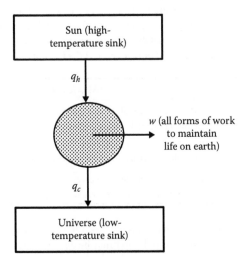

EXAMPLE FIGURE 3.8.1 Carnot analogy model of hurricane.

The efficiency of a Carnot engine is $\eta_{carnot\,cycle} = w/q_h$, then we may calculate the maximum work produced by the hurricane as $W = (\eta_{carnot\,cycle})(Q_h)$. When $q_h = 10^8$ J/m^2 and the heat transfer area is 800,000 km^2, for example, about the same area of Turkey, or two times the area of California, the work produced by this hurricane will be

$$w = (\eta_{carnot\,cycle})(q_h) = (0.3)(10^8\,\text{J/m}^2)(800,000\,\text{km}^2)\left(\frac{1,000\,\text{m}}{1\,\text{km}}\right)^2$$

$$= (2.4\times10^{19}\,\text{J})\left(\frac{1\,\text{GJ}}{10^9\,\text{J}}\right) = 10^{10}\,\text{GJ}$$

This is the maximum amount of the work produced by the hurricane, even if only 1% of it should be employed to devastate the environment, it will be about 10^8 GJ.

Sorgüven and Özilgen (2015) proposed a Carnot analogy for muscular work production as described in the following Example 3.9.

Example 3.9: Carnot Cycle Analogy of Work Production by the Muscles

Energy extracted from high-energy nutrients via metabolic pathways is used to fuel the cellular work. Sorgüven and Özilgen (2015) established an analogy between the work production by the muscle cells and Carnot engine (Example Figure 3.9.1).

In this analogy, one muscle cell is considered similar to the Carnot engine. This cell extracts energy from the high chemical potential reservoir, for example, O$_2$, glucose, and fatty acid–rich blood in the arteries. The *waste* energy is released in the form of chemical energy to the low chemical potential reservoir, that is, the blood in veins, which is rich in metabolic waste and CO$_2$. The following table lists the similar properties in the *ideal muscle cell* and the Carnot engine.

Carnot Engine	Ideal Muscle Cell
T_H	Chemical potential of the nutrient-rich blood in the artery, μ_H
T_L	Chemical potential of the nutrient-poor blood in the vein, μ_L
q_H	The Gibbs free energy extracted from the artery, ΔG_{in}
q_L	The Gibbs free energy rejected to the vein, ΔG_{out}
w	Muscle work performed via contraction, $w = \int F\,dl$, here F = fiber tension, dl = contraction length

The primary nutrient of a muscle cell is glucose. Oxidation of glucose via metabolic activity to carbon dioxide and water may be described as

$$\text{Glucose} + 6O_2 \rightarrow 6CO_2 + 6H_2O \qquad (E3.9.1)$$

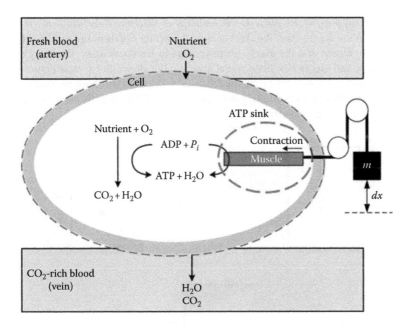

EXAMPLE FIGURE 3.9.1 Carnot engine analogy of work-performing muscle cell.

EXAMPLE TABLE 3.9.1 Enthalpy and the Gibbs Free Energy Change of the Chemicals Contributing to Reaction Given in (E3.9.1) at 310.15 K, pH = 7 with $I = 0.18$

	Glucose	Oxygen	Water	Carbon Dioxide
$\Delta h^T_{f,i}$ (kJ/mol)	−1267	−12	−287	−701
$\Delta g^T_{f,i}$ (kJ/mol)	2080	22	−149	−127

The enthalpy and the Gibbs free energy change of the chemicals contributing to this reaction under the conditions prevailing in the cell are given in Example Table 3.9.1 (Genç et al., 2013a, b).

Then we may calculate the enthalpy and the Gibbs free energy change upon this reaction as

$$\Delta h^T_R = \sum_{i=1}^{i=4} \alpha_i \Delta h^T_{f,i} = -4586 \text{ kJ/mol} \tag{E3.9.2}$$

and

$$\Delta g^T_R = \sum_{i=1}^{i=4} \alpha_i \Delta g^T_{f,i} = -3862 \text{ kJ/mol} \tag{E3.9.3}$$

An actual muscle cell converts the enthalpy of reaction, Δh_R^T, into heat and work. Accordingly, the Gibbs free energy of reaction equals to the sum of the work produced and the exergy destroyed due to irreversibilities. The energy, entropy, and exergy flow in an actual muscle cell is shown in Example Figures 3.9.2 and 3.9.3. Note that in a muscle cell $T_{system} = T_{body} = T_0$; therefore, heat transfer occurs through an infinitely small temperature difference. This means that there is no exergy transfer due to heat transfer.

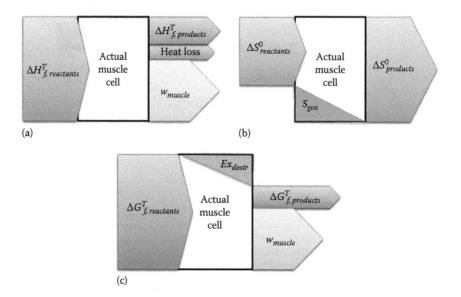

EXAMPLE FIGURE 3.9.2   (a) Energy, (b) entropy, and (c) exergy flow diagrams for a steady flow ideal muscle cell.

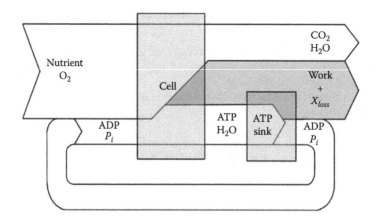

EXAMPLE FIGURE 3.9.3   Exergy chart of the muscle cell describing the inputs, outputs, and the destruction.

An *ideal muscle cell* produces the maximum work by eliminating all the internal and external irreversibilities. If all the irreversibilities can be eliminated, then the maximum work and the minimum heat release in a muscle cell via reversible processes are calculated as

$$w_{max} = \sum (mb)_{in} - \sum (mb)_{out} = 3862 \text{ kJ/mol glucose}$$

$$q_{min} = \sum (mh)_{out} - \sum (mh)_{in} + w = -724 \text{ kJ/mol glucose}$$

where $w_{max} = \Delta g$ and $q_{min} = \Delta h - \Delta g$ of reaction (E3.9.1).

In Example Table 3.9.2 the maximum work calculated based on the Carnot analogy is compared with the experimentally measured muscle efficiencies as reported by Smith et al. (2008).

The first law efficiency of the muscle work is defined as

$$\eta_I = \frac{W}{\Delta h} \tag{E3.9.4}$$

The maximum first law efficiency for glycolysis may be calculated after substituting Δg of the reaction (E3.9.1) for w_{max} in Equation E3.9.4:

$$\eta_{I, rev} = \frac{3862}{4586} = 84\%$$

The second law efficiency is defined as the ratio of the actually produced work to the maximum available work:

$$\eta_{II} = \frac{w}{w_{max}} = \frac{w}{\Delta g} = \frac{w}{\sum (m_j b_j)_{in} - \sum (m_j b_j)_{out}} \tag{E3.9.5}$$

EXAMPLE TABLE 3.9.2 First and Second Law Efficiencies of the Muscle Work Production

	Theoretical Limit of Maximum Work Production in a Reversible Process	Theoretical Limit of Maximum Heat Production when No Work Is Done	Measured Minimum Heat Release (Smith et al., 2008)	Measured Maximum Heat Release (Smith et al., 2008)
ex_{loss}	0	3862		
w	3862	0	3707	569
q	724	4586	879	4017
η_I	0.84	0	0.81	0.12
η_{II}	1	0	0.96	0.14

Source: Adapted from Sorgüven, E. and Özilgen, M., *Int. J. Exergy,* 18, 142, 2015.

Equation E3.9.4 implies that $w = \eta_I \Delta h$; therefore, the second law efficiency can be rewritten as

$$\eta_{II} = \eta_I \frac{\Delta h}{\Delta g} \qquad (E3.9.6)$$

The calculated maximum muscle efficiency of an *ideal muscle cell* compares well with the experimental measurements. The efficiency of the mechanical power output is estimated to be about 0.36 in rabbit psoas muscle fiber segments (He et al., 1999) and about 0.68 in the human first dorsal interosseous muscle (Jubrias et al., 2008).

Between ages 20 and 80, humans lose approximately 20%–30% of their muscles. The age-related loss of muscles is also accompanied with the loss of the efficiency and related with the mitochondrial functioning and the cellular energetics (Carmeli et al., 2002). The loss of efficiency in the process summarized by Equations E3.9.1 and E3.9.2 may be associated with the exergy loss and irreversibility in the cellular processes (Ketzer and de Meis, 2008; Silva and Annamalai, 2008; Lems et al., 2009; The and Lutz, 2011; Mady et al., 2012; Mady and de Olivera, 2013; Genç et al., 2013a; Sorgüven and Özilgen, 2013). Yaniv et al. (2013) argue that the myocardial ATP supply and demand mechanisms are age dependent. Although a vast number of papers are available in the literature, the complexity of the topic implies that there is still need for more research to understand the details in the cellular systems, which are not studied in detail yet. The decline of muscle work efficiency is associated with heart failure (Mariunas et al., 2008; Ribeiro et al., 2012). Inefficient calcium ion transportation in the muscles (Barclay, 2008), restriction of the diffusion of ADP from the site that it is produced (Vendelin et al., 2004; Barclay, 2008), and formation of the localized pools of adenine nucleosides around the sites of ATP use and generation (Joubert et al., 2002, 2008) are among the reasons of the failure. The use of partial circulatory support devices is hypothesized for treatment of the inefficient muscle work–related heart problems (Morley et al., 2007).

Numerous experimental and analytical studies are performed to determine the heat released and work produced via muscle contractions since the pioneer work of Hill (Hill, 1938; Reggiani et al., 1997; Holmes, 2006). Until recently, the heat released during the glycolysis and ATP hydrolysis was assumed to be constant. But recent studies show that heat released during ATP hydrolysis may vary depending on metabolic conditions such as Ca^{2+} concentration (Ketzer and de Meis, 2008). This observation is consistent with the model suggested by Genç et al. (2013a), where the Gibbs free energy and enthalpy difference of the reactions leading to Equation E3.9.2 and E3.9.3 were reported to vary substantially depending on the pH, ionic strength, and metabolite concentration in the cell. When 1 mol of glucose synthesizes 30 mol of ATP, the measured heat release varies between 879 and 4017 kJ for each mole of consumed glucose. When we compare the theoretical maximum limits with the literature values given in Example Table 3.9.3; the measured first law efficiencies appear

EXAMPLE TABLE 3.9.3   First and Second Law Efficiencies of
Isolated Muscles Undergoing Full Contraction Cycles

Muscle	T (°C)	η_I	η_{II}
Tortoise rectus femoris	15	0.35	0.42
Frog sartorius	12	0.25	0.30
Rat soleus	20	0.17	0.20
Mouse soleus	35	0.15	0.18
Mouse EDL	35	0.14	0.17

First law efficiencies, η_I, are taken from Smith et al. (2008) which present
the average values reported by Gibbs and Chapman (1974), Heglund and
Cavagna (1987), and Barclay (1996). The second law efficiencies, η_{II}, are
calculated with $\Delta h = -4586$ kJ/mol of glucose and $\Delta g = -3862$ kJ/mol of
glucose (Genç et al., 2013a).

dramatically lower. Among the factors complicating the experimental measurements, we may account the fraction of the energy that may be stored in the
elastic connections between the myofibrils (Linari et al., 2003) and the energy
that is absorbed during lengthening of the muscle and converted into mechanical work during subsequent shortening (Constable et al., 1997).

In the following, we will enhance the Carnot–*ideal muscle cell* analogy to derive a general model for biological systems. If we consider an evolved biological system, like the
human body, instead of one cell, then the thermodynamic system can be sketched as
shown in Figure 3.19.

For this macroscopic system, the reservoirs would be the nutrients and the atmospheric air. Blood would remain inside the system boundaries and it would be circulated

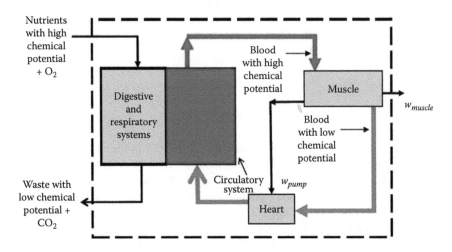

FIGURE 3.19   Schematic description of work production by the muscles.

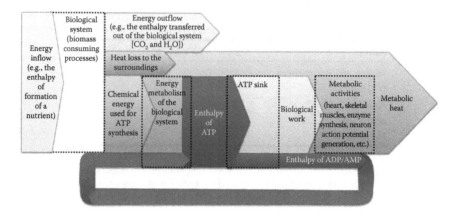

FIGURE 3.20   Energy flow in a Rankine cycle like biological system.

in a closed loop, much like the working fluid in the Rankine cycle. The *energy carrier* in the blood is the adenosine compound ATP, which enables muscles to perform work. Figure 3.20 shows the energy flow in such a system.

This model reduces the whole energy metabolism to two apparent chemical reactions: nutrient catabolism and ATP production. Comparing Figure 3.20 with the energy flow diagram of the Rankine cycle (Figure 3.6) shows that the energy metabolism acts like the boiler, which increases the *work production* potential of the working fluid via increasing its enthalpy by transferring Q_H. Similarly, the energy metabolism increases the *work production* potential of the adenosine compounds by increasing the number of the P-bonds by transferring Δh_{rxn}. This additional energy is consumed in various metabolic activities by breaking the P-bond and releasing the bond energy. Work production restores the *working fluid* into its original thermodynamic state, that is, ADP.

The overall coupled reactions that make the metabolic activities possible are (Demirel, 2010; Nelson and Cox, 2013)

$$ADP + P_i \rightarrow ATP$$

$$ATP \rightarrow ADP + P_i$$

One of these reactions describes the production of the ATP in the energy metabolism, which is then consumed with the following model reaction to do work as described in Example Figures 3.9.1 through 3.9.3. The ATP production reactions can be considered as reversible, which is a valid assumption, since the cellular biochemical reactions have efficiencies higher than 99% (Genç et al., 2013a). Even though the elementary reactions occurring are almost reversible, there are a number of irreversibilities that can occur in a biological system. For example, a cell has to transfer the nutrient and oxygen into the system boundaries. External irreversibility will occur if the cellular concentration of the reactants is larger than the concentrations in the reservoir. Similarly, depending on the physiological conditions, internal irreversibilities occur and a lesser number of ATP can be synthesized per consumed nutrient. It is reported that oxidation of 1 mol of glucose can result in 30–38 mol of ATP (Nelson and Cox, 2013).

3.3 Questions for Discussion

Q3.1 A power plant produces 300 MJ/h work for electric power generation. Steam enters to the turbine at 700°C and 8 MPa. It is discharged isentropically at 20 kPa.
(a) What should be the flow rate of the steam?
(b) Calculate the work done by an actual turbine that has the same inlet conditions but operates with an isentropic efficiency of 80%.

Q3.2 (a) In Example 3.6 operating conditions of a cascade refrigeration system was given. If this system operates with actual compressors that have isentropic efficiencies of 85%, then calculate the coefficient of performance of the system.
(b) Calculate the total entropy generated at the compressors.

Q3.3 Assuming that the cellular metabolism may be summarized with the reactions described as
Energy metabolism (simulated as oxidation of glucose):

$$C_6H_{12}O_6 + 6O_2 \rightarrow 6CO_2 + 6H_2O + q_{EM} \qquad \text{(Q3.3.1)}$$

with

$$q_{EM} = 6\Delta h^f_{CO_2} + 6\Delta h^f_{H_2O} - \Delta h^f_{C_6H_{12}O_6} - \Delta h^f_{O_2} \qquad \text{(Q3.3.2)}$$

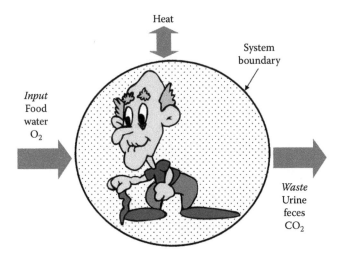

QUESTION FIGURE 3.3.1 Description of the *system* around which the mass and the entropy balances will be performed.

Lipid metabolism (simulated as oxidation of palmitic acid):

$$C_{16}H_{32}O_2 + 23O_2 \rightarrow 16CO_2 + 16H_2O + q_{LM} \qquad \text{(Q3.3.3)}$$

with

$$q_{LM} = 16\Delta h^f_{CO_2} + 16\Delta h^f_{H_2O} - \Delta h^f_{C_{16}H_{32}O_2} - 23\Delta h^f_{O_2} \qquad \text{(Q3.3.4)}$$

The conversion of amino acids to urea:

$$C_6H_{12}N_2O_4S_2 + 7.5O_2 \rightarrow 5CO_2 + 4H_2O + CH_4N_2O + 2SO_2 + q_{AAM} \qquad \text{(Q3.3.5)}$$

$$q_{AAM} = 5\Delta h^f_{CO_2} + 4\Delta h^f_{H_2O} + \Delta h^f_{CH_4N_2O} + 2\Delta h^f_{SO_2} - \Delta h^f_{C_6H_{12}N_{24}O_4S_2} - 7.5\Delta h^f_{O_2} \quad \text{(Q3.3.6)}$$

Estimate the amount of the entropy generation by the person depicted in Question Figure 3.3.1, after consuming 1 kg of food consisting of 20% carbohydrate, 10% lipid, 5% amino acid, and 65% water, if the entropy change of the body is negligibly small.

4

Thermodynamic Aspects of Biological Processes

4.1 Biological System and Biological Processes

In the previous chapter, we categorized mechanical cycles as work-producing (heat engines) and work-consuming (heat pumps) systems. Metabolic activities performed by biological systems can be categorized similarly as *biomass-consuming* and *biomass-producing* processes. The driving force of the biological systems is the internal energy of chemical substances. Biomass-consuming processes catabolize high-energy-level chemical substances (i.e., nutrients) and produce work to sustain cellular activities. Biomass-producing processes employ metabolic or solar energy to produce long chains of hydrocarbons (Figure 4.1).

Like the mechanical systems, all biological systems are also governed by mass, energy, entropy, and exergy balances. Figure 4.2 shows a general biological system with multiple inlet and outlet ports, which are denoted with the subscripts *in* and *out*, respectively. Chemical species flow in or out of the system boundaries through the inlet/outlet ports. Organisms usually control the flow rate and chemical composition at each inlet/outlet port either via passive semipermeable membranes (like the epidermis of human skin) or via actively controlled gates (like the Na^+ gates in axons). In general, a biological system can be in contact with multiple heat reservoirs. The temperature of the boundary segment through which the heat q_j is transferred is denoted as T_j.

The following equations summarize the governing equations for the system depicted in Figure 4.2:

$$\sum_{in} \dot{m}_{in} - \sum_{out} \dot{m}_{out} = \frac{dm_{system}}{dt} \tag{4.1}$$

$$\sum_{in} [\dot{m}(h + e_p + e_k)]_{in} - \sum_{out} [\dot{m}(h + e_p + e_k)]_{out} + \sum_i \dot{Q}_i - \dot{W} = \frac{d[m(u + e_p + e_k)]_{system}}{dt} \tag{4.2}$$

$$\sum_{in} [\dot{m}s]_{in} - \sum_{out} [\dot{m}s]_{out} + \sum_i \frac{\dot{Q}_i}{T_{b,i}} + \dot{S}_{gen} = \frac{d[ms]_{system}}{dt} \tag{4.3}$$

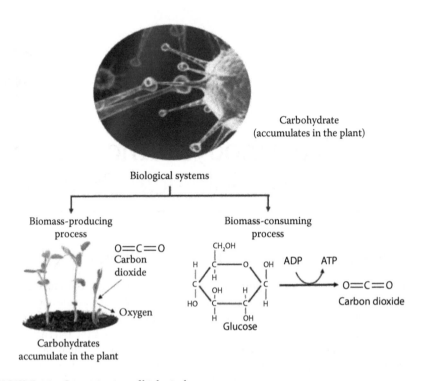

FIGURE 4.1 Categorization of biological processes.

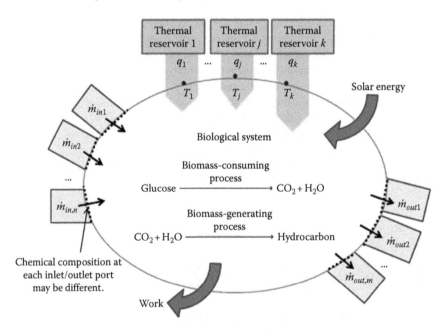

FIGURE 4.2 Schematic description of a general biological system.

$$\sum_{in} [\dot{m}ex]_{in} - \sum_{out} [\dot{m}ex]_{out} + \sum_{i} \left(1 - \frac{T_0}{T_{b,i}}\right) \dot{Q}_i - W - \dot{Ex}_{destr} = \frac{d[mex]_{system}}{dt} \quad (4.4)$$

Even in the simplest biological system, hundreds of biochemical reactions occur simultaneously, and at any given time, hundreds of biochemical species are present within the system boundaries. The mass of each chemical species may change with respect to time, depending on the net influx of the species through the ports and the rates of the reactions that produce or generate the species. Thus, the mass balance for the ith chemical species in a biological system can be written as

$$\sum_{in} \dot{m}_{i,\,in} - \sum_{out} \dot{m}_{i,\,out} + \sum_{rxn} \dot{m}_{i,\,generated} - \sum_{rxn} \dot{m}_{i,\,consumed} = \left(\frac{dm_i}{dt}\right)_{system} \quad (4.5)$$

Chemical species that flow into the system or that is generated via the biochemical reactions in the system may be excreted out or accumulate in the system. Organisms tend to excrete waste but accumulate the useful substances. For example, biomass-consuming species, like animals and humans, accumulate the unused nutrients in the form of fat. If fat is stored in a biological system, then both the total mass of the system $(dm/dt)_{system}$ and the total internal energy of the system $(dmu/dt)_{system}$ increase.

4.1.1 Biomass-Consuming Process

In biomass-consuming processes, long-chain hydrocarbons, such as fats, proteins, and carbohydrates are oxidized via energy metabolism. The main purpose of the energy metabolism is to produce ATP. ATP is used in a cell to perform work, such as muscle contraction. In a complete energy metabolism, nutrients are reduced to carbon dioxide and water. In the absence of oxygen, energy metabolism may not be completed, and the end product becomes lactic acid instead of carbon dioxide. Figure 4.3 shows the energy flow of a steady-state energy metabolism. The difference in the energy levels of the inflowing and outflowing chemicals is converted into metabolic heat and work. The amount of the work produced depends on the efficiency of the system.

Energy metabolism consists of three stages: (1) glycolytic pathway, (2) tricarboxylic acid cycle, and (3) the electron transport chain (ETC) as described in Figure 4.4. Glucose and oxygen are taken up by the cell through the extracellular matrix, and immediately glucose enters glycolysis, producing pyruvate and NADH. This pyruvate enters the citric

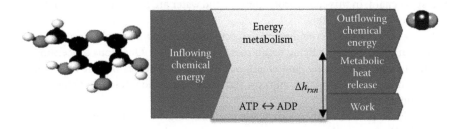

FIGURE 4.3 A biomass-consuming process.

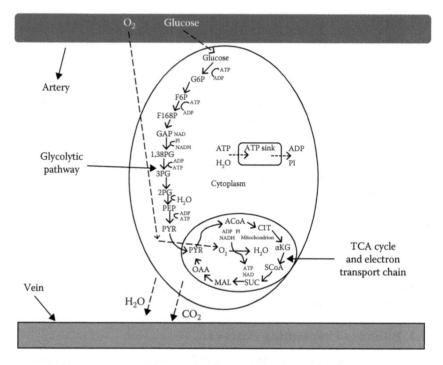

FIGURE 4.4 Schematic description of the cellular energy metabolism model.

acid cycle in the mitochondria, and gets fully oxidized with the help of ETC on the inner mitochondrial membrane. The overall output of the system is carbon dioxide and water, which are transported to blood. Any work that consumes energy (ATP) in the cell is implicitly shown as an ATP sink. Transport events are shown with dashed lines, whereas biochemical reactions are shown with solid lines.

There are models in the literature where the energy metabolism is described with a set of representative reactions (Tables 4.1 and 4.2).

Under anaerobic conditions, after following the glycolytic pathway, pyruvate may be converted into either lactate or ethanol. Under the presence of normal levels of oxygen, pyruvate and NADH are translocated to the mitochondria and used to produce 28 mol of additional ATP through the citric acid cycle and then to the electron transport chain; in this case, carbon dioxide and water are produced instead of lactate or ethanol (Berg et al., 2002).

Understanding the thermodynamic aspects of the energy metabolism would enable us to gain insight on how living organisms are functioning. Questions concerning the health state of a cell, aging, and emergence of cellular diseases may be answered by understanding how the energy, entropy, and exergy of metabolites are manipulated by intracellular reactions.

In order to perform a thermodynamic analysis, first, we need to define our system precisely. Here, we will explain the thermodynamic analysis of the energy metabolism based on the *model cell* proposed by Genç et al. (2013a,b). This model cell is in contact

TABLE 4.1 Glycolytic Steps of the Energy Metabolism That Occur in the Cytoplasm

$Glu_c + ATP_c \rightarrow G6P_c + ADP_c$

$G6P_c \rightarrow F6P_c$

$F6P_c + ATP_c \rightarrow F16BP_c + ADP_c$

$F16BP_c \rightarrow 2GAP_c$

$2GAP_c + 2Pi_c + 2NAD_c \rightarrow 2(13BPG)_c + 2NADH_c$

$2(13BPG)_c + 2ADP_c \rightarrow 2(3PG)_c + 2ATP_c$

$2(3PG)_c \rightarrow 2(2PG)_c$

$2(2PG)_c \rightarrow 2(PEP)_c + 2(H_2O)_c$

$2PEP_c + 2ADP_c \rightarrow 2PYR_c + 2ATP_c$

The following model reactions are employed by Genç et al. (2013a,b) to represent the glycolytic steps of the energy metabolism.

TABLE 4.2 Steps of the Energy Metabolism That Occur in the Mitochondria

$2(PYR_m + CoA_m + NAD_m \rightarrow ACoA_m + NADH_m + CO_{2m})$

$2(ACoA_m + OAA_m + H_2O_m \rightarrow CIT_m + CoA_m)$

$2(CIT_m \rightarrow ISOCIT_m)$

$2(ISOCIT_m + NAD_m \rightarrow \alpha KG_m + CO_{2m} + NADH_m)$

$2(\alpha KG_m + CoA_m + NAD_m \rightarrow SCoA_m + CO_{2m} + NADH_m)$

$2(SCoA_m + ADP_m + Pi_m \rightarrow SUC_m + CoA_m + ATP_m)$

$2(SUC_m + (2/3)NAD_m \rightarrow FUM_m + (2/3)NADH_m)$

$2(FUM_m + H_2O_m \rightarrow MAL_m)$

$2(MAL_m + NAD_m \rightarrow OAA_m + NADH_m)$

$O_{2m} + 26ADP_m + 26Pi_m + (34/3)NADH_m \rightarrow 28H_2O_m + 26ATP_m + (34/3)NAD_m$

The following model reactions are employed by Genç et al. (2013a,b) to represent the tricarboxylic acid steps and the electron transport chain steps of the energy metabolism.

with the extracellular fluid as shown in Figure 4.5. The overall system *model cell* is divided into six subsystems:

1. *Mitochondrion*: Here, pyruvate is degraded via the citric acid cycle and the ETC to produce ATP, that is, reactions listed in Table 4.2 are occurring here. Mitochondrial fluid is assumed to be an ideal solution with uniform T_m, p_m, and $c_{i,m}$.

2. *Mitochondrial membrane* (T1): One boundary of this system is inside the cytoplasm, and the other is inside the mitochondrion (Figure 4.5). No reactions occur here. The system T1 transfers the reactants of the mitochondrial reactions from cytoplasm to mitochondrion and the products from mitochondrion to cytoplasm. The system has uniform temperature and pressure. The concentrations of the metabolites vary linearly between the two boundaries.

3. *Cytoplasm*: Reactions listed in Table 4.3 occur here. Cytoplasmic fluid is assumed to be an ideal solution with uniform T_c, p_c, and $c_{i,c}$.

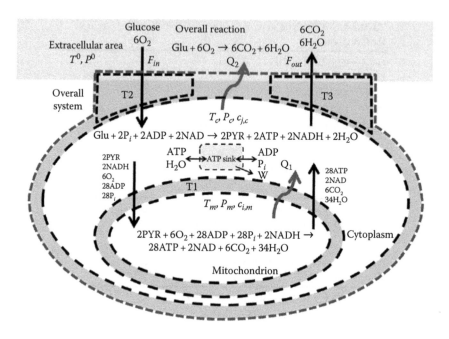

FIGURE 4.5 The *model* cell.

4. *ATP sink*: The net ATP produced flows into the ATP sink. A stoichiometric amount of water flows also in. The subsystem *ATP sink* represents all the cellular activities that need ATP hydrolysis. These may change from cell to cell. For example, a neuron may need ATP hydrolysis to generate an electrical signal, whereas a muscle cell may need it for contraction. Products of the ATP hydrolysis (i.e., ADP and P_i) flow out of the *ATP sink*, as well as work. This system is also an ideal solution with constant temperature and pressure, and a linear concentration gradient for ATP, ADP, H_2O, and p_i.

5. *Cellular membrane for influx* (T2): In an actual cell, mass transfer between the surroundings and the cytoplasm occurs through gates, which are distributed nearly homogenously along the cellular membrane. Here, we have simplified the mass transfer process by defining the subsystem T2 (Figure 4.5) as the only part of the membrane that allows influx of metabolites. Through T2, glucose and oxygen are transported into the cell. One boundary of T2 is inside the extracellular region, and on that boundary, the metabolite concentrations are equal to the concentrations in the extracellular region ($c_{i,0}$). The other boundary is inside the cytoplasm, where the metabolite concentrations are equal to $c_{i,c}$.

6. *Cellular membrane for outflux* (T3): This subsystem is similar to T2, except that only the outflux of H_2O and CO_2 is allowed here (Figure 4.5).

Once the system (in this case six subsystems) is defined, the next step of the thermodynamic analysis is to perform the mass balance. For each subsystem and each metabolite, Equation 4.5 is to be solved. Since our model is at steady state, there is no accumulation

TABLE 4.3 Thermodynamic Properties of Each Metabolite in the System as Calculated Using the Equations Given in Section 2.6

Metabolite	$[c]$ (mM)	Δm_f^T (kJ/mol)	Δk_f^T (kJ/mol)	ex_{ch}^0 (kJ/mol)	$RT \ln [c]$ (kJ/mol)	ex (kJ/mol)
Acetyl coenzyme A$_c$	1.00E+03	−1.30	−69.07	12,672.50	0.00	12,603.43
ADP$_c$	2.70E+02	−2644.19	−1376.62	5,851.38	−3.38	4,471.38
ADP$_m$	2.50E+02	−2644.19	−1376.62	5,851.38	−3.57	4,471.18
ADP$_{bl,in}$	3.00E+02	−2644.19	−1376.62	5,851.38	−3.10	4,471.65
ADP$_{bl,out}$	2.70E+02	−2644.19	−1376.62	5,851.38	−3.38	4,471.38
ATP$_c$	2.00E+02	−3632.27	−2239.05	6,720.93	−4.15	4,477.73
ATP$_m$	2.20E+02	−3632.27	−2239.05	6,720.93	−3.90	4,477.98
1,3-bisphosphoglycerate$_c$	4.00E+03	1.14	−2295.63	2,977.83	3.57	685.77
Citrate$_m$	1.00E−05	−1524.00	−943.91	2,475.46	−47.50	1,484.04
Isocitrate$_m$	3.39E+00	−3.97	−997.98	2,475.46	−14.67	1,462.81
Coenzyme A$_m$	5.00E−08	−1.82	−7.60	11,850.04	−61.17	11,781.27
ATP$_{bl,in}$	1.00E+00	−3632.27	−2239.05	6,720.93	−17.81	4,464.07
ATP$_{bl,out}$	3.10E+01	−3632.27	−2239.05	6,720.93	−8.96	4,472.92
CO$_{2,m}$	3.80E+01	−700.59	−541.25	414.23	−8.43	−135.45
Dihydroxy acetone phosphate$_c$	1.40E−02	−3.23	−1139.87	2,106.29	−28.82	937.60
Fructose6phos$_c$	5.00E+00	−5.84	−1369.12	3,343.03	−13.66	1,960.24
Fructose16phos$_c$	5.00E+00	−4.45	−2295.25	4,212.58	−13.66	1,903.67
Fumarate$_m$	6.00E−08	−779.05	−513.24	1,648.98	−60.70	1,075.04
CO$_{2,c}$	3.70E+01	−700.59	−541.25	414.23	−8.50	−135.52
CO$_{2,bl,in}$	1.00E+00	−700.59	−541.25	414.23	−17.81	−144.83
CO$_{2,bl,out}$	7.00E+00	−700.59	−541.25	414.23	−12.80	−139.82
Glucose$_c$	1.20E+00	−1267.40	−393.80	2,473.47	−17.34	2,062.33
G6P$_c$	1.70E−01	−2281.24	−1280.83	3,343.03	−22.38	2,039.81
GAP$_c$	1.40E−02	−3.23	−1131.90	2,106.29	−28.82	945.57
Glucose$_{bl,in}$	7.00E+00	−1267.40	−393.80	2,473.47	−12.80	2,066.87
Glucose$_{bl,out}$	6.00E+00	−1267.40	−393.80	2,473.47	−13.19	2,066.48
H$_2$O$_c$	1.80E+02	−286.70	−150.53	1.99	−4.42	−152.97
H$_2$O$_m$	2.00E+02	−286.70	−150.54	1.99	−4.15	−152.71
α-Ketoglutarate$_m$	1.70E−01	0.00	−659.09	2,061.23	−22.38	1,379.76
Malate$_m$	9.40E−03	−2.80	−710.33	1,650.97	−29.85	910.79
H$_2$O$_{bl,in}$	4.00E−04	−286.70	−150.53	1.99	−37.99	−186.53
H$_2$O$_{bl,out}$	2.60E+01	−286.70	−150.53	1.99	−9.41	−157.96
NAD$_c$	5.70E−02	−10.86	1100.24	10,372.87	−25.20	11,447.90
NAD$_m$	2.12E+00	−10.86	1100.24	10,372.87	−15.88	11,457.23
NADH$_c$	6.40E−03	−41.93	1165.09	10,372.87	−30.84	11,507.12
NADH$_m$	2.40E−01	−41.93	1165.09	10,372.87	−21.49	11,516.46
O$_{2,m}$	1.00E−04	−11.70	17.53	3.97	−41.56	−20.06
Oxaloacetate$_m$	1.70E−07	0.87	−743.65	1,650.97	−58.01	849.31

(Continued)

TABLE 4.3 (*Continued*) Thermodynamic Properties of Each Metabolite in the System as Calculated Using the Equations Given in Section 2.6

Metabolite	$[c]$ (mM)	Δm_f^T (kJ/mol)	Δk_f^T (kJ/mol)	ex_{ch}^0 (kJ/mol)	$RT \ln [c]$ (kJ/mol)	ex (kJ/mol)
Phosphenol pyruvate$_c$	2.50E−04	−0.37	−1237.25	2,106.29	−39.20	829.84
2-phospho-D-glycerate$_c$	1.00E−03	−1.23	−1395.64	2,108.28	−35.63	677.01
3-phospho-D-glycerate$_c$	2.00E−02	−1.23	−1401.80	2,108.28	−27.90	678.57
$O_{2,c}$	1.00E−04	−11.70	17.53	3.97	−41.56	−20.06
$O_{2,bl,in}$	7.00E+00	−11.70	17.53	3.97	−12.80	8.70
$O_{2,bl,out}$	6.00E+00	−11.70	17.53	3.97	−13.19	8.31
Pi_c	2.00E+02	−1303.34	−1049.70	871.54	−4.15	−182.31
Pi_m	1.89E+02	−1303.34	−1049.71	871.54	−4.30	−182.47
$Pi_{bl,in}$	2.00E+02	−1303.34	−1049.70	871.54	−4.15	−182.31
$Pi_{bl,out}$	1.70E+02	−1303.34	−1049.70	871.54	−4.57	−182.73
Pyruvate$_c$	1.00E−10	−597.09	−341.02	1,236.74	−77.19	818.52
Pyruvate$_m$	1.00E−10	−597.09	−341.02	1,236.74	−77.19	818.52
Succinate$_m$	6.50E−01	−915.46	−515.43	1,648.98	−18.92	1,114.63
SuccinylcoA$_m$	8.00E−01	−3.56	−361.63	13,497.00	−18.39	13,116.98
Lactate$_c$	1.00E+00	−688.38	−298.93	1,236.74	−17.81	920.00
Lactate$_{bl,in}$	1.40E+01	−688.38	−298.93	1,236.74	−11.01	926.80
Lactate$_{bl,out}$	1.20E+01	−688.38	−298.93	1,236.74	−11.41	926.40

within the system boundaries. Accordingly, the net consumption of the metabolite i via chemical reactions inside the system boundaries $\left(-\sum_{rxn} \dot{m}_{i,generated} + \sum_{rxn} \dot{m}_{i,consumed}\right)$ has to be equal to the net influx of the metabolite $\left(\sum_{in} \dot{m}_{i,in} - \sum_{out} \dot{m}_{i,out}\right)$ into this system. Accordingly, mass fluxes of the metabolites can be calculated based on the reaction rates. Since the system is at steady state, all of the mass fluxes can be calculated as a function of only one reaction rate, for example, the oxygen consumption rate. Note that, if the system would undergo an unsteady process, then the ordinary differential equations that describe the mass balance of each species have to be solved simultaneously. Interested reader can find the details of such an analysis in Genç et al. (2013b).

Concentrations of the species in the cytoplasm, mitochondria, and blood are either taken from the literature (Aubert and Costalat, 2005; Zhou et al., 2005) or estimated via a kinetic model (Genç et al., 2011). Mass balance for the overall system dictates

$$c_{i,bl,out} = \frac{(c_{i,bl,in}\dot{V}_{in} + N_i b_{O_2})}{\dot{V}_{out}}$$

where
    N_i is the stoichiometric constant of the species i
    b_{O_2} is the consumption rate of oxygen (mmol/h)
    $\dot{V}_{in}$ and $\dot{V}_{out}$ are the volumetric blood flow rate at the inlet and the outlet of the neuron, respectively

Here, $\dot{V}_{in} = \dot{V}_{out} = 1\,\text{L/h}$, and b_{O_2} is varied between 0.5 and 4 mmol/h according to the results of the kinetic analysis (Genç et al., 2011).

Now that the mass fluxes and concentrations for each metabolite in each subsystem are calculated, energy and exergy analyses can be performed. For this, we need the thermodynamic properties of the metabolites in the cell. Reference conditions for standard Gibbs free energy and enthalpy of formation are 298.15 K and 101.0 kPa (Szargut et al., 1998). Physiological conditions for the cell are $T = 310.15$ K, pH = 7, with ionic strength $I = 0.18$ M, and ion concentrations of $[\text{Mg}^{2+}] = 0.8$ mM, $[\text{K}^+] = 140$ mM, $[\text{Na}^+] = 10$ mM, and $[\text{Ca}^{2+}] = 0.00001$ mM (Li et al., 2010). In Chapter 2, we have mentioned that in biochemical systems under physiological conditions, where pH and ionic strength are controlled and transformed, thermodynamic properties have to be employed. Examples in Section 2.7 demonstrate the calculation of the transformed Gibbs free energy and the transformed enthalpy of formation under physiological conditions. Table 4.3 lists the thermodynamic properties of each metabolite in the system as calculated using the equations given in Section 2.6.

The first reaction in Table 4.1, for example, phosphorylation of glucose in the cytoplasm is

$$\text{Glu}_c + \text{ATP}_c \rightarrow \text{G6P}_c + \text{ADP}_c$$

Heat released in this reaction is

$$\sum \left(n\Delta h_f^T \right)_{i,in} - \sum \left(n\Delta h_f^T \right)_{i,out} = q$$

where

$n_{i,in}$ and $n_{i,out}$ are the number of moles of each chemical species i entering (reactants) or leaving (products) the system
Δh_f^T is the enthalpy of each chemical species

Accordingly, the heat released and exergy destruction in the phosphorylation reaction are calculated as (Genç et al., 2013a)

$$q = 25.8 \text{ kJ/mol}$$

The exergy destruction, $ex_{dstrctn}$, of the phosphorylation reaction of glucose is computed as

$$ex_{dstrctn} = ex_{\text{Glu}} + ex_{\text{ATP}} - ex_{\text{G6P}} - ex_{\text{ADP}} = 28.9 \text{ kJ/mol}$$

Note that the exergy transfer via heat transfer is zero, since the boundary temperature and the reference temperature are the same. In other words, the term $\sum_j (1 - (T_0 - T_j))q_j$ in Equation 4.4 is equal to $(1-(T_0 - T_j))q = 0$, since $T_0 = T_b = 310.15$ K.

The exergetic efficiency of a reaction may be calculated as

$$\eta_{rxn} = \frac{\sum ex_{products}}{\sum ex_{reactants}}$$

where

$\sum ex_{products}$ is the sum of the exergy of products

$\sum ex_{reactants}$ is the sum of reactants

Calculations show that the exergetic efficiency of all of the metabolic reactions listed in Tables 4.1 and 4.2 are greater than 99% (Genç et al., 2013a,b).

The exergy destruction in each subsystem of Figure 4.5 is listed in Table 4.4.

If we interpret the total exergy destruction values as listed in Table 4.4, we would conclude that most of the exergy is destructed in the mitochondrion. Note that, most of the ATP is produced in the mitochondrion. To perform a fair comparison between the anaerobic reaction in the cytoplasm and the aerobic reactions in the mitochondrion, Table 4.5 shows the exergy destroyed per 1 mol of ATP produced. Cytoplasmic reactions destroy nearly three times more exergy that produce 1 mol of ATP. Thus, the efficiency of the anaerobic energy metabolism in the cytoplasm is lower than the aerobic metabolism in the mitochondrion.

In order to understand how the biochemical reactions occur, we can analyze the exergy flow diagrams of each subsystem. Figure 4.6 shows the exergy flow in the cytoplasm.

In the cytoplasm, the main exergy carrier is glucose. One mol of glucose brings 2062.33 kJ of exergy into the cytoplasm. Of this exergy, 118.4 kJ is used to

TABLE 4.4 Exergy Destruction in Each Subsystem in Figure 4.5

Subsystem	Exergy Destruction (kJ)
Cytoplasm	262.3
ATP sink	669.0
Mitochondrion	1247.6
T1	17.1
T2	32.2
T3	22.1

TABLE 4.5 Exergy Destroyed Per 1 mol of ATP Produced in the System Described in Figure 4.5

Subsystem	Exergy Destruction (kJ)	Number of ATP Produced	Exergy Destroyed/mol ATP (kJ/mol)
Cytoplasm	262.3	2	131.15
Mitochondrion	1247.6	28	44.55

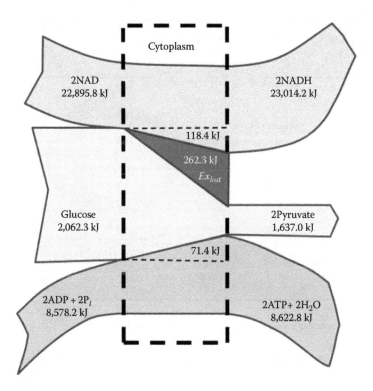

FIGURE 4.6 Exergy flow in the cytoplasm.

produce NADH from NAD and 71.4 kJ is used for the conversion of $2ADP + 2P_i$ into $2ATP + 2H_2O$. In other words, almost 8% of the exergy from glucose is consumed to drive production reactions (NADH, ATP), while 12.7% of it is lost due to irreversibilities. The main output, pyruvate, carries the rest of the exergy. Considering that glucose is used up to produce pyruvate, the cumulative degree of perfection of pyruvate can be defined as

$$CDP_{ATP, cytoplasm} = \frac{\sum (mex)_{ATP}}{\sum (mex)_{in}} = 0.13$$

Studies of Genç et al. (2013a,b) demonstrate how thermodynamic analyses may be utilized to understand how a cell switches between different metabolic pathways to reduce the entropy generation rate or react under varying conditions, such as starvation or hypoxia.

4.1.2 Thermodynamic Assessment of the Work Generated via Biomass-Consuming Processes

In the previous section, the energy metabolism of a model cell is analyzed in detail by considering the biochemical reactions occurring in the cell. We have now a detailed information about the biochemical reactions, which produce ATP. The main purpose

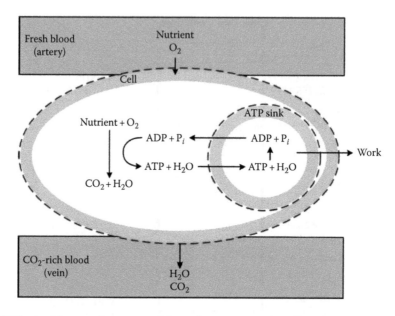

FIGURE 4.7　Schematic description of the work generation in the cell.

of the energy metabolism is to generate ATP, which is used by the cell to perform work. Work generated via biomass-consuming processes can be used for several purposes, for example, transporting of a vital chemical in to the cell, via activating gates on the cellular membrane. Another example can be expressing enzymes that are required for the survival of an organism. Since the role of each cell in an organism is different, the way that ATP is consumed and, hence, the definition of the work performed are different. In this section, we will discuss what happens inside the ATP sink and how work is generated (Figure 4.7).

The following examples aim to demonstrate different aspects of the biological work and provide methods to estimate it.

Example 4.1: Work Performance during Building a Spider Web by the European Garden Spider, *Araneus diadematus*

Orb-web-building spiders have developed highly energy-efficient lifestyles. They build their webs with the silk they produce by themselves and then wait at its center without moving for the prey to come. Internal energy of the prey sustains all the activities (Example Figure 4.1.1). The silk has a very strong structure (Qin et al., 2015) and some parts of it are covered with very effective biological glue (Choresh et al., 2009). Spiders may wrap and preserve their prey in their silk until eating them. Spider web inspired the production of numerous products. For example, Jin et al. (2013) report fabrication of a chitin nanofiber silk

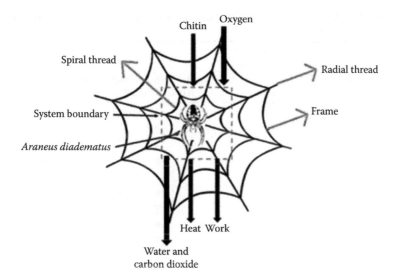

EXAMPLE FIGURE 4.1.1   Schematic drawing of the web of European garden spider.

biocomposite made of chitin nanofibers embedded in a silk-like protein matrix. The energy efficiency of the silken web-making process determines the sustainability of the life of the spiders. They may eat their web and rebuilt it at a different location at a very modest energy expenditure. A web may be recycled by using only 3.8% of the energy that would be needed to rebuild it from the very beginning, for example, synthesis of the silk for web making (Peakall and Witt, 1976).

In this example, we will consider building a web with 22 radial threads. There are pieces of silk named as *spirals* between the radiuses (Example Figure 4.1.1). Peakall and Witt (1976) calculated the energy utilized by *Araneus diadematus* to carry its body for one step as 1.254×10^{-3} J/[(g body weight)(step)] and expressed the energy utilization in *number of steps* equivalent. The number of the steps of spider's work employed for making the radial and spiral threads spirals are given in Example Table 4.1.1:

(Total number of the steps involved) = (Number of the steps employed in making the radial threads) + (Number of the steps employed in making the spiral threads) + (The pendulum work) + (Number of the steps needed to build the site where the web will be placed in) + (Steps of the nonweb making activities involved)

In our analysis, the number of the steps needed to build the site is referred to as the *provisional work*. Spider swings around like a pendulum in some stages of the web-building activity, and this is referred to as the *pendulum work* and expressed in equivalent *steps* in MATLAB® code E4.1. Mass of the spider is taken as $m_{spider} = 0.1$ g.

EXAMPLE TABLE 4.1.1   Number of the Steps of Spider's Work Employed for Making the Radial and Spiral Threads

Radial Thread Number	Number of Steps of Spider's Work Employed while Manufacturing the Radial Thread	Circle Number (Center Is 0, Frame Is 20)	Number of Steps of Spider's Work Employed while Manufacturing Each Spiral
1	22	1	1
2	21	2	1
3	23	3	1
4	18	4	1
5	17	5	2
6	13	6	2
7	17	7	2
8	19	8	2
9	20	9	2
10	21	10	3
11	22	11	3
12	21	12	3
13	19	13	3
14	17	14	3
15	15	15	3
16	15	16	3
17	17	17	3
18	19	18	3
19	19	19	3
20	20	20	3
21	21		
22	22		

Source: Data adapted from Peakall, D.B. and Witt, P.N., *Comp. Biochem. Physiol.*, 54, 187, 1976.

MATLAB CODE E4.1

Command Window

```
clear all
close all

% ENTER THE DATA
circleNo=14; % number the spiral circles
radiusNo=22; % number the radial threads
energyForOneStep = 1.254e-4; % (J/[(g body weight) (step)])
massOfSpider = 0.100; % (g)
nonWebMakingW = 240; % non-web making work (steps)

% number of the steps needed to build the spirals in the 1st
  to 4th circles from the center is 1
```

```
% number of the steps needed to build the spirals in the 5st
  to 9th circles from the center is 2

% number of the steps needed to build the spirals in
  the 10th to 14th circles from the center is 3

% vector of the number of the steps of work done to produce
  each spiral thread
  stepsS = [1,1,1,1,2,2,2,2,2,3,3,3,3,3];
  spirals = 22; % number of the spirals in each circle

% vector of the number the steps of work done to produce
  each radial thread
      stepsR =
[22,21,20,18,17,15,17,19,20,21,22,21,14,17,15,15,17,19,19,20,21,22];

% calculate the number of the steps employed while producing
  the radial threads
  sumR = 0;
     for i =1: radiusNo
         sumR= sumR+ stepsR(i);
     end

% calculate the number of the steps employed while producing
  the spiral threads
  sumS = 0;
     for j =1: circleNo
       sumS= sumS+ spirals*stepsS(j);
    end

% calculate the pendulum and the provional works
  PendulumWork=3*7*5; % steps equivalent of the pendulum work
  ProvionalWork=7*3*radiusNo; % steps equivalent of the
    provisional work

% add the steps of the pendulum and provional works to the sum
       sumT = sumR+sumS+PendulumWork + ProvionalWork +
  nonWebMakingW;

% display the total number of steps of work done to build
  the web
  disp('Total number of steps of work done to build the web ')
  totalNumberOfSteps =sumT

% display the total work done
  disp('Total work done to build the web (J)')
  totalW = totalNumberOfSteps*energyForOneStep*massOfSpider

%
%
```

When we run the code the following lines will appear in the screen

```
Total number of steps of work done to build the web
totalNumberOfSteps =
    1857
Total work done to build the web (J)
totalW =
                0.0233
```

$$W_{web\,building} = (1857\ \text{steps})(1.254 \times 10^{-4}\ \text{J/steps g spider})(0.1\ \text{g spider}) = 0.0233\ \text{J}.$$

The web had a very small weight, 0.1 mg, and the spider performed (0.0233 J)/ (0.1 × 10⁻³ g) = 233 J/g of work to synthesize it. If the spider would not use the readily produced raw material for web making, the cost of the web-building process would be (233 J/g)/(0.038) = 6131.6 J/g. In addition to the spider's use of energy sparingly because of the long waiting stages for the prey at the basal energy utilization rates, and its ability of recycling the already synthesized chemicals, the sustainability of the process is also supported by the strength of the web, which made it possible for the spider to produce the thread in the form of very thin fibers of light weight.

Example 4.2: Nectar Collection Energy Efficiency of the Honeybees

Wasps and bees have very complicated energy utilization schemes. They usually do not regulate their body temperature, when they are in the shade. Specific heat production rate of the honeybees increases with the ambient temperature (Schmolz et al., 2002). They benefit from the solar energy to maintain their body temperature a few degrees over that of the environment, when they are foraging, for example, collecting nectar from the plants, under the sunshine (Stabentheiner et al., 2012). The weight of a honeybee may be 75–165 mg (Stabentheiner et al., 2012). It has a very high capability to do muscle work. Thorax, a division of the body that lies between the head and the abdomen, accounts for 20%–40% of the honeybee's body mass, and the flight muscle occupies approximately 75% of the thorax (Harrison and Fewell, 2002).

Harrison and Fewell (2002) report that a honeybee may carry a nectal load of 30 μL (energy content 9 J/μL). Each trip for nectar collection may take on the average 30 min, and the metabolic energy cost of a foraging trip at 30°C is 2.5 J/bee min. Honeybees may make 12 foraging trips/day. Only 10% of the bees in a hive are foraging, while the others remain in the hive. In hive, metabolism of a honeybee may be assumed to be 0.16 J/bee min.

Energy of the product is (9 J/μL) × (30 μL/trip) × (12 trips/day) = 3240 J/day. Energy cost of the production is the sum of the energy utilized by the foragers and the energy utilization by the bees staying in the hive.

Energy utilization by the foraging bees is (metabolic energy utilization rate/ trip) × (duration of a trip) × (number of trips/day):

$$= (2.5\ \text{J/min}) \times (30\ \text{min/trip}) \times (12\ \text{trips/day}) = 30.4\ \text{J/day}$$

A foraging bee spends 6 h for foaraging and spends the rest of the day in the hive; 10% of the bees forage, and the rest stays in the hive. Energy expenditure of the bees staying in the hive is

$$[(18/24)\times(0.10)+(24/24)\times(0.90)]\times(0.16\,J/min)$$

$$\times(60\,min/1\,h)\times(24\,h/day)=224.6\,J/day$$

$$\frac{\text{Energy of the product}}{\text{Energy cost of production}}=\frac{3240\,J/day}{30.4\,(J/day)+224.6\,(J/day)}=12.7$$

These results imply that under the given conditions, the honeybees provide 12.7 times of the energy they utilize.

Example 4.3: Thermodynamic Assessment of the Argument That the Oriental Wasp, *Vespa orientalis*, Flight May Be Achieved Employing the Sunshine as the Exclusive Energy Source

Recent experimental studies suggest that the solar energy absorbed by the cuticle at the outer body surface of the oriental wasp may be converted to electricity and then the electric power may be employed to perform muscle work (Ishay, 2004; Ishay et al., 2004). Biological details of this phenomenon are not known yet. In this example, we will assess whether it is thermodynamically possible for a hornet to fly with an energy taken from the sunlight without consuming any organic matter. System boundaries for the mass and energy balances of hornet flight are given in Example Figure 4.3.1.

The hornet that is subject to this assessment has an average metabolic rate of 120 J/kg s, maintains stable body temperature, has a surface area of $2.5\times10^{-3}\,m^2$ to absorb the sunshine, and receives 941 J/m^2 s of solar energy when the atmospheric temperature is 25°C.

In this case, we can formulate the first law of thermodynamics as

$$\sum_{in}\dot{E}_{in}-\sum_{out}\dot{E}_{out}+\sum_{i}\dot{Q}_{i}-\dot{W}=\frac{dE_{system}}{dt}$$

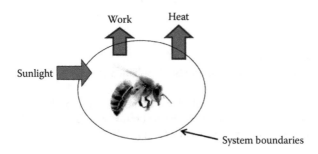

EXAMPLE FIGURE 4.3.1 System boundaries and energy flow of a hornet in flight.

The nutrient consumed in the metabolic pathway will be neglected in this analysis. The only form of energy that is consumed by the hornet is assumed to be solar energy, E_s. When the wasp is flying under the steady-state conditions with $m_{in} = m_{out} = 0$, the first law expression is

$$\dot{E}_s - \dot{Q} - \dot{W} = 0$$

The solar energy input to the system is

$$\dot{E}_s = (941\,\text{J/s}\,\text{m}^2)(2.3 \times 10^{-3}\,\text{m}^2) = 2.35 \times 10^{-3}\,\text{J/s}$$

The maximum voltage and current measured on the cuticles of the wasp is 180 mV and 6 µA (Ishay, 2004); if the wasp can achieve to convert the voltage times current into the muscle work, then we can calculate the work as

$$\dot{W} = (180\,\text{mV})(6\,\mu\text{A})\left(\frac{\text{J/s}}{(\text{V})(\text{A})}\right)\left(\frac{1\,\text{V}}{1000\,\text{mV}}\right)\left(\frac{1\,\text{A}}{10^6\,\mu\text{A}}\right) = 1.1 \times 10^{-6}\,\text{J/s}$$

The muscle work done by the wasp appears to be three orders of magnitude smaller than the solar energy received, and then the heat released from the system via convection and radiation may be calculated as

$$\dot{Q} = \dot{E}_s - \dot{W} = 2.35 \times 10^{-3}\,\text{J/s}$$

The entropy balance around the wasp is

$$\sum \dot{S}_{in} - \sum \dot{S}_{out} + \frac{\delta \dot{Q}}{T_b} + \dot{S}_{gen} = \frac{dS_{system}}{dt}$$

Since there is no mass flow, the inflowing entropy is the entropy of the solar radiation, which is defined as (Petela, 2010)

$$s_s = \varepsilon \frac{4}{3} \sigma T^3$$

where
   ε is the emissivity of surface
   σ is the Boltzmann constant for black radiation ($\sigma = 5.6693 \cdot 10^{-8}\,\text{W/(m}^2\,\text{K}^4))$

Entropy of solar radiation is calculated as the same as s_s of the earth, that is, 1.272 J/s m² K (Wu and Liu, 2010). The heat transfer surface area of the wasp is $(1.272\,\text{J/s}\,\text{m}^2\,\text{K})(2.5 \times 10^{-3}\,\text{m}^2) = 3.18 \times 10^{-3}\,\text{J/s K}$. If

$$\frac{\dot{Q}}{T} = \frac{(2.35 \times 10^{-3})\,\text{J/s}}{(273 + 25)\,\text{K}} = 7.9 \times 10^{-6}\,\text{J/s K},$$

then we may calculate the entropy generation rate as

$$\dot{S}_{gen} = \frac{\dot{Q}}{T} - \dot{S}_s = 3.17 \times 10^{-3} \text{ J/s K}$$

Under these conditions, the rate of exergy destruction is

$$\dot{Ex}_{destr} = \dot{S}_{gen}(T_{atmosphere}) = (3.17 \times 10^{-3} \text{ J/s K})(273 + 25) \text{ K} = 0.95 \text{ J/s}$$

The results obtained here do not conflict with the physical and biological facts; therefore, we may say that the argument claiming the oriental wasp converting the absorbed solar energy to electric current and then using that for performing the muscle work may be correct, but we need more data regarding the biological details including the metabolic pathways to accept the argument.

Example 4.4: Propulsion Work Done by the Squid

Squid is an animal that attracts attention of scientists working in different fields. Scientists working on hydrodynamics study squids to understand how it fills its mantle cavity with water and ejects it at a very high speed forming a vortex ring and accelerating the squid rapidly. Neuroscientists prefer to work on squids because of its giant axon that allows easy experimentation. In this example, we chose squid to demonstrate how the calculation of work changes with respect to the choice of the system boundaries.

Squid anatomy consists of a prominent head, eight arms, two tentacles, and a mantle. The mantle consists of muscles in a conical shape, encloses vital organs and the mantle cavity. Openings on each side of the prominent head are used to draw water into the mantle cavity, and then water is ejected from the mantle cavity via contractions of mantle muscles at different speeds to provide the squid with the necessary momentum for propulsion (Example Figure 4.4.1). Squid has two modes of motion. Normally, its mantle cavity contractions are weak and swimming speed is low. In case of emergency, powerful contractions occur, which is referred to as jet propulsion.

Contraction of the mantle cavity wall to discharge water from the squid's nozzle requires mechanical work. This mechanical work may be calculated with the integral

$$w = \int p \, dv$$

where
  p is the water pressure inside the mantle cavity
  dv is the change in the mantle volume

The volume of the mantle cavity is $V = \int_0^L \pi r^2 dx$, where r is the radius of the mantle cavity at the point x and L is squid's dorsal mantle length (Example Figure 4.4.2).

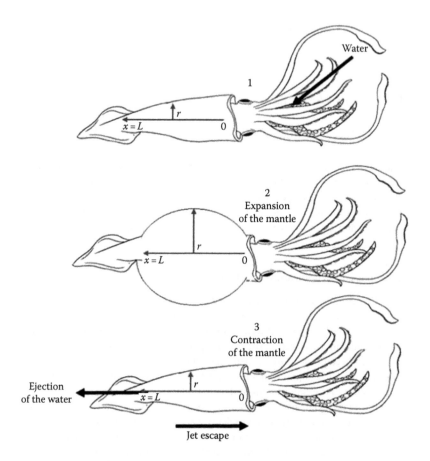

EXAMPLE FIGURE 4.4.1 Schematic description of the jet escape of the squid.

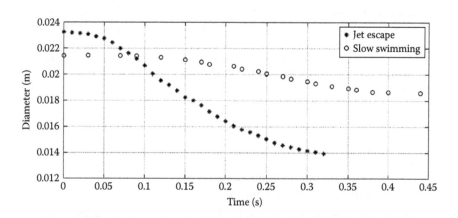

EXAMPLE FIGURE 4.4.2 Change of a mantle cavity diameter with time during ejection of water.

Volume change of the squid's mantle cavity can be modeled as a sinusoidal motion with respect to time as

$$v = \int \pi \left[\sin\left(\frac{x\pi}{2L}\right) r_{max} \right]^2 dx$$

Geometrical details and the pressure inside the mantle cavity are found in the literature as a time-dependent data set. Furthermore, stress on the mantle wall fibers from the muscle contraction was measured via electromyography. The work done by the mantle cavity can also be also calculated as a function of the muscle contraction force and displacement as

$$w = \int F dl$$

MATLAB code E4.4 calculates the work performed by the squid by using both $w = \int p dv$ and $w = \int F dl$ during both slow swimming and jet escape modes:

MATLAB CODE E4.4

Command Window

```
clear all
close all

% Fast Swiming ****************************
% Fast Swiming ****************************

% Volume Part
syms x

t=0.004;          %thickness
% L = 15 cm DML in our squid example.
% L=0.150;         %Length

% Bartol et al. 2001 examines squid lenght of 4.4 cm DML
L=0.044;             %Length

counter=1;
res=150;

%Pressure Part
% A: relative mantle diameter information from Gosine and
   DeMont (1985)_Biomimetics_jet propelled swimming in Squids

A=[1.069; 1.069; 1.069; 1.068; 1.063; 1.059; 1.048; 1.032;
   1.019; 1.004; 0.985; 0.962; 0.944; 0.933; 0.917; 0.898;
   0.890; 0.877; 0.860; 0.846; 0.834; 0.821; 0.812; 0.806;
   0.798; 0.789; 0.779; 0.773; 0.769; 0.765; 0.762; 0.760;
   0.757];
```

```
A = transpose(A);

%Z: pressure inside the mantle cavity information from
   Gosine and DeMont (1985)_Biomimetics_jet propelled
   swimming in Squids

Z=[0; 13.037; 20.067; 23.401; 24.060; 24.481; 24.796;
   24.890; 24.748; 24.665; 24.432; 23.648; 22.200; 21.651;
   20.510; 18.886; 17.605; 15.625; 13.919; 12.195; 10.435;
   9.002; 6.784; 5.266; 4.364; 3.146; 2.001; 1.055; 0.521;
   0.126; -0.168; -0.650; -1.269];

Z = transpose(Z);

% Z=1000*Z; % to convert the pressure from kPa to Pascal unit
Z=10*Z*0.8; % Finke et al. (1996) pressure in the mantle
   cavity was correlated with the body length.
% Therefore, 250 Pa mean pressure was obtained by
   multiplying 10 from kPa to Pascal unit

A=A*(0.025); % Radius is given as 25mm. So, by multiplying
   with 0.025 relative diameter
% became radius of this specific squid. The unit is in meters.

Thick=linspace(0.0035,0.005,33);
T=Thick;
A=A-Thick;

for i=1:length(A)
   V(i)=double(int(pi*(A(i)*sin(x*pi/(2*L)))^2,0,L)); %
   Volume calculation from radius.
end

% figure, plot(V,Z,'+'); grid on; title('Pressure vs Volume
   (Pa vs m³)')
W_PdV_FAST_in_joule = trapz(V,Z)

% 65 gram squid mean mass and 0.32 second is the period for
   fast mode.
W_PdV_FAST_in_joule_per_second_per_gram = trapz(V,Z)/0.32/65
% Result is negative because pressure (Z) rises when volume
   (V) decreases.

% Relative mantle diameter values as adapted from Gosline
   and DeMont (1985)

X = [1.098417558; 1.094866014; 1.087940502; 1.078622472;
   1.0651056; 1.053471428; 1.043710409; 1.03213352;
   1.022405917; 1.005712704; 0.990258712; 0.978758201;
   0.960325495; 0.949458915; 0.935580205; 0.922614204;
   0.911175749; 0.900635682; 0.883192062; 0.863081919;
   0.848107196; 0.836144601; 0.825406908; 0.81498618;
   0.805761712; 0.798045696; 0.791525939];

X = transpose(X);
```

```
% Curve interpolation stress values as adapted from from
  Gosline and DeMont (1985)

Y = [102.6272199; 137.2659155; 154.8113719; 170.4225733;
     183.4372648; 192.0682177; 196.9461811; 201.7926388;
     204.4629801; 207.3279243; 208.3216248; 208.1220891;
     205.9095343; 203.8282529; 200.7483384; 197.3688021;
     193.0693967; 189.4164934; 181.8583233; 170.4683997;
     159.7984912; 150.1272159; 139.5308206; 127.9935525;
     117.4234119; 107.1949005; 97.93351674];

Y = transpose(Y);

% figure, plot(X,Y,'+'); grid on;
% Y = Y*1000; % to convert the stress from kPa to Pascal unit
Y = Y*10*0.8; % Finke et al. (1996) only 10 is used because
  pressure inside the
% mantle cavity is proportional to the stress at the
  circular muscle.

X1=X;
X=X*(0.025);
Thick=linspace(0.0035,0.005,27);
X=X-(Thick/2);

for k=1:length(X)
    for i=1:res
        pn(k,i)=2*pi*X(k)*sin(i*pi/(2*res)); %Volume
    end
end

% Obtaining force from pressure times area. Y is the
  pressure.
for i=1:length(X)
    F(i)=Y(i)*Thick(i)*(L/res);
end

for c=1:res
    W_Fdl_m(c)=trapz(pn(:,c),F);
end

W_Fdl_FAST_in_joule = sum(W_Fdl_m)
% 65 gram squid mean mass and 0.32 second is the period for
  fast mode.
W_Fdl_FAST_in_joule_per_second_per_gram = sum(W_Fdl_m)/0.32/65

time= [0.000; 0.010; 0.020; 0.030; 0.040; 0.050; 0.060;
       0.070; 0.080; 0.090; 0.100; 0.110; 0.120; 0.130; 0.140;
       0.150; 0.160; 0.170; 0.180; 0.190; 0.200; 0.210; 0.220;
       0.230; 0.240; 0.250; 0.260; 0.270; 0.280; 0.290; 0.300;
       0.310; 0.320];

% Slow Swimming ************************
% Slow Swimming ************************
```

```
t1=0.004;           %thickness

% L1=0.150;          %Length
L1=0.044;           %Length

counter1=1;
res1=150;

%Pressure Part

% A1: relative mantle diameter information from Gosline and
  DeMont (1985)_Biomimetics_jet propelled swimming in Squids

A1=[0.997; 0.998; 0.998; 0.997; 0.993; 0.986; 0.980; 0.973;
  0.968; 0.960; 0.953; 0.946; 0.939; 0.932; 0.925; 0.919;
  0.911; 0.905; 0.901; 0.895; 0.894; 0.892];

A1 = transpose(A1);

%Z1: pressure inside the mantle cavity information from
  Gosline and DeMont (1985)_Biomimetics_jet propelled
  swimming in Squids

Z1=[0.049; 0.752; 1.331; 1.656; 1.919; 2.182; 2.379; 2.513;
  2.647; 2.718; 2.662; 2.668; 2.738; 2.745; 2.751; 2.634;
  2.515; 2.272; 2.029; 1.724; 1.358; 0.500];

Z1 = transpose(Z1);

% Z1=1000*Z1;  % to convert the pressure from kPa to Pascal
  unit
Z1=10*Z1*0.8; % Finke et al. (1996) pressure in the mantle
  cavity was correlated with the body length.
% Therefore, 250 Pa mean pressure was obtained by multiplying
  10 from kPa to Pascal unit

A1=A1*(0.025); % Radius is given as 25mm. So, by multiplying
  with 0.025 relative diameter
% became radius of this specific squid. The unit is in meters.

Thick1=linspace(0.0035,0.00371,22);
T1=Thick1;
A1=A1-Thick1;

for i=1:length(A1)
   V1(i)=double(int(pi*(A1(i)*sin(x*pi/(2*L1)))^2,0,L1));
   % Volume
end

% figure, plot(V1,Z1,'+'); grid on; title('Pressure vs
  Volume (Pa vs m³)')
W_PdV_SLOW_in_joule = trapz(V1,Z1)
% 65 gram squid mean mass and 0.32 second is the period for
  fast mode.
W_PdV_SLOW_in_joule_per_second_per_gram =
  trapz(V1,Z1)/0.44/65

%Curve interpolation Stress Values from article by Gosline
  and DeMont (1985)
```

```
% Read the sample data with 11 points.

X1 = [1.005712704; 0.990258712; 0.978758201; 0.960325495;
  0.949458915; 0.935580205; 0.922614204; 0.911175749;
  0.900635682; 0.883192062];
X1 = transpose(X1);

Y1 = [207.3279243; 208.3216248; 208.1220891; 205.9095343;
  203.8282529; 200.7483384; 197.3688021; 193.0693967;
  189.4164934; 181.8583233];

Y1 = transpose(Y1);

% figure, plot(X1,Y1,'+'); grid on;
% Y1 = Y1*1000;
Y1 = Y1*10*0.8; % Finke et al. (1996) only 10 is used
  because pressure inside the
% mantle cavity is proportional to the stress at the
  circular muscle.

X2=X1;
X1=X1*(0.025);
Thick1=linspace(0.0035,0.00371,10);
X1=X1-(Thick1/2);

for k=1:length(X1)
    for i=1:res1
        pn1(k,i)=2*pi*X1(k)*sin(i*pi/(2*res1)); %Volume
    end
end

% Obtaining force from pressure times area. Y1 is the
  pressure.
for i=1:length(X1)
    F1(i)=Y1(i)*Thick1(i)*(L1/res1);
end

for c=1:res1
    W_Fdl_m1(c)=trapz(pn1(:,c),F1);
end

W_Fdl_SLOW_in_joule = sum(W_Fdl_m1)

W_Fdl_SLOW_in_joule_per_second_per_gram =
  sum(W_Fdl_m1)/0.44/65

time1= [0.00; 0.03; 0.07; 0.09; 0.12; 0.15; 0.17; 0.18;
  0.21; 0.22; 0.24; 0.25; 0.27; 0.28; 0.30; 0.31; 0.33;
  0.35; 0.36; 0.38; 0.40; 0.44];

% Example Figure 4.3.2
figure, plot(time,A,'*',time1,A1,'o', 'MarkerSize',12,...
  'LineWidth',2,'Color',[0 0 0]); xlabel('\sl Time (sec)',
  'FontSize',35,'FontName','Times New Roman');
ylabel('\sl Diameter (m)','FontSize',35,'FontName','Times
  New Roman'); grid on;
size_legend = legend('Jet Escape','Slow
  Swimming','Location', 'northeast');
set(size_legend,'FontSize',35,'FontName','Times New Roman');
```

```
% Example Figure 4.3.3
figure, plot(time,T,'*',time1,T1,'o', 'MarkerSize',12,...
 'LineWidth',2,'Color',[0 0 0]); xlabel('\sl Time (sec)',
 'FontSize',35,'FontName','Times New Roman');
ylabel('\sl Thickness (m)','FontSize',35,'FontName','Times
 New Roman'); grid on;
size_legend = legend('Jet Escape','Slow Swimming','Location',
 'northeast');
set(size_legend,'FontSize',35,'FontName','Times New Roman');

% Example Figure 4.3.4
figure, plot(time,V,'*',time1,V1,'o', 'MarkerSize',12,...
 'LineWidth',2,'Color',[0 0 0]); xlabel('\sl Time (sec)',
 'FontSize',35,'FontName','Times New Roman');
ylabel('\sl Volume (m^3)','FontSize',35,'FontName','Times
 New Roman'); grid on;
size_legend = legend('Jet Escape','Slow
 Swimming','Location', 'northeast');
set(size_legend,'FontSize',35,'FontName','Times New Roman');

% Example Figure 4.3.5
figure, plot(time,Z,'*',time1,Z1,'o', 'MarkerSize',12,...
 'LineWidth',2,'Color',[0 0 0]); xlabel('\sl Time (sec)',
 'FontSize',35,'FontName','Times New Roman');
ylabel('\sl Pressure (Pa)','FontSize',35,'FontName','Times
 New Roman'); grid on;

size_legend = legend('Jet Escape','Slow
 Swimming','Location', 'northeast');
set(size_legend,'FontSize',35,'FontName','Times New Roman');

% Example Figure 4.3.6
figure, plot(V,Z,'*',V1,Z1,'o', 'MarkerSize',12,...
 'LineWidth',2,'Color',[0 0 0]); xlabel('\sl Volume (m^3)',
 'FontSize',35,'FontName','Times New Roman');
ylabel('\sl Pressure (Pa)','FontSize',35,'FontName','Times
 New Roman'); grid on;
size_legend = legend('Jet Escape','Slow Swimming','Location',
 'northwest');
set(size_legend,'FontSize',35,'FontName','Times New Roman');
```

When we run the code, the following lines and figures will appear in the screen:

```
W_PdV_FAST_in_joule =
 -0.0033


W_PdV_FAST_in_joule_per_second_per_gram =
 -1.6035e-04


W_Fdl_FAST_in_joule =
 -0.0090


W_Fdl_FAST_in_joule_per_second_per_gram =
 -4.3460e-04
```

```
W_PdV_SLOW_in_joule =
 -1.5204e-04

W_PdV_SLOW_in_joule_per_second_per_gram =
 -5.3162e-06

W_Fdl_SLOW_in_joule =
 -0.0032

W_Fdl_SLOW_in_joule_per_second_per_gram =
 -1.1301e-04
```

The following MATLAB output figures describe the change of a mantle cavity diameter (Example Figure 4.4.2), mantle cavity wall thickness (Example Figure 4.4.3), mantle cavity volume (Example Figure 4.4.4), mantle cavity pressure (Example Figure 4.4.5) with time, and variation of the mantle cavity pressure with volume (Example Figure 4.4.6) during ejection of water.

The force–muscle elongation and the pressure–volume change works calculated during slow swimming and jet escape are summarized in Example Table 4.4.1.

Comparing the calculated pressure–volume change work $w = \int p dv$ and force–muscle elongation work $w = \int Fdl$ shows a large difference for both of the swimming modes. The reason for that becomes clear when we investigate the muscle structure closely. Microscopic visualization shows that squid mantle is made of two types of fibers, that is, superficial-mitochondria-rich (SMR) and central-mitochondria-poor (CMP) fibers. CMP fibers make up ca. 88%, whereas SMR fibers make up ca. 4% of the total mantle volume. During the steady-state swimming, SMR fibers are activated, and ca. 95% of the mantle volume remains passive (Example Figure 4.4.7).

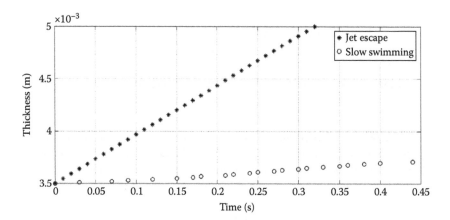

EXAMPLE FIGURE 4.4.3 Change of a mantle cavity wall thickness with time during ejection of water.

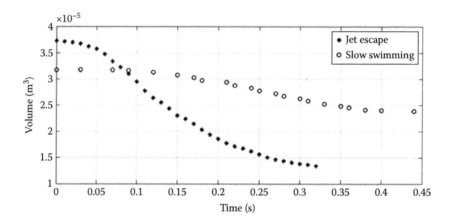

EXAMPLE FIGURE 4.4.4 Variation of the mantle cavity volume with time during ejection of water.

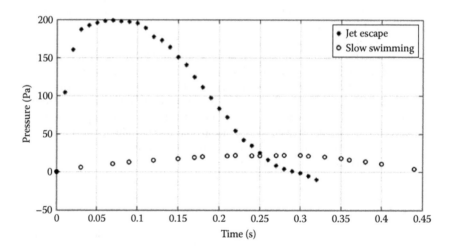

EXAMPLE FIGURE 4.4.5 Variation of the mantle cavity pressure with time during ejection of water.

The contraction work done by the muscle fibers is consumed to contract the passive tissue and to compress the water filled inside the cavity. Thus, only a small fraction of the force–muscle elongation work $w = \int F dl$ is transferred to pressure–volume change work $w = \int p dv$. The contraction efficiency is

$$\eta = \frac{\int p dv}{\int F dl} = 4.7\%.$$

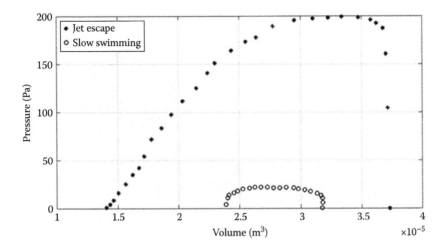

EXAMPLE FIGURE 4.4.6 Variation of the mantle cavity pressure with volume during ejection of water.

EXAMPLE TABLE 4.4.1 Summary of the Force–Muscle Elongation Work and Pressure–Volume Change Work during Slow Swimming and Jet Escape

	Slow Swimming	Jet Escape
Force–muscle elongation work (J)	−0.0032	−0.0090
Pressure–volume change work (J)	−0.0002	−0.0033

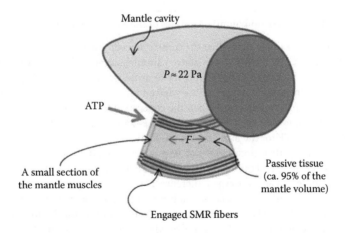

EXAMPLE FIGURE 4.4.7 Microscopic visualization of the squid mantle and the muscles contributing to steady-state slow swimming.

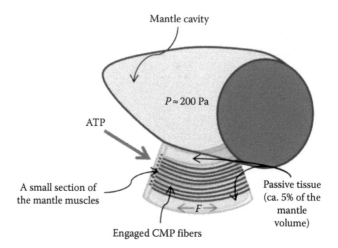

EXAMPLE FIGURE 4.4.8   Microscopic visualization of the squid mantle and the muscles contributing to the fast powerful contractions during the jet escape.

The ratio of the force–elongation work, used to compress the water inside the cavity, is larger during the jet escape, since a larger volume of muscle fibers are engaged and the passive tissue makes up only about 5% of the total mantle volume. The contraction efficiency during jet escape (Example Figure 4.4.8) is

$$\eta = \frac{\int p\,dv}{\int F\,dl} = 36.8\%.$$

Example 4.5: Thermodynamic Analyses of the Energy Metabolism of Squid

The previous example shows that the contraction efficiency of the squid mantle muscle is very low during steady-state swimming compared to jet escape. We know from observations that squids try to avoid jet escape swimming unless there is an immediate danger or they have to hunt. But why does the squid prefer to swim at a low speed when the contraction efficiency is that low? To answer that question, we have to model the energy metabolism that produces the required ATP for the muscle work production.

Squid has neither circulatory nor respiratory systems. Oxygen is taken in through the inner and the outer surfaces of the mantle. These superficial regions of the mantle can access oxygen, therefore here aerobic respiration is possible. Squids are evolved to have a large number of mitochondria on both surfaces of the mantle (SMR). Oxygen concentration decreases dramatically toward the center of the mantle. Therefore, mitochondria would be useless for the cells in the center of the mantle. Evolution eliminated any unused cellular metabolites, and the central cells are poor in mitochondria (CMP).

Finke et al. (1996) measured the O_2 consumption rate of $9.72 \times 10^{-4}\,\mu mol/(g\ s)$ for slow swimming. During slow swimming, steady-state aerobic respiration occurs, so glucose is consumed in a stoichiometric amount with the intake of oxygen, and the products (CO_2 and H_2O) are transported out of the system. For the jet escape, Finke et al. (1996) measured the O_2 consumption rate ($16.67 \times 10^{-4}\ \mu mol/(g\ s)$) as well as the changes in the concentration levels for several metabolites. Measurements show that during jet escape, some metabolites like ATP and arginine are used up and some metabolites like octopine are accumulated. The experimental data of Finke et al. are taken as a basis for the thermodynamic analysis. First, the metabolic reactions during steady-state and jet escape swimming are modeled. Then, reaction rates are estimated based on the experimental concentration changes. Example Figure 4.5.1 shows the mass balance for the jet escape mode.

Thermodynamic properties of the metabolites are calculated via the equations given in Tables 4.1 and 4.2 by taking the physiological condition in a squid cell as $I = 0.25$ M, pH $= 7$, $T = 298.15$ K (Example Table 4.5.1).

Example Table 4.5.2 shows the reactions occurring in the metabolism, reaction rates, and the Δh_{rxn} and Δg_{rxn}.

Thermodynamic analysis of the steady-state slow swimming shows that oxidation of glucose within the squid mantle releases an exergy of 3.82 J/(kg s). Depending whether 30 or 38 mol of ATP is produced per 1 mol of glucose, 0.029–0.037 mmol/(kg s) of ATP is produced. We assume that during slow swimming, ATP concentration remains steady. Accordingly, an exergy of 1.05–1.34 J/(kg s) is stored in the produced ATP, which can be converted into mechanical work via muscle fiber contraction (Fdl). If we define the exergetic efficiency of the respiration as

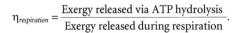

$$\eta_{respiration} = \frac{\text{Exergy released via ATP hydrolysis}}{\text{Exergy released during respiration}}.$$

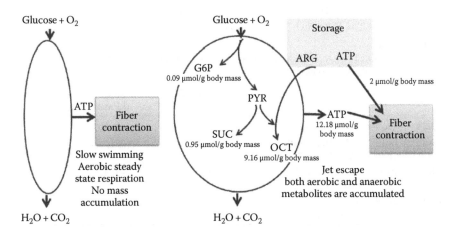

EXAMPLE FIGURE 4.5.1 The mass balance for the jet escape mode.

EXAMPLE TABLE 4.5.1 Thermodynamic Properties of the Metabolites Involving in the Metabolic Activity

Metabolite	Concentration (mM)	Δh at $I =$ 0.25 M, pH = 7, $T = 298.15$ K (kJ/mol)	Δg at $I =$ 0.25 M, pH = 7, $T = 298.15$ K (kJ/mol)	ex_{ch} (kJ/mol)	$RT \ln (c)$ (kJ/mol)	ex (kJ/mol)
Glu	1.20	−1267.12	−426.71	2,473.47	0.45	2,047.21
G6P	0.10	−2279.30	−1318.92	3,343.03	−5.71	2,018.40
ATP	0.36	−3616.92	−2292.50	6,720.93	−2.53	4,425.90
ADP	0.17	−2627.24	−1424.70	5,851.38	−4.39	4,422.29
NAD	1.27	−41.83	1120.09	10,372.87	0.59	11,493.55
NADH	0.24	−10.26	1059.11	10,372.87	−3.54	11,428.44
P_i	1.89	−1299.36	−1059.49	871.54	1.58	−186.37
H_2O	1.00	−286.75	−155.66	1.99	0.00	−153.67
PYR	0.55	−597.04	−350.79	1,236.74	−1.48	884.47
ARG	0.20	−277.38	25.20	2,466.91	−3.99	2,488.12
OCT	0.20	−608.52	−272.85	3,701.63	−3.99	3,424.79
SUC	0.065	−908.70	−530.64	1,648.94	−6.78	1,111.52
CO_2	20.00	−692.86	−547.10	414.32	7.43	−125.35
O_2	0.96	−11.70	16.40	3.97	−0.10	20.27

EXAMPLE TABLE 4.5.2 Reactions Occurring in the Metabolism, Reaction Rates, and the Δh_{rxn} and Δg_{rxn}

Slow Swimming Mode
➤ ATP produced through the aerobic energy metabolism.
➤ SMR fibers are engaged.

		Reaction Rate (μmol/(g s))	Δh_{rxn} (kJ/mol)	Δg_{rxn} (kJ/mol)
SMR1	$Glu + 6O_2 \Rightarrow 6CO_2 + 6H_2O$	$6.48 \times 10^{-4} – 1.85 \times 10^{-3}$	−4540.34	−3930.39
SMR2	$ADP + P_i \Rightarrow ATP + H_2O$	$1.94 \times 10^{-2} – 7.04 \times 10^{-2}$	−22.93	−36.31

Jet Escape Mode
➤ ATP produced both through the aerobic and anaerobic energy metabolisms.
➤ Mostly CMP fibers are engaged.

		Reaction Rate (μmol/(g s))	Δh_{rxn} (kJ/mol)	Δg_{rxn} (kJ/mol)
CMP1	$Glu + ATP \Rightarrow G6P + ADP$	1.00×10^{-4}	−22.50	−21.00
CMP2	$Glu + 2NAD + 2ADP + 2P_i \Rightarrow$ $2PYR + 2NADH + 2ATP + 2H_2O$	6.14×10^{-3}	182.04	−381.73
CMP3	$PYR + ARG + NADH \Rightarrow$ $OCT + NAD + H_2O$	1.02×10^{-2}	−52.42	−18.19
CMP4	$2PYR + 2H_2O + 3NAD \Rightarrow$ $SUC + 2CO_2 + 3NADH$	1.06×10^{-3}	−432.13	−779.68

Then, the exergetic efficiency of the squid aerobic respiration is

$$28\% = \frac{1.05 \, J/(kg \, s)}{3.82 \, J/(kg \, s)} \leq \eta_{respiration, \, aero} \leq \frac{1.34 \, J/(kg \, s)}{3.82 \, J/(kg \, s)} = 35\%.$$

Exergy chart for the slow swimming mode is given in Example Figure 4.5.2.

During jet escape, energy comes from both the aerobic and anaerobic respiration. Oxygen consumption rate increases nearly linearly between 0.5 and 4 mantle length/s. Therefore, at a speed of 3 mantle length/s, squid consumes a large amount of oxygen, and about 6.55 J/(kg s) of exergy is released via aerobic respiration, which is nearly twice the amount of exergy as the previous case. The chemical exergy released via the anaerobic pathway is 3.35 J/(kg s). About 0.072 J/(kg s) of the chemical exergy comes via depleting the cellular ATP reserves.

The exergetic efficiency of the squid anaerobic respiration is

$$\eta_{respiration, \, anaero} = \frac{0.44 \, J/(kg \, s)}{3.35 \, J/(kg \, s)} = 13.2\%.$$

A squid swimming with a speed of 3 mantle length/s produces an exergy of 2.33 J/(kg s) via ATP hydrolysis. 0.44 J/(kg s) of this comes from ATP produced via anaerobic respiration, 0.07 J/(kg s) comes from the ATP used up from cellular reserves, and 1.82 J/(kg s) comes from ATP produced via aerobic respiration. Since the jet escape swimming speed (3 mantle length/s) is very close to the critical speed (2 mantle length/s) at which anaerobic respiration begins; the squid still uses the aerobic pathway extensively to produce the required ATP. One can speculate that if the swimming speed would be higher, then the ATP production rate from the anaerobic pathway would increase dramatically, so that a much higher percent of the consumed ATP would come from the anaerobic pathway. Here, at this low jet escape speed, still 78% of the consumed ATP comes from the aerobic pathway. Therefore, the respiration efficiency of the overall system, which is 23.37%, is close to the aerobic respiration efficiency.

The animals in the nature generally prefer to stay in steady-state conditions. The squids also do not depart from this rule. Squid generally prefers the steady-state swimming mechanism in their lifetime unless they are in danger or when they are hunting. The squid switches its swimming behavior for survival or for capturing the prey. When the squid switches to the jet escape swimming mode, it means a drop in the respiration efficiency. At very high speeds, squids may be forced to use the anaerobic pathway exclusively, which would drop the respiration efficiency to 13.2%. The exergy chart for the jet escape swimming mode is given in Example Figure 4.5.3.

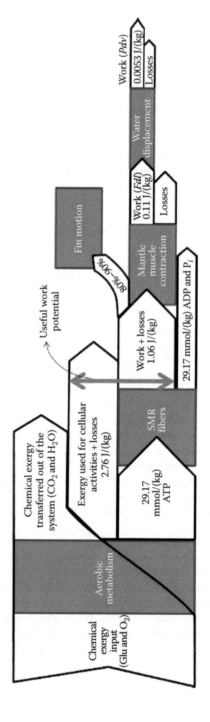

EXAMPLE FIGURE 4.5.2   Exergy chart for the slow swimming mode.

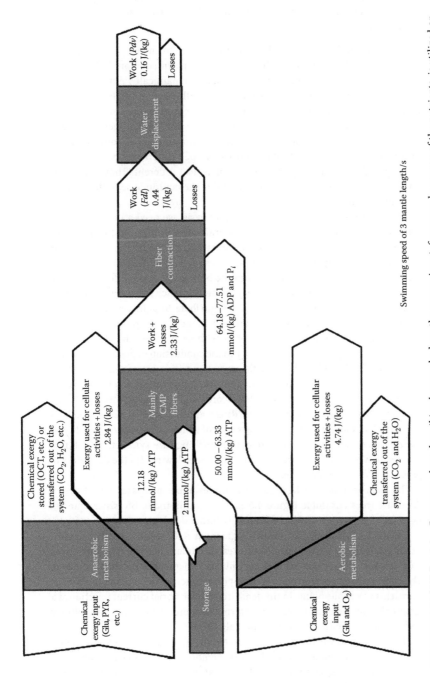

EXAMPLE FIGURE 4.5.3 Grassman chart describing how and where the exergy input, for example, exergy of the nutrients, is utilized or destroyed during the jet escape mode.

Example 4.6: ATP Consumption, Entropy Generation, and Exergy Destruction during a Single Action Potential Generation in the Squid Neuron

Squid has been a model animal for neuron studies for many years because of the large size of its giant axon that allow easy experimentation. In this example, we will analyze the action potential thermodynamically. Action potential is the second largest energy consumer in overall usage in the brain (Attwell and Laughlin, 2001). So it is very important to calculate the consumption of ATP in action potential to be able to estimate the total entropy generation and exergy destruction in a neuron. During the two different swimming mechanisms of the squid (steady state and jet escape), different types of muscles are engaged. The intensity of contractions is also different. Thus, a different amount of ATP has to be consumed to enable the neuron to send the contraction signal to muscles.

In order to begin the thermodynamic analysis, we need a model to represent the squid neuron. All neurons in the brain are directly or indirectly linked to each other. While transmitting information, some energy is dissipated. Dissipation of energy is an irreversible process and quantified via entropy generation. The entropy generation within the brain may cause different types of diseases in the central nervous system like multiple sclerosis, epilepsy, cancer, and Alzheimer's disease and eventually contribute to aging. Neuroscience is interested in two questions: how synaptic input signals propagate through the axon/dendrite and how they interact with each other. The Hodgkin and Rushton (1946) applied the cable theory to the conduction of potentials in axon. Later, Rall (1977) applied the same theory to dendritic trees of neurons.

In a typical neuron, thousands of synaptic input spread across its surface (Example Figure 4.6.1). Using the Rall's core conductor cable theory, the dissipation in a neuron can be calculated. Rinzel and Rall's (1974) neuron model is composed of N identical dendritic trees each of which exhibits M orders of symmetric branching. They assume that all branchings are symmetrically bifurcated satisfying the 3/2-power law. The 3/2-power law relates the diameters of parent and daughter branches.

The neuron model investigated in this example possesses five different dendritic trees (Example Figure 4.6.2); BI (input branch), BS (sister branch), BC-1, BC-2 (two cousin branches), and OT (other dendritic trees).

The model considers each branch segment as a dimensional, infinite cylindrical electronic cable with isopotential extracellular medium. At all branches, core conductor is conserved and continuously membrane potential received. The cable model uses convolution of response function to compute the voltage transients at various points in the dendritic tree for a brief current injection at a terminal branch and its Laplace transform space product domain for arbitrary and boundary conditions (Rinzel and Rall, 1974).

Since the system is linear, electric potential $V(t,x)$ at time t and location x transient current applied outward across membrane $I(t)$ at time t computed in the form of $K(t,x; L)$ as a function of t at location x. For the response function in the time domain, we have two types of response functions: one of which is

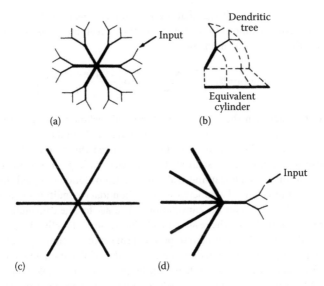

(a)   (b)

(c)   (d)

EXAMPLE FIGURE 4.6.1 Rinzel and Rall's (1974) idealized neuron model. (a) represents the neuron model composed of six identical dendritic trees. (b) indicates the relation of a dendritic tree to its equivalent cylinder. (c) represents the same model as (a), with each dendritic tree replaced by an equivalent cylinder. (d) represents the same model as (a) and (c), with dendritic branching shown explicitly only for the tree that receives input current injected into the terminal of one branch; the five other trees of the model are represented by their equivalent cylinders, here shown gathered together.

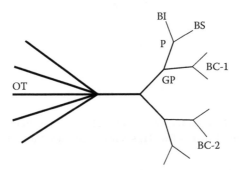

EXAMPLE FIGURE 4.6.2 Five different dendritic trees of a model neuron; BI, input branch; BS, sister branch; BC-1 and BC-2, two cousin branches; and OT, other dendritic trees, Rall core conductor.

$K_{ins}(t, x, L)$, for example, response at time t and location x in a cylinder of length L insulated ($\partial V/\partial x = 0$) at the origin for instantaneous point charge placed at the end $x = L$ demonstrated as

$$K_{ins}(x, t, L) = R_{\infty} \frac{e^{-t}}{(\pi t)^{1/2}} \sum_{-\infty}^{\infty} \exp\left\{ -\frac{[L(2n-1)+x]^2}{4t} \right\}$$

The second one is $K K_{clp}(t,x,L)$ that is a response at time t and location x in a cylinder length L clamped ($V = 0$) at the origin for instantaneous point charge placed at the end $x = L$ demonstrated as

$$K_{clp}(x,t,L) = R_{\infty} \frac{e^{-t}}{(\pi t)^{1/2}} \sum_{-\infty}^{\infty} (-1)^n \exp\left\{-\frac{[L(2n-1)+x]^2}{4t}\right\}$$

The relation between the response function and instantaneous point charge is represented in terms of the Dirac delta functions, $\delta(t)$ and $\delta(x)$. For infinite cable model after the response functions calculated the general Dirac delta response function of the system at different electronic positions was calculated for the dimensionless infinite cable model. Convolution of signals $I(t)$ and K calculated to ensure the adherence to the boundary conditions defined by Rinzel and Rall (1974) for the convolution, all values for the convolution after time t were deleted from the result convolution vectors. Finally dissipation of neuron calculated with respect to trapezoidal rule with multiplying membrane capacitance (C_m) and characteristic length (λ). The cable parameters for squid taken from Rall core conductor theory are listed in Example Table 4.6.1.

The ATP consumption during a single instantaneous deviation from the resting membrane potential was calculated from the Na⁺ influx (Sengupta et al., 2010). The parameters are adjusted and sodium channel activation and deactivation rates α_m, β_m, α_h, and β_h are calculated as $\overline{g_{Na}} = 120$ m mho/cm^2 and $E_{Na} = -115$ mV:

$$\alpha_m = \frac{0.1(V_m + 25)}{e^{[(V_m+25)/10]} - 1} \tag{E4.6.1}$$

$$\beta_m = 4e^{V_m/18} \tag{E4.6.2}$$

$$\alpha_h = \frac{0.07}{e^{V_m/20}} \tag{E4.6.3}$$

$$\beta_h = \frac{1}{e^{[(V_m+30)/10]} + 1} \tag{E4.6.4}$$

The total Na⁺ load for an individual action potential is integrated and then multiplied with $1/3F$ where F is Faraday's constant to reach the ATP consumption.

EXAMPLE TABLE 4.6.1   Cable Parameters for Squid

Axon diameter, $d = 500$ μm
Length constant, $\lambda = 0.65$ cm
Core resistance per unit length, $r_i = 15 \times 10^3$ Ω/cm
Membrane resistance for unit length, $r_m = 6.5 \times 10^3$ Ω cm
Intracellular resistivity, $R_i = 30$ Ω cm
Membrane resistivity, $R_m = 1 \times 10^3$ Ω cm^2
Membrane time constant, $\tau = 1$ ms
Membrane capacitance, $C_m = 1$ μF/cm^2

Source: Adapted from Rinzel, J. and Rall, W., *Biophys. J.*, 14, 759, 1974.

The exergy destruction per 1 mol of ATP (kJ/mol) is adapted from Genç et al. (2013a) for hypoxic and normoxic conditions. It is known that neurons can emit action potential constantly at a rate of 10–100 per second (Barnett and Larkman, 2007). We assumed that in normoxic conditions, neurons emit action potential at the rate of 10 s^{-1} and in hypoxic at the rate of 30 s^{-1}. Entropy generation is calculated with the MATLAB code E4.6.

MATLAB CODE E4.6

Command Window

```
clear all
close all

% IMPORTANT! Commented out for testing with parameters from
  papers to
% ensure that the desired results are obtained. Once the
  code has been
% verified, uncomment the defined parameters and delete the
  test
% parameters.

% % First, define the parameters:
%
I_p = 3.75E-10; %[=] A; Peak value of transient current
  (Turker & Powers, 2001)
% rise_time = 0.5; % [=] ms; (Turker & Powers, 2001)

% Note: Divide the rise time by the (dimensional) time to
  get 1/alpha. Then
% find the reciprocal of this to find alpha. Make sure
  dimensions of time
% agree!

% Parameters FROM HODGKIN HUXLKEY CORE CONDUCTOR:

a = 20;
T = 0:0.01:1.25; % Dimensionless time
I_T = I_p .*a.*T.*exp(1-a.*T); %(Rinzell & Rall, 1974)
L = 0.6; % length of cable [=] cm
N = 6; % Number of trees
M = 3; % Degree of branching

% The following parameters were found for the squid
  motoneuron from *Core
% Conductor Theory* (Rall,1977)

X = 0:L/((M+1).*200):L; % Start with potential at injection
  site of cable.
Xk_vals = L/(M+1):(1/(M+1))*L:L*(1-(1/(M+1)));
% Electrotonic distance to branching.
```

```
R_m = 1000; % [=] ohm*cm^2; Membrane resistivity.
R_N = 5E3; % [=] ohm; Neuron input resistance.
R_i = 30; % [=] ohm*cm; Intracellular resistivity.
C_m = 1E-6; % [=] F*cm^-2; Membrane capacitance.
d = (500E-4); % [=] cm; diameter of membrane cylinder.
rho = 10; % dimensionless; conductance ratio (dendrites/soma).


% Cable parameters: Calculated via definitions by Rall (Core
  Conductor
% Theory)


c_m = pi.*d.*C_m; % [=] *Farad*cm^-1; Membrane capacity per
  unit
% length.
%r_i and r_m from FROM HODGKIN HUXLKEY CORE CONDUCTOR
r_i = 15E3; % [=] ohm*cm^-1; Core resistance per unit length.
r_m = 6.5E3; % [=] ohm*cm; Resistance across unit length of
% membrane.
lambda = 0.65; % [=] cm; characteristic length.
tau = 1; % [=] millisecond; Time constant.
R_inf = lambda.*r_i; % [=] ohm; Input resistance of semi-
  infinite cable.


% Finding Response Functions in time domain.
% Finding K_ins, dirac-delta response to charge placed at
  origin for
% insulated cylinder (initial condition dV/dX = 0);
% Finding K_clp, dirac-delta response to charge placed at
  origin for
% voltage-clamped cylinder (initial condition V = 0)
% From Rinzel & Rall (1974)


K_ins = zeros(length(X),length(T)); % Empty vector for storage
K_clp = zeros(length(X),length(T)); % Empty vector for storage
K_clp_delta = zeros(length(X),length(T)); % Empty vector for
  storage
trunk_end = find(X == Xk_vals(1)) - 1; % Cutoff index of the
  trunk.
bgp_end = 2.* trunk_end; % Cutoff index of the grand-parent
  branch.
bp_end = 3.* trunk_end; % Cutoff index of the parent branch.
bin_end = length(X); % Cutoff index of input branch


end_inds = [1, trunk_end, bgp_end, bp_end, bin_end];


for i = 1:length(X)

    for j = 1:length(T)

            sum1_neg = 0;
            sum1_pos = 0;
```

```
n_neg = 0;
n_pos = 1;
perc_pos = 100;
perc_neg = 100;

while perc_neg > 0.00001
    sum1_neg_prev = sum1_neg;
    sum1_neg = sum1_neg_prev + ...
        (-1).^n_neg.*exp(-((L.*(2.*n_neg-1) +
            X(i)).^2)/(4*T(j))));
    n_neg = n_neg - 1;
    perc_neg = (sum1_neg - sum1_neg_prev)./sum1_neg;
end

while perc_pos > 0.00001
    sum1_pos_prev = sum1_pos;
    sum1_pos = sum1_pos_prev + ...
        (-1).^n_pos.*exp((-(L.*(2.*n_pos-1) +
            X(i)).^2)/(4*T(j))));
    n_pos = n_pos + 1;
    perc_pos = (sum1_pos - sum1_pos_prev)./sum1_pos;
end

K_clp(i,j) = R_inf.*exp(-T(j))./(pi.*T(j)).^(0.5) .* ...
    (sum1_pos + sum1_neg);

sum2_neg = 0;
sum2_pos = 0;
n2_neg = 0;
n2_pos = 1;
perc2_pos = 100;
perc2_neg = 100;

while perc2_neg > 0.00001
    sum2_neg_prev = sum2_neg;
    sum2_neg = sum2_neg_prev + ...
        exp((-(L*(2.*n2_neg - 1) + X(i)).^2)/(4*T(j))));
    n2_neg = n2_neg - 1;
    perc2_neg = (sum2_neg - sum2_neg_prev)./sum2_neg;
end

while perc2_pos > 0.00001
    sum2_pos_prev = sum2_pos;
    sum2_pos = sum2_pos_prev + ...
        exp((-(L*(2.*n2_pos - 1) + X(i)).^2)/(4*T(j))));
    n2_pos = n2_pos + 1;
    perc2_pos = (sum2_pos - sum2_pos_prev)./sum2_pos;
end

K_ins(i,j) = R_inf.*exp(-T(j))./(pi.*T(j)).^(0.5) .* ...
    (sum2_pos + sum2_neg);

    end
end
```

```
K_ins(isnan(K_ins)) = 0;
K_clp(isnan(K_clp)) = 0;


% Calculating the general dirac-delta response function of
  the system at
% different times and electrotonic positions.
% From Rinzel & Rall (1974)


% B_k parameters for all branches which are are simple
  constants whose values are specified according to
  location:

Bk0_vals = [0,0,0]; % Trunk
Bk1_vals = [1,0,0; -1,0,0]; % BGP, BC2
Bk2_vals = [1,1,0; 1,-1,0; -1,0,0]; % BP, BC1, BC2
Bk3_vals = [1,1,1; 1,1,-1; 1,-1,0; -1,0,0]; % BI, BS, BC1, BC2


% The branches of B_k are trunk, BGP, BP, BI, BS, BC1, BC2
  respectively.

K_M0 = zeros(trunk_end,length(T),size(Bk0_vals,1)); % Empty
  vector for storage
K_M1 = zeros(trunk_end,length(T),size(Bk1_vals,1)); % Empty
  vector for storage
K_M2 = zeros(trunk_end,length(T),size(Bk2_vals,1)); % Empty
  vector for storage
K_M3 = zeros(trunk_end + 1,length(T),size(Bk3_vals,1)); %
  Empty vector for storage


A = N - 1; % Pre-defined parameter.


for i = 1:length(X)

    if X(i) < Xk_vals(1)
        for j = 1:length(T)
            sum_K = 0;
            K_M0(i,j) = N.^(-1).*K_ins(i,j) +...
                A.*N.^(-1).*K_clp(i,j) + sum_K;
        end

    elseif X(i) >= Xk_vals(1) && X(i) < Xk_vals(2)
        for j = 1:length(T)
            for m = 1: size(Bk1_vals,1);
                sum_K = 0;
                for k = 1:M
                    sum1_neg = 0;
                    sum1_pos = 0;
                    n_neg = 0;
                    n_pos = 1;
                    perc_pos = 100;
                    perc_neg = 100;
```

```
                        while perc_neg > 0.001
                            sum1_neg_prev = sum1_neg;
                            sum1_neg = sum1_neg_prev + ...
                                (-1).^n_neg.
                                   *exp(-((((L-Xk_vals(k))*(2.*n_
                                   neg-1)...
                                   + (X(i)-Xk_vals(k))).^2)/(4*T(j)));
                            n_neg = n_neg - 1;
                            perc_neg = (sum1_neg - sum1_neg_
                                prev)./sum1_neg;
                        end

                        while perc_pos > 0.001
                            sum1_pos_prev = sum1_pos;
                            sum1_pos = sum1_pos_prev + ...
                                (-1).^n_pos.*exp((-((L-Xk_
                                   vals(k))*(2.*n_pos-1)...
                                   + (X(i)-Xk_vals(k))).^2)/(4*T(j)));
                            n_pos = n_pos + 1;
                            perc_pos = (sum1_pos - sum1_pos_
                                prev)./sum1_pos;
                        end

                        B_k = Bk1_vals(m,:);
                        K_clp_delta(i,j) = R_inf.*exp(-T(j))./
                            (pi.*T(j)).^(0.5)... .*
                            (sum1_pos + sum1_neg);
                        sum_K = sum_K + 2.^(k-1).*B_k(k).
                            *K_clp_delta(i,j);
                        KM1_index = i - (trunk_end).*(size(Bk1_
                            vals,1)-1);
                        K_M1(KM1_index,j,m) = ...
                            N.^(-1).*K_ins(i,j) + A.*N^(-1).
                                *K_clp(i,j) + sum_K;
                    end
                end
            end

    elseif X(i) >= Xk_vals(2) && X(i) < Xk_vals(3)
        for j = 1:length(T)
            for m = 1: size(Bk2_vals,1);
                sum_K = 0;
                for k = 1:M
                    sum1_neg = 0;
                    sum1_pos = 0;
                    n_neg = 0;
                    n_pos = 1;
                    perc_pos = 100;
                    perc_neg = 100;

                    while perc_neg > 0.001
                        sum1_neg_prev = sum1_neg;
```

```
                        sum1_neg = sum1_neg_prev + ...
                            (-1).^n_neg.
                                *exp(-(((L-Xk_vals(k))*(2.*n_
                                neg-1)...
                                + (X(i)-Xk_vals(k))).^2)/(4*T(j)));
                        n_neg = n_neg - 1;
                        perc_neg = (sum1_neg - sum1_neg_
                            prev)./sum1_neg;
                    end

                    while perc_pos > 0.001
                        sum1_pos_prev = sum1_pos;
                        sum1_pos = sum1_pos_prev + ...
                            (-1).^n_pos.*exp((-((L-Xk_
                            vals(k))*(2.*n_pos-1)...
                            + (X(i)-Xk_vals(k))).^2)/(4*T(j)));
                        n_pos = n_pos + 1;
                        perc_pos = (sum1_pos - sum1_
                            pos_prev)./sum1_pos;
                    end

                    B_k = Bk2_vals(m,:);
                    K_clp_delta(i,j) = R_inf.*exp(-T(j))./
                      (pi.*T(j)).^(0.5)...
                        .* (sum1_pos + sum1_neg);
                    sum_K = sum_K +
                      2.^(k-1).*B_k(k).*K_clp_delta(i,j);
                    KM2_index
                      = i - (trunk_end).*(size(Bk2_vals,1)-1);
                    K_M2(KM2_index,j,m) =...
                        N.^(-1).*K_ins(i,j)+A.*N^(-1).*K_
                            clp(i,j) + sum_K;
                end
            end
        end
    else
        for j = 1:length(T)
            for m = 1: size(Bk3_vals,1);
                sum_K = 0;
                for k = 1:M
                    sum1_neg = 0;
                    sum1_pos = 0;
                    n_neg = 0;
                    n_pos = 1;
                    perc_pos = 100;
                    perc_neg = 100;

                    while perc_neg > 0.00001
                        sum1_neg_prev = sum1_neg;
                        sum1_neg = sum1_neg_prev + ...
                            (-1).^n_neg.*exp(-(((L-Xk_
                            vals(k))*(2.*n_neg-1)...
                            + (X(i)-Xk_vals(k))).^2)/(4*T(j)));
```

```
                        n_neg = n_neg - 1;
                        perc_neg = (sum1_neg - sum1_neg_
                           prev)./sum1_neg;
                  end

                  while perc_pos > 0.00001
                        sum1_pos_prev = sum1_pos;
                        sum1_pos = sum1_pos_prev + ...
                           (-1).^n_pos.*exp((-((L-Xk_
                           vals(k))*(2.*n_pos-1)...
                           + (X(i)-Xk_vals(k))).^2)/(4*T(j)));
                        n_pos = n_pos + 1;
                        perc_pos = (sum1_pos - sum1_pos_
                           prev)./sum1_pos;
                  end

                  B_k = Bk3_vals(m,:);
                  K_clp_delta(i,j) = R_inf.*exp(-T(j))./
                     (pi.*T(j)).^(0.5)...
                           .* (sum1_pos + sum1_neg);
                  sum_K = sum_K + 2.^(k-1).
                     *B_k(k).*K_clp_delta(i,j);
                  KM3_index = i - (trunk_end).*
                     (size(Bk3_vals,1)-1);
                  K_M3(KM3_index,j,m) =...
                           N.^(-1).*K_ins(i,j)+A.*N^(-1).*K_
                           clp(i,j) + sum_K;
               end
            end
         end
      end
end

% Forming the matrix K such that each dimension represents
  the impulse
% response of the path leading to BI, BS, BC1, BC2
  respectively.

K = zeros(length(X), length(T), M + 1);
K(:,:,1) = cat(1,K_M0(:,:), K_M1(:,:,1), K_M2(:,:,1),
   K_M3(:,:,1)); % BI
K(:,:,2) = cat(1,K_M0(:,:), K_M1(:,:,1), K_M2(:,:,1),
   K_M3(:,:,2)); % BS
K(:,:,3) = cat(1,K_M0(:,:), K_M1(:,:,1), K_M2(:,:,2),
   K_M3(:,:,3)); % BC1
K(:,:,4) = cat(1,K_M0(:,:), K_M1(:,:,2), K_M2(:,:,3),
   K_M3(:,:,4)); % BC2

K(isnan(K)) = 0;
K(K<0) = 0;
neuron_volts = zeros(size(K));
```

```
% Convolution of signals I(t) and K as defined by Rinzel and
  Rall (1974) =
% To ensure adherence to the boundary conditions defined by
  Rinzel and Rall
% for the convolution, all values for the convolution after
  time T were
% deleted from the result convolution vectors.

K_rows = size(K,1);
K_dims = size(K,3);

for k = 1:K_dims
    for i = 1:K_rows
        row = K(i,:,k);
        K_involt = conv(I_T, row);
        K_involt((length(T) + 1) : end) = [];
        neuron_volts(i,:,k) = K_involt;
    end
end

% Calculating charge dissipation with respect to position
  for all time.

dissipation = c_m.*lambda.*trapz(T, neuron_volts, 2);

branch_dissipation = zeros(4,4); % This creates a vector for
  the storage
% of dissipation values for the
% Now finding the dissipation that occurs over an entire
  branch over time.

for i = 1:size(dissipation, 3)
    for j = 1:(length(end_inds)-1)
        branch_interval = (end_inds(j):end_inds(j+1));
        branch_pos = X(branch_interval);
        branch = dissipation(branch_interval, 1, i);
        branch_dissipation(i,j) = trapz(branch_pos, branch);
    end
end

% Producing corresponding bar graph:

bar(end_inds(1:end-1),branch_dissipation')
legend('BI', 'BS', 'BC1', 'BC2')

% xlabel('Trunk', 'BGP', 'BP', 'BI')
% === Set the initial states FROM Hodgkin Huxley 1952 === %
% equations taken from table 1
% Alpha and beta are variables that determine gating kinetics

gbar_K=17.5; gbar_Na=120; gbar_L=0.1;%mmho/cm2=msiemens/cm2
  (giant axon of squid)
```

```
E_K = -5.0; E_Na=-115; E_L=0;%mV (Squid)
load_vector = zeros(length(X),size(neuron_volts,3)); %
  storage vector

for j = 1:size(neuron_volts,3)
    V_0 = neuron_volts(:,1,j).*10^3;
    alpha_m0 =0.1.*(V_0+25)./(exp((V_0 + 25)./ 10)-1);
    beta_m0 = 4.*exp(V_0./18);
    alpha_h0 = 0.07.*exp(V_0./20);
    beta_h0 = 1./(exp((V_0+30)./10)+1);
    alpha_n0 = 0.01.*(V_0+10)./(exp((V_0+10)./10)-1);
    beta_n0 = 0.125*exp(V_0./80);

    n_0 = alpha_n0./(alpha_n0 + beta_n0); %Equation 9
    m_0 = alpha_m0./(alpha_m0 + beta_m0); %Equation 18
    h_0 = alpha_h0./(alpha_h0 + beta_h0); %Equation 18

    if all(m_0 == m_0(1))
        m_0 = m_0(1);
    end

    if all(n_0 == n_0(1))
        n_0 = n_0(1);
    end

    if all(h_0 == h_0(1))
        h_0 = h_0(1);
    end

    for i = 1:length(X)
        V = neuron_volts(i,:,j)*10^3;
        alpha_m =0.1.*(V_0(i)+25)./(exp((V_0(i)+25)./10)-1);
        beta_m = 4.*exp(V_0(i)./18);
        alpha_h = 0.07.*exp(V_0(i)./20);
        beta_h = 1./(exp((V_0(i)+30)./10)+1);
        alpha_n = 0.01.*(V_0(i)+10)./(exp((V_0(i)+10)./10)-1);
        beta_n = 0.125.*exp(V_0(i)./80);

        n_inf = alpha_n./(alpha_n+beta_n); %Equation 9
        m_inf = alpha_m./(alpha_m+beta_m); %Equation 18
        h_inf = alpha_h./(alpha_h+beta_h); %Equation 18

        tau_m = m_inf ./ alpha_m;
        tau_n = n_inf ./ alpha_n;
        tau_h = h_inf ./ alpha_h;
        tau_h(isnan(tau_h)) = 0;

        n = n_inf - (n_inf - n_0).*exp(-T./tau_n);
        m = m_inf - (m_inf - m_0).*exp(-T./tau_m);
        h = h_inf - (h_inf - h_0).*exp(-T./tau_h);
```

```
                I_Na = (gbar_Na.*m.^3.*h.*(V - E_Na))*10^-6;
                Na_load = trapz(T, I_Na); % Area under the Na curve
                load_vector(i,j) = Na_load;
                %      I_K = gbar_K.*n.^4.*(V-E_K);
                %      I_l = gbar_L.*(V-E_L);
            end
        end


        total_ATP = zeros(1,size(load_vector,2));
        Faraday = 9.64870E4;

        for i = 1:length(total_ATP)
            total_ATP(i) = sum(load_vector(:,i))./(3.*Faraday);
            total_Exnormoxic = (total_ATP .* 131.3) %131.3 overall
              exergy destruction rate (kj/mol) for normoxic
              condition Seda Genc
            total_Exhypoxic = (total_ATP .* 129.8) %129.8 overall
              exergy destruction rate (kj/mol) for normoxic
              condition Seda Genc
            total_Snormoxic = ((sum(total_Exnormoxic))*10)/298.15
              %10 represent action potential rate 10 per second
            total_Shypoxic = ((sum( total_Exhypoxic))*30)/298.15
              %30 represent action potential rate 10 per second
        end
```

When we run the code to calculate the total ATP utilization per one action potential, the following line and Example Figure 4.6.3 will appear on the screen:

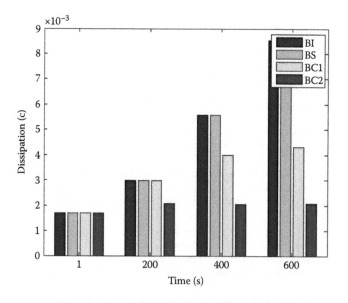

EXAMPLE FIGURE 4.6.3 Schematic diagram of the dissipation in the input (BI), sister (BS), cousin 1 (BC1), and cousin 2 (BC2) branches.

```
total_Exnormoxic =
    1.0e-06 *
    0.5542          0          0          0

total_Exhypoxic =
    1.0e-06 *
    0.5479          0          0          0

total_Snormoxic =
    1.8589e-08

total_Shypoxic =
    5.5130e-08

total_Exnormoxic =
    1.0e-06 *
    0.5542          0.5542          0          0

total_Exhypoxic =
    1.0e-06 *
    0.5479          0.5479          0          0

total_Snormoxic =
    3.7178e-08

total_Shypoxic =
    1.1026e-07

total_Exnormoxic =
    1.0e-06 *
    0.5542          0.5542          0.5542          0

total_Exhypoxic =
    1.0e-06 *
    0.5479          0.5479          0.5479          0

total_Snormoxic =
    5.5766e-08

total_Shypoxic =
    1.6539e-07

total_Exnormoxic =
    1.0e-06 *
    0.5542          0.5542          0.5542          0.5542

total_Exhypoxic =
    1.0e-06 *
    0.5479          0.5479          0.5479          0.5478

total_Snormoxic =
    7.4352e-08

total_Shypoxic =
    2.2051e-07
```

Example 4.7: Light Production Efficiency of Firefly,
***Photinus pyralis*, Bioluminescence**

The detailed chemical nature of the process bioluminescence is described by Gomi and Kajiyama (2001). Luciferin reacts with ATP and oxygen to generate carbon dioxide, AMP, and oxyluciferin. There are different species of oxyluciferin. One of them is first formed in the excited state and is then decayed to the ground state with the emission of green light (Nakamura et al., 2005; Vieira et al., 2012). Oxyluciferin then reacts with water, and it is converted into 2-cyano-6-hydroxybenzothiazole and thioglycolic acid. 2-Cyano-6-hydroxybenzothiazole reacts with D-cysteine to produce luciferin and ammonia. In this sequence of reactions, luciferin, oxyluciferin, and 2-cyano-6-hydroxybenzothiazole remain in the tissue, within the system boundaries; O_2, ATP, H_2O, and D-cysteine enter the system boundaries, while CO_2, AMP, thioglycolic acid, and ammonia flow out of the system boundaries (Example Figure 4.7.1). These luminescence reactions are employed in numerous assays (Branchini et al., 2011).

The chemical structures of the species that circulate in the firefly bioluminescence cycle are given in Example Table 4.7.1.

Normally, yellow-green light with a wavelength, for example, λ, of about 560 nm is produced by fireflies; detailed information about the spectra of the firefly bioluminescence reaction is given by Wang et al. (2011). When excited chemicals go to the ground state, emission of visible light appears. It is the transduction of chemical energy Δh into radiant energy $h\nu$. It is possible when

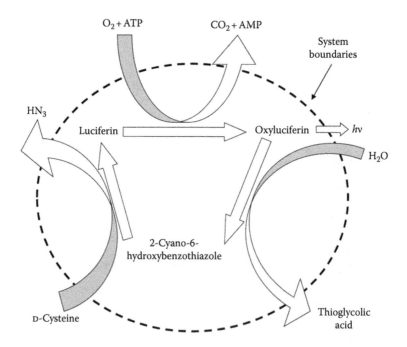

EXAMPLE FIGURE 4.7.1   The firefly bioluminescence cycle.

EXAMPLE TABLE 4.7.1 Chemical Structures of the Species That Circulate in the Firefly Bioluminescence Cycle

Chemical Species	Chemical Structure
D-Luciferin	
Oxyluciferin	
Thioglycolic acid	
2-Cyano-6-hydroxybenzothiazole	
D-Cysteine	

the Δh of the reaction is large enough to allow for production of at least one of the products in excited state. Exergy of the firefly bioluminescence, $h\nu$, may be estimated as

$$h\nu = h\frac{c}{\lambda} = \frac{(6.63\times10^{-34}\ \text{J s})(3.00\times10^{8}\ \text{m/s})}{(560\ \text{nm})(1\ \text{m}/10^{-9}\ \text{nm})} = 3.5\times10^{-19}\ \text{J}$$

where $\nu = c/\lambda$, for example, speed of the light, c, divided with the wavelength λ. Exergies of formation of O_2, CO_2, H_2O, NH_3, ATP, and AMP are readily available in the literature. Chemical exergies of D-cysteine and thioglycolic are estimated with the group contribution method.

Chemical exergies of D-cysteine and thioglycolic are estimated with the expression

$$ex_{ch} = \Delta g_f + \sum n_e ex_{ch,e}$$

where
ex_{ch} is the standard chemical exergy of the compound
Δg_f is the Gibbs free energy of formation at 298.15 K, kJ/mol
n_e is the number of moles of the element e
$ex_{ch,e}$ is the standard chemical exergy (Example Table 4.7.2) of each contributing element (Değerli et al., 2015)

EXAMPLE TABLE 4.7.2 Standard Chemical Exergy of the Elements

Molecule	$ex_{ch,e}$/molecule (kJ/mol)
C	410.3
H_2	236.1
O_2	4.0
N_2	0.7
S	607.1

Source: Exergoecology Portal, Exergy calculator, available at http://www.exergoecology.com/excalc/exergo/refenv/#equation1, accessed August 16, 2013.

EXAMPLE TABLE 4.7.3 Calculation of the Exergies of the Firefly Bioluminescence Cycle Chemicals

Chemical Species	Δg^0_{f298} (kJ/mol)	n_e	$ex_{ch} = \Delta g^0_{f298} + \sum n_e ex_{ch,e}$ (kJ/mol)
D-Luciferin $C_{11}H_8N_2O_3S_2$	25.5	$n_{e,C} = 11, n_{e,H_2} = 4, n_{e,N_2} = 1,$ $n_{e,O_2} = 3/2, n_{e,S} = 2$	6703.4
Oxyluciferin $C_{10}H_6N_2O_2S_2$	252.2	$n_{e,C} = 10, n_{e,H_2} = 3, n_{e,N_2} = 1,$ $n_{e,O_2} = 1, n_{e,S} = 2$	6281.8
Thioglycolic acid $C_2H_4O_2S$	−348.6	$n_{e,C} = 2, n_{e,H_2} = 2, n_{e,O_2} = 1,$ $n_{e,S} = 1$	1555.2
2-Cyano-6-hydroxybenzothiazole $C_8H_4N_2OS$	303.5	$n_{e,C} = 8, n_{e,H_2} = 2, n_{e,N_2} = 1,$ $n_{e,O_2} = 1/2, n_{e,S} = 1$	4667.5
D-Cysteine $C_3H_7NO_2S$	−276.1	$n_{e,C} = 3, n_{e,H_2} = 7/2,$ $n_{e,N_2} = 1/2, n_{e,O_2} = 1, n_{e,S} = 1$	2392.4

Details of the calculations of the chemical exergies are depicted in Example Table 4.7.3.

Gibbs free energy of formation of a molecule, Δg_f, is estimated via atomic group contributions at 298.15 K and 1 atm (101.325 kPa):

$$\Delta g^0_{f298} = 53.88 + \sum n_i \Delta g_i$$

where

Δg^0_{f298} is the Gibbs free energy of formation in kJ/mol at 298.15 K
Δg_i is the numerical value of the group contribution of the atomic group

The details of the calculations, for the estimation of the chemical exergies of D-cysteine and thioglycolic acid, are given in Example Table 4.7.4.
Results of these calculations are listed in Example Table 4.7.5.

EXAMPLE TABLE 4.7.4 Estimation of the Chemical Exergies of D-Cysteine and Thioglycolic Acid

Group	Occurrence	$\Delta g_{f,i}^0$ (kJ/mol)
i. Calculation of the contribution of the atomic groups of D-cysteine		
—NH$_2$	1	14.1
—COOH (acid)	1	−387.9
—CH	1	58.4
—CH$_2$—	1	8.4
—SH	1	−23.0
		$\sum n_i \Delta g_{f,i}^0 = -330.0$ kJ/mol
		$\Delta g_{f298}^0 = -276.1$ kJ/mol
ii. Calculation of the contribution of the atomic groups of thioglycolic acid		
—COOH (acid)	1	−387.9
—CH$_2$—	1	8.4
—SH	1	−23.0
		$\sum n_i \Delta g_{f,i}^0 = -404.4$ kJ/mol
		$\Delta g_{f298}^0 = -348.6$ kJ/mol
iii. Atomic group contributions employed for the estimation of Δg_{f298}^0 of D-luciferin		
—COOH (acid)	1	−387.9
—OH (phenol)	1	−197.4
—CH$_2$—	1	−3.7
—CH	1	41.0
—S— (ring)	2	27.8
—N= (ring)	2	79.9
=C—	5	54.1
=CH	3	11.3
		$\sum n_i \Delta g_{f,i}^0 = -28.4$ kJ/mol
D-Luciferin		$\Delta g_{f298}^0 = 25.5$ kJ/mol
iv. Atomic group contributions employed for the estimation of Δg_{f298}^0 of oxyluciferin		
—OH (alcohol)	1	−189.2
—OH (phenol)	1	−197.4

(Continued)

EXAMPLE TABLE 4.7.4 (Continued)   Estimation of the Chemical Exergies of D-Cysteine and Thioglycolic Acid

Group	Occurrence	$\Delta g_{f,i}^0$ (kJ/mol)
$=C-$	6	54.1
$=CH$	4	11.3
$-S-$ (ring)	2	27.8
$-N=$ (ring)	2	79.9
		$\sum n_i \Delta g_{f,i}^0 = -198.3$ kJ/mol
Oxyluciferin		$\Delta g_{f298}^0 = 252.2$ kJ/mol

v. Atomic group contributions employed for the estimation of Δg_{f298}^0 of 2-cyano-6-hydroxybenzothiazole

$-OH$ (phenol)	1	-197.4
$=C-$	4	54.1
$=CH$	3	11.3
$-N=$ (ring)	1	79.9
$-S-$ (ring)	1	27.8
$-CN$	1	89.2
		$\sum n_i g_{f,i}^0 = 249.6$ kJ/mol
		$\Delta g_{f298}^0 = 303.5$ kJ/mol

EXAMPLE TABLE 4.7.5   Exergies of the Inputs and the Outputs of the Firefly Bioluminescence Cycle

Inputs	Exergy of Formation (kJ/mol)	Outputs	Exergy of Formation (kJ/mol)
O_2	2,955.0 (Değerli et al., 2015)	CO_2	19.5 (Değerli et al., 2015)
ATP	6,720.9 (Genç et al., 2013a)	AMP[a]	4981.8
H_2O	0.9 (Değerli et al., 2015)	Thioglycolic acid	1555.2
D-Cysteine	2,392.4	NH_3	337.9 (Szargut, 2007)
Total	12,069.2	Total	6894.4

[a] Calculated by ATP → ADP + P and ATP → AMP + 2P.

Since the purpose of running this chemical cycle is producing light, the exergetic efficiency of the process may be calculated as

$$\eta = \frac{h\nu\, N_A}{\sum ex_{inputs} - \sum ex_{outputs}} = \frac{(3.5\times10^{-22})(6.02\times10^{23})\,\text{kJ}/\text{mol}}{(12,069.2 - 6,894.4)\,\text{kJ}/\text{mol}}\times100 = 4.1\%$$

Example 4.8: Work Done to Develop an Immune System

In the immune-challenged poultry, the molecules, which are involved in the immune response development, make up less than 1%, and their daily turnover corresponds to less than 0.05% of the body mass, but the resting metabolic rate of these birds is 5%–15% higher than the average (Klasing, 1998; Hasselquist and Nilsson, 2012). When the energy cost of developing the immune response is too high, animals may tend to increase the consumption of the nutrients or reduce the amount of the work they perform. Amat et al. (2007) argue that the antibody production and *Brucella abortus* increase the energy expenditure of the small bird *Carduelis chloris* by about 4.7%, and some individuals pay for this cost by reducing the energy expenditure for growing hair. Raberg et al. (2000), after inducing immune response in the small bird *Parus caeruleus*, observed that the vaccinated female birds reduced their feeding workload by 4 trips/h. It was estimated from the respiration data that the energy cost of each feeding trip was 0.047 kJ; therefore, in a 14 h of feeding period, the reduced workload of an individual female bird was (4 trips/h)(14 h/day)(0.047 kJ/trip) = 2.63 kJ/day (Hasselquist and Nilsson, 2012).

4.1.3 Biomass-Producing Process

Living organisms produce biomass to store energy for future, as described in Figure 4.8. Animals produce lipids and plants produce starch or cellulose from the excess glucose. Biological species that perform photosynthesis employ solar energy to create bonds between carbon atoms to generate long-chain hydrocarbons. Energy and exergy of the chemicals that flow in (mostly CO_2 and H_2O and some inorganic substances) are lower than the energy and exergy levels of the products of photosynthesis. In other words,

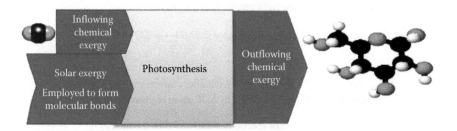

FIGURE 4.8 Schematic description of the exergy flow during photosynthesis.

high-entropy chemicals are flowing into the photosynthesis process, and their entropy is decreased by organizing them into low-entropy long-chain hydrocarbons. The work required to increase the exergy (i.e., decrease entropy) is extracted from solar energy.

Example 4.9: Exergetic Efficiency of Glucose Synthesis via Photosynthesis

Photosynthesis occurs in the *chloroplast*, a specialized organelle of the green plants. Some of the reactions involved in photosynthesis consume photon's energy, and therefore, these are called the *light reactions*. Exergy of the photons reaching the earth's surface from the sun is expressed as (Lems, 2009)

$$ex_{photons} = \frac{hc}{\lambda}\left(1 - \frac{T_{earth}}{T_{sun}}\right) - k_B T_{earth} \ln\left(\frac{d_{earth}^2}{r_{sun}^2}\right)$$

where
   c is the speed of light (2.9998×10^8 m/s)
   d_{earth} is the distance between the sun and the earth
   h is Planck's constant (6.626×10^{-34} J s)
   k_B is Boltzmann constant (1.38×10^{23} J/K)
   r_{sun} is the radius of the sun
   T_{earth} is the temperature of the surface of the earth (estimated as 295 K)
   T_{sun} is the temperature of the surface of the sun (estimated as 5762 K)
   λ is the wavelength of the radiation (nm)

Solar photons are in the range of 380 nm $< \lambda <$ 750 nm. The second term in this equation refers to the exergy of the photons lost to dispersion:

$$ex_{lost} = k_B T_{earth} \ln\left(\frac{d_{earth}^2}{r_{sun}^2}\right)$$

where $(d_{earth}^2 / r_{sun}^2) = 46,448$. Exergy of the direct solar radiation arriving in the Earth's atmosphere and the fraction available for the photosynthesis are calculated by Petela (2008) as 1.2835 and 0.5155 kW/m², respectively. ATP produced with the light reactions is consumed in the dark reactions, mainly in the Calvin cycle where CO_2 is converted into glucose. Production of glucose by green plants via photosynthesis is described schematically in Example Figure 4.9.1.
   The overall process of the light reactions is (Lems, 2009)

$$6CO_2 + 24H_{NADHP} + 18P_{ATP} \rightarrow C_6H_{12}O_6 + 6H_2O + 18P$$

where
   H_{NADHP} represents the H added to NADP after conversion into $NADPH^+$
   P_{ATP} is the organic phosphate added to ADP upon conversion to ATP

The overall process of the dark reactions is expressed as (Lems, 2009)

$$12H_2O + 24hv_{680} + 24hv_{700} \rightarrow 6O_2 + 24H_{NADHP} + 24P_{ATP}$$

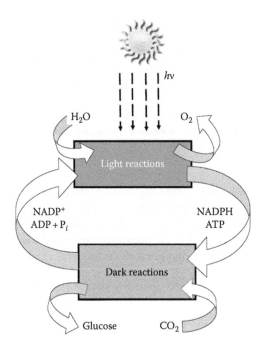

EXAMPLE FIGURE 4.9.1 Schematic description of photosynthesis in the green plants.

The sum of the light and the dark reactions gives the overall reaction

$$6CO_2 + 6H_2O + 6P + 24hv_{680} + 24hv_{700} \rightarrow C_6H_{12}O_6 + 6O_2 + 6P_{ATP}$$

The overall exergy balance of photosynthesis is presented in Example Table 4.9.1. Then the total exergetic efficiency of photosynthesis is calculated as

$$\eta = \frac{\text{Total exergy output}}{\text{Total exergy input}} = \frac{3261\,\text{kJ}}{7896\,\text{kJ}} = 0.41$$

EXAMPLE TABLE 4.9.1 Inputs and Outputs and the Exergy Destruction of Photosynthesis

Exergy Input		Exergy Output		Exergy Loses	
$6CO_2$	0 kJ	$1C_6H_{12}O_6$	2955 kJ	Photon absorption	1278 kJ
$6H_2O$	0 kJ	$6O_2$	0 kJ	Cellular processes	3357 kJ
6P	0 kJ	$6P_{ATP}$	306 kJ		
$24hv_{680}$	4005 kJ				
$24hv_{700}$	3891 kJ				
TOTAL	7896 kJ	TOTAL	3261 kJ	TOTAL	4635 kJ

Source: Adapted from Lems, S., *Thermodynamic explorations into sustainable energy conversion, learning from living systems*, PhD dissertation, Technical University of Delft, Rotterdam, the Netherlands, 2009.

This efficiency indicates that in photosynthesis about 59% of the exergy is lost. The yield potential of a grain crop is determined by the availability of light, CO_2 and H_2O, and the genetically determined plant properties. Among the efficiency of light capture, the efficiency of conversion of the intercepted light into biomass is related to the exergetic efficiency of photosynthesis. Long et al. (2006) explored the potential routes of increasing the plant yield by considering the possibilities ranging from the altered canopy architecture to improved regeneration of the acceptor molecule for CO_2 and argued that these changes may improve the yield by 50%.

Cells store their internal energy in the high-energy interatomic bonds of their carbohydrate or lipid reserves (Stryer, 1981; Nelson and Cox, 2013). For short-term storage, glucose is converted into glucose polymers (glycogen in animal cells, and starch or cellulose in plant cells) and stored in the immediate tissues (Morgenthaler et al., 2006). A glucose unit may be added to glycogen with the following reactions (Stryer, 1981):

$$\text{Glucose} + \text{ATP} \rightarrow \text{Glucose-6-phoshate} + \text{ADP} + P_i$$

$$\text{Glucose-6-phosphate} + \text{ATP} + (\text{Glycogen})_{n\,residues} + H_2O \rightarrow (\text{Glycogen})_{n+1 residues} + \text{ADP} + P_i$$

Cleavage of glycogen into glucose-1-phosphate may be expressed as

$$(\text{Glycogen})_{n\,residues} + P_i \rightarrow \text{Glucose-1-phoshate} + (\text{Glucogen})_{n-1 residues}$$

The glucose-1-phosphate formed with the cleavage of glycogen is converted into glucose-6-phosphate by the enzyme phosphoglucomutase. Glucose-6-phosphate is processed further in the glycolytic pathway of the energy metabolism to produce ATP. Glycogen stored in the muscles provides the energy needed for the muscular activity. A major function of the liver is to maintain a constant level of glucose in the blood. Phosphorylated glucose cannot diffuse out of the cells. The liver contains an enzyme glucose 6-phosphatase that removes the P_i and enables glucose to leave the organ:

$$\text{Glucose-6-phosphate} + H_2O \rightarrow \text{Glucose} + P_i$$

The storage consumes slightly more than one ATP per glucose 6-phosphate (Stryer, 1981). In case of an urgent energy demand, a series of reactions occur to oxidize glucose and produce ATP. Glucose oxidation can be summarized as

$$\text{Glucose} + 6O_2 \rightarrow 6CO_2 + 6H_2O$$

Complete oxidation of glucose is mainly achieved in three steps. The first step is the conversion of 1 mol of glucose to 2 mol of pyruvate and 2 mol of ATP by glycolysis in the cytoplasm, followed by conversion of pyruvate to acetyl CoA and full oxidation in the mitochondria through the Krebs cycle and the ETC (Berg et al., 2002). During these reactions, 30–38 mol of ATP is produced, which is the main energy source for cellular activities.

For nonurgent long-term consumption, glucose is converted into triglycerides in the animal cells (Stryer, 1981). Conversion of glucose into fat is a biomass-producing process, where molecular bonds are formed to generate a long-chain lipid. Forming molecular bonds can only happen at an expenditure of metabolic energy. Therefore, it decreases the metabolic energy efficiency (Flatt, 1970; Fell and Small, 1986; Sorgüven and Özilgen, 2013). Note that polymers made of glucose (i.e., glycogen, starch) do not undergo such an extensive structural change, and therefore, their storage is energetically very efficient (Stryer, 1981; Nelson and Cox, 2012).

The next example demonstrates how glucose is produced via photosynthesis and then used to synthesize lipids and how the stored lipids are converted into work.

Example 4.10: Exergetic Efficiency of Synthesis of Palmitic Acid and Its Reconversion into Work by Photosynthetic Algae

For modeling purposes, the alga is considered as the mixture of two parts: the lipids and the rest of the cell (Sorgüven and Özilgen, 2010). The so-called rest of the cell may be assumed to have a similar elemental composition with bacteria. The elemental composition of a bacterial cell was determined by Luria et al. (1960) as

$$C_{4.17}H_8O_{1.25}N_1P_{0.10}S_{0.03}K_{0.03}Na_{0.04}Ca_{0.01}Mg_{0.02}C_{10.01}Fe_{0.004}$$

Lipid content of algae varies in a large range: the lipid content of *Scenedesmus quadricauda* is only 1.9% of its dry weight, whereas *Scenedesmus dimorphus* can have a lipid content of 40%. There are cases reported in the literature where the oil content in microalgae exceeds 80% by weight of dry biomass (Metting, 1996; Spolaore et al., 2006). Algal lipids consist of different unsaturated fatty acids. For example, oleic acid is about 45% of the total fatty acid in microalgae *Botryococcus braunii* (Yoo et al., 2010).

Photosynthesis done by microalgae may be summarized with the following equation (Sorgüven and Özilgen, 2010):

$$CO_2 + H_2O + \text{Inorganics} \xrightarrow[\substack{\text{Calvin} \\ \text{cycle}}]{} \text{Cell debris}$$

$$+\text{Glucose} + O_2 \xrightarrow[\substack{\text{Fatty acid} \\ \text{and triglyceride} \\ \text{synthesis}}]{} \underbrace{\text{Cell debris} + \text{Lipid}}_{\text{Microalgae}} + O_2$$

Mass balance during production of 100 kg of microalgae (dry weight) in an open pond and exergy of the chemical species are listed in Example Table 4.10.1.

During the lipid synthesis by the algae, first glucose is synthesized and then converted into lipids; and then the lipids are stored in the cell. In such a process, we may define the exergetic efficiency of triglycerol synthesis from glucose as

$$\eta = \frac{(\text{Exergy of the triglycerol produced})}{\text{Exergy of the glucose consumed}}$$

EXAMPLE TABLE 4.10.1 Mass Balance and Exergy of the Chemical Species during Production of 100 kg of Microalgae (Dry Weight) in an Open Pond

	Total Mass to Produce 100 kg of Microlgae (kg)	Chemical Exergy (kJ/mol)	Total Exergy (kJ/m² Day)
Input of CO₂ and water			
CO_2	249.79	0.00	0.0
H_2O	4894.84	0.77	125.0
Input of inorganics			
$Ca(OH)_2$	0.22	103.33	
KCl	0.67	14.16	
$MgSO_4$	0.72	32.67	
Na_2SO_4	0.85	19.14	
$FeCl_2$	0.15	159.64	
NH_4OH	10.54	328.80	
H_3PO_4	2.95	80.60	
Total inorganics			57.5
Input of sunlight			
Total sunlight			44,539
Algal output			
Debris	30.00	2,753.94	509
Lipid	70.00	33,998.09	1,694
Algae total			2,203
Chemical outputs			
O_2	249.23	0.00	0
HCl	0.31	48.90	
H_2SO_4	0.29	160.53	
H_2O	4810.91	0.77	
Total inorganics			0.4
Exergy destroyed			42,396

Source: Adapted from Sorgüven, E. and Özilgen, M., *Energy*, 58, 679, 2013.

Chemical exergy of glucose is 4,675 kJ/mol; 214.47 kg of glucose with total exergy of (214.47 kg)(1 mol/0.18 kg)(4,675 kJ/mol) = 5,570,262 was used to produce 70 kg of tripalmitoylglycerol with exergy of [(70 kg)/(807,000 mol/kg)] (33,998 kJ/mol) = 2,949,000 kJ; the exergetic efficiency of triglycerol production in this process is calculated as

$$\eta = \left(\frac{2,949,000}{5,570,262}\right)(100) = 53\%$$

When the organism needs to utilize the energy stored in the chemical bonds of triglycerides, palmitoylglycerol ($C_{51}H_{98}O_6$) with a chemical exergy of 33,998 kJ/mol is first converted into 3 mol of palmitic acid with a chemical exergy of 10,819 kJ/mol and then oxidized. During these reactions, 1 mol of ATP is converted into AMP, which costs an exergy consumption of 40.6 kJ/mol (Rastogi, 2007).

Triglycerides undergo a series of enzymatic reactions to enter the mitochondrial matrix, where fatty acid oxidation occurs. A fatty acid can enter into the mitochondria only in the form of a fatty-acyl–S–CoA:

$$\text{Palmitic acid} + \text{ATP} + \text{CoA} \rightarrow \text{Palmitoyl S CoA} + \text{AMP} + \text{PP}_i$$

Nelson and Cox (2012) present the accounting of the ATP production with the oxidation of CoA ester of palmitic acid (16 carbons) as

$$\text{Palmitoyl} - \text{S} - \text{CoA} + 23\text{O}_2 + 131\text{P}_i + 131\text{ADP} \rightarrow \text{CoA} + 131\text{ATP} + 16\text{CO}_2 + 146\text{H}_2\text{O}$$

Each ATP molecule can produce 36.8 kJ/mol of work, when one phosphate bond is broken and ADP is formed. We may calculate the work potential of the ATP produced from these reactions as

$$\text{Net work potential} = \text{Work potential stored in ATP}$$

$$- \text{Work potential consumed for AMP}$$

$$= (131\,\text{mol ATP})(36.8\,\text{kJ/mol}) - (40.6\,\text{kJ/mol})$$

$$= 4708\,\text{kJ/mol of palmitic acid}$$

Those exergy values are subject to change slightly, based on the physiological conditions, such as temperature, pH, ionic strength, and metabolic concentrations (Genç et al., 2013a). If we assume that ATP is the only agent that a cell can employ to produce work, then the total work potential stored as a result of triglyceride oxidation can be calculated as

$$\eta = \frac{\text{Net work potential}}{\text{Chemical exergy of the palmitoyl glycerol}} = \frac{(3)(4{,}708)}{(33{,}998)} = 41.5\%$$

Examples 4.9 and 4.10 provide analysis for the synthesis reactions occurring in a biomass-producing system. Actual systems also have to perform a set of biochemical reactions for the survival of the organism. The most basic of these reactions is respiration. Some of the biomass produced via photosynthesis has to be consumed by the organism via respiration to perform cellular activities, such as expressing enzymes and performing work to take necessary chemicals inside the cell against an unfavorable concentration gradient.

Example 4.11 demonstrates the complex thermodynamic processes occurring in a biomass-producing system by modeling the photosynthetic algae *Chlamydomonas reinhardtii*.

Example 4.11: Thermodynamic Processes Occurring in the Biomass-Producing System of the Photosynthetic Algae *Chlamydomonas reinhardtii*

In a laboratory experiment, this photosynthetic alga is produced in a bioreactor under varying environmental conditions, and concentrations of several chemicals are measured as a function of time. We will begin our analysis by defining the system in accordance with the available experimental data. Here, our system is an isothermal, open system that contains 1 L trisphosphate medium. The chemical species CO_2, H_2O, O_2, NH_4, and H_2O_4P may flow in or out of the system boundaries. Light also enters the system and allows photosynthesis. Hundreds of biochemical reactions occur in a *C. reinhardtii* cell. For the sake of simplicity, elementary reactions can be grouped and basic metabolic mechanisms can be represented with one overall reaction. Küçük et al. (2015) modeled the bioreactions occurring in a *C. reinhardtii* cell as a three-step mechanism involving photosynthesis, respiration, and lipid production:

   a. Photosynthesis

$$8CO_2 + 7.37H_2O + 0.22NH_4^+ + 0.02H_2O_4P^-$$

$$+14.71\,\text{photon} \rightarrow 2CO_{0.48}H_{1.83}N_{0.11}P_{0.01} + C_6H_{12}O_6 + 8.25O_2$$

   b. Respiration

$$C_6H_{12}O_6 + 6O_2 \rightarrow 6CO_2 + 6H_2O$$

   c. Lipid production

$$11.5C_6H_{12}O_6 \rightarrow C_{69}H_{98}O_6 + 21.5O_2 + 20H_2O$$

The first step of the thermodynamic analysis is to solve the mass balance for each species. For this purpose, the experimentally measured weights of dry biomass and lipid are employed. Based on the experimental data, a kinetic model employing the logistic equation is solved. The growth rate of the *C. reinhardtii* is modeled as

$$\frac{dx}{dt} = \mu_m x \left(1 - \frac{x}{x_{max}} \right)$$

where x is the size of the algal population. And the lipid production rate (dP/dt) is modeled as

$$\frac{dP}{dt} = \alpha x + \beta \frac{dx}{dt}$$

where α, β, and μ_m are constants. MATLAB code E4.11.1 provides the solutions to the biomass and lipid production equations:

MATLAB CODE E4.11.1

Command Window

```
clear all
close all
format compact
global xmax

% enter the constants

xmax=8.60;

% enter the data
tData1=[0:14];
a=[0 0.10 1.00 2.10 3.20 3.20 3.30 4.18 5.00 5.20 6.20 6.80
   7.60 7.70 8.60];
x1=[0; 0.10; 1.00; 2.10; 3.20; 3.20; 3.30; 4.18; 5.00; 5.20;
   6.20; 6.80; 7.60; 7.70; 8.60];
lipid=[0 0.0001 0.00325 0.00625 0.00975 0.0121 0.0125 0.0135
   0.014 0.0125 0.01525 0.0145 0.014 0.014 0.015];

[t,x]=ode45('odetotalkinetics_first', [0 14], [0.1
   0.00000001]);

hold on

[ax, h1, h2]=plotyy(t, x(:,1), t, x(:,2));

set(ax(1),'ylim',[0 10],'ytick',[0:1:10],'ycolor','black');
set(ax(2),'ylim',[0 0.02],'ytick',[0:0.002:0.02],'ycolor',
   'black');
set(h1,'LineStyle',':','color','black','LineWidth',2);
set(h2,'LineStyle','-','color','black','LineWidth',2);
legend ('biomass','lipid','Location', 'southEast');
[ax, h3, h4]=plotyy(tData1, a, tData1, lipid);
set(h3, 'LineStyle', 'x', 'LineWidth', 2.0, 'Color', 'black');
set(h4, 'LineStyle', 'o', 'LineWidth', 2.0, 'Color', 'black');
set(get(ax(1),'Ylabel'),'String','dry biomass (g/L)');
set(get(ax(2),'Ylabel'),'String','lipid (g/L)');
grid on
xlabel('t(h)')
title ('Kinetic Model - 18.7 mM')
set(ax(1),'YTick',[0:1:10])
set(ax(1),'YColor','black')
set(ax(2),'YColor','black')
```

M-File:

```
function dx = odetotalkinetics_first(t,x);
% This function models product formation with Luedeking-
   Piret model
```

```
mu=1;
xmax=8.60;

if t<=4;
    mu=0.978;
    xmax=8.6;
    alpha=0.001;
    beta=0.002;
    dx1=mu*x(1)*(1-(x(1)/xmax));
    dx2=alpha*x(1) + beta*dx1;
end

if 4<=t;
    alpha=0.0005;
    beta=0.002;
    dx1=0;
    dx2=alpha*x(1) + beta*dx1;
end

if t>=6;
    mu=0.38;
    xmax=8.6;
    alpha=0.01;
    beta=0.0001;
    dx1=mu*x(1)*(1-(x(1)/xmax));
    dx2=0;
end

    dx = [dx1; dx2];
```

When we run this code Example Figure 4.11.1 appears on the screen.

Mass balances for the rest of the chemical species are solved based on the time-dependent results of the lipid and biomass concentrations. The mass fluxes through the system boundaries are described in Example Figure 4.11.2, when the biomass production rate is 57 mM/s. The reactants and the products of these reactions describe the situation where 2 M of algae is produced (apparent molar mass 23.4 g/mol). Note that the cell produces glucose via photosynthesis. Then, some of this glucose is consumed during respiration to maintain cellular activities. The remaining glucose is converted into lipid and stored in the cell. Respiration and lipid production are competitive reactions. The amount of lipid produced depends on the cellular conditions and the work produced in the system. Lipid and biomass are stored in the cell, and all of the other products are flowing out of the cell.

Once the mass balance equations are solved, we get the mass of each chemical species inside the system boundaries at any given time ($m_i(t)$) and the mass flow rate of each chemical species through the system boundaries ($\dot{m}_i(t)$) at any given time.

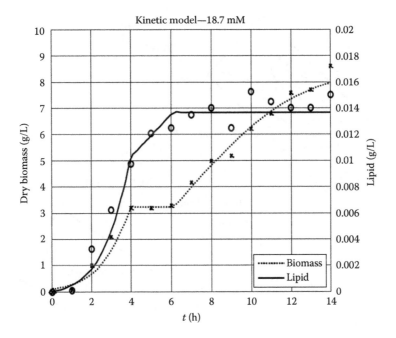

EXAMPLE FIGURE 4.11.1 Variation of the dry biomass and the lipid contents with time during cultivation of *Chlamydomonas reinhardtii* with the presence of 18.7 mM of NH_4^+.

For the *C. reinhardtii* system, the energy balance is

$$\sum_{in} \dot{m}_{in}h_{in} - \sum_{out} \dot{m}_{out}h_{out} - \dot{w} + \dot{q} + \dot{e}_{light} = \left(\frac{d(mu)}{dt}\right)_{system}$$

Since this is a time-dependent process, we must evaluate each term at each time t. To solve the governing equations numerically, we can discretize them with a time step size of Δt:

$$\sum_{in} \dot{m}_{in}h_{in} = (\dot{m}h)_{in,\,CO_2}\big|_t + (\dot{m}h)_{in,\,H_2O}\big|_t + (\dot{m}h)_{in,\,NH_4}\big|_t + (\dot{m}h)_{in,\,H_2O_4P}\big|_t$$

$$\sum_{out} \dot{m}_{out}h_{out} = (\dot{m}h)_{out,\,CO_2}\big|_t + (\dot{m}h)_{out,\,H_2O}\big|_t + (\dot{m}h)_{out,\,O_2}\big|_t$$

where enthalpies of the species are evaluated at the thermodynamic state of the corresponding inlet/outlet boundary at the time t

$$\left(\frac{d(mu)}{dt}\right)_{system} = \frac{1}{\Delta t}\left[(mu)_{biomass}\big|_{t+\Delta t} + (mu)_{lipid}\big|_{t+\Delta t} - (mu)_{biomass}\big|_t - (mu)_{lipid}\big|_t\right]$$

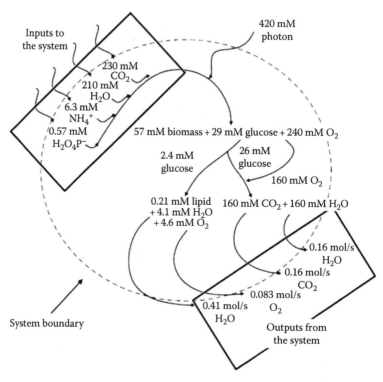

EXAMPLE FIGURE 4.11.2 Schematic description of the mass fluxes around a cell when the biomass production rate is 57 mM/s.

Since for this system $d(pv) = 0$, $du = dh$:

$$\left(\frac{d(mu)}{dt}\right)_{system} = \frac{1}{\Delta t}\left[\left((mh)\big|_{t+\Delta t} - (mh)\big|_{t}\right)_{biomass} + \left((mh)\big|_{t+\Delta t} - (mh)\big|_{t}\right)_{lipid}\right]$$

where enthalpies of the species are evaluated at the thermodynamic state of the system at time t. In this system, specific enthalpy h can be calculated as a function of temperature and pressure. For this system, both pressure and temperature are homogenously distributed and remain constant at 1 atm and 298 K, respectively. Example Table 4.11.1 that lists the thermodynamic data of the chemicals is adapted from Küçük et al. (2015).

During cultivation, 85 W superhigh-output fluorescent lights are used to provide 200 µmol photons/m² s (Tevatia et al., 2012). The energy of light is

$$E = \frac{hc}{\lambda}$$

where

    h is Planck's constant (6.626×10^{-34} J s)
    c is the speed of the light (2.9998×10^{8} m/s)
    λ is the wavelength of the radiation (555 nm)

EXAMPLE TABLE 4.11.1 Enthalpy of Formation and Exergy of the Chemical Compounds

Chemical Compounds	MW (g/mol)	$\Delta h^0_{f\,298K}$ (kJ/mol)
Glucose	180.2	−1274.5
Oxygen	32.0	0
$H_2O_4P^-$	97.0	−1302.5
NH_4^+	18.0	−132.8
Water (l)	18.0	−285.8
Carbon dioxide (g)	44	−393.5
Lipid	1026.6	−1056
Biomass	23.4	−201.8

The bulb has a color temperature of 6500 K (Lights of America, 2014) that corresponds to daylight (Amour, 1955); hence, exergy of sunlight is taken as 0.5155 kW/m² (Petela, 2008).

With these calculations, the only unknowns in the energy balance become the heat and work transfer rates. The only form of work that is transferred out of the system boundaries is the work done by the *C. reinhardtii* to swim. *C. reinhardtii* uses flagella for motility. To calculate the flagella work, we multiplied the drag force that the *C. reinhardtii* has to overcome to move with its average speed. Hydrodynamic measurements report that the drag force is nearly constant under the given circumstances. So we can calculate the flagellar work also as a function of time.

MATLAB code E4.11.2 employs the thermodynamic data given in Example Table 4.11.1 and calculates flagellar work and solves the energy balance to calculate the heat generated.

MATLAB CODE E4.11.2

Command Window

```
clear all
close all

MW_co2=44;
MW_wa=18.02;
MW_o2=31.99;
MW_glu=180.16;
MW_nh=18.04;
MW_hpo=96.99;
MW_biomass=23.36;
MW_tga=1023.6;
no_cell=1.3*10^11;
W=(no_cell*1.03*10^(-14))*(-1);

H_data=[-394; -285.8; 0; -1274.5; -132.80; -1302.5; -204;
  -1056; 215.9];
```

```
[t,x] = ode45('odetotalkinetics_first', [0 14], [0.1
  0.00000001]);


% biomass&photsynthesis rxn
for i=1:60;
    n_biomass(i,:)=(((x(i+1,1)-x(i,1))/MW_biomass));
    Nh_CO2_BP=n_biomass*4*H_data(1,:);
    Nh_water_BP=n_biomass*3.685*H_data(2,:);
    Nh_nh_BP=n_biomass*0.11*H_data(5,:);
    Nh_hpo_BP=n_biomass*0.01*H_data(6,:);
    Nh_biomass_BP=n_biomass*H_data(7,:);
    Nh_glu_BP=n_biomass*0.5*H_data(4,:);
    Nh_oxygen_BP=n_biomass*4.2125*H_data(3,:);
    Nh_photon=n_biomass*7.353*H_data(9,:);
end


%lipid rxn
for i=1:60;
    n_lipid(i,:)=(((x(i+1,2)-x(i,2))/MW_tga));
    Nh_glu_TGA=n_lipid*11.5*H_data(4,:);
    Nh_oxygen_TGA_in=n_lipid*H_data(3,:);
    Nh_lipid_TGA=n_lipid*H_data(8,:);
    Nh_oxygen_TGA=n_lipid*22.5*H_data(3,:);
    Nh_water_TGA=n_lipid*20*H_data(2,:);
end

% respiration rxn
for i=1:60;
    n_glu_R=((n_biomass*0.5)-(n_lipid*11.5));
    Nh_glu_R=n_glu_R*H_data(4,:);
    Nh_oxygen_R=n_glu_R*6*H_data(3,:);
    Nh_CO2_R=n_glu_R*6*H_data(1,:);
    Nh_water_R=n_glu_R*6*H_data(2,:);
end


for i=60;
    Nh_oxygen_ex= ((n_biomass*4.2125)-(n_lipid*22.5)
      -(n_glu_R*6))* H_data(3,:);
    Nb_oxygen_ex=((n_biomass*4.2125)-(n_lipid*22.5)-
      (n_glu_R*6))*H_data(3,:);
end


for i=1:60;
    n_biomass(i,:)=(((x(i+1,1)-x(i,1))/MW_biomass));
    wb=cumsum(n_biomass*W);
end


Qin=Nh_CO2_BP+Nh_water_BP+Nh_nh_BP+Nh_hpo_BP+Nh_photon;
Qout=Nh_CO2_R+Nh_water_R+Nh_oxygen_TGA+Nh_water_TGA+Nh_
  oxygen_ex;
```

```
Qacc=Nh_lipid_TGA+Nh_biomass_BP;
Q_total=cumsum((Qout-Qin-Qacc));

for i=1:60;
    time(i)=t(i+1);
end

[ax, h1, h2]=plotyy(time, Q_total, time, wb)

set(ax,'xlim',[0 14],'xtick',[0:1:14]);
set(get(ax(1),'Ylabel'),'String','q(kJ/L)')
set(get(ax(2),'Ylabel'),'String','Work(kJ/L)')
xlabel ('time (h)')
title ('heat & flagella work generation - 18.7 mM')
set(ax(1),'ylim',[-400 0],'ytick',
  [-400:40:0],'ycolor','black');
set(ax(2),'ylim',[-0.0005 0],'ytick',[-0.0005:0.00005:0],
  'ycolor','black');
set(h1,'LineStyle','-.','color','black','LineWidth',2);
set(h2,'LineStyle',':','color','black','LineWidth',2);
legend ('q', 'work','Location', 'Southwest');
grid on
```

When we run this code, a figure, described in Example Figure 4.11.3, appeared on the screen.

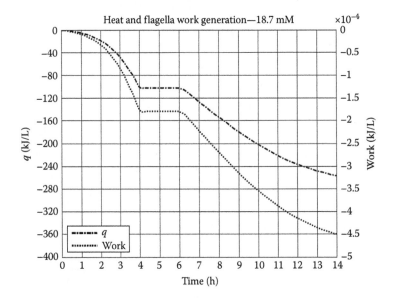

EXAMPLE FIGURE 4.11.3 Variation of the heat and flagella work generation with time during cultivation of *Chlamydomonas reinhardtii* with the presence of 18.7 mM of NH_4^+.

EXAMPLE TABLE 4.11.2   Chemical Exergy of the Chemical Compounds

Chemical Compounds	MW (g/mol)	ex_{ch}^0 (kJ/mol)
Glucose	180.2	2961
Oxygen	32.0	4.0
$H_2O_4P^-$	97.0	11.3
NH_4^+	18.0	393.1
Water (l)	18.0	0.9
Carbon dioxide (g)	44	0 (dead state)
Lipid	1026.6	40,542
Biomass	23.4	628.2

Next, we will solve the entropy or the exergy balance equations to calculate either the entropy generation or the exergy destruction. Note that $\dot{ex}_{destr} = T_0 \dot{s}_{gen}$; therefore, solving one of these equations provides the necessary information on the irreversibilities of the system. Here, we choose to solve for the exergy equation because calculating the exergy cost of a living cell and the exergy destructed provides more insight.

In order to solve the exergy equation, we need the work, the exergy of light, and the chemical exergy of the species inside the system. Flagella work and exergy of light have already been calculated. The chemical exergy of some species are found in the literature (Szargut et al., 1988). But the chemical exergy of the lipid and the biomass are not available in the literature. We can estimate the chemical exergy based on a hypothesis that the chemical exergy content of

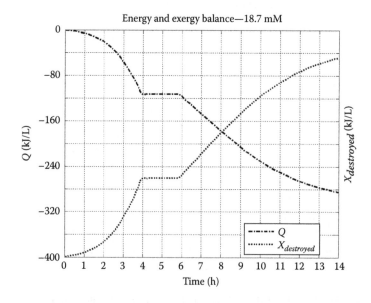

EXAMPLE FIGURE 4.11.4   Variation of the heat released and the exergy destroyed during cultivation of *Chlamydomonas reinhardtii*.

a substance is related to its lower heating value (LHV) as explained in Example 2.13. Both correlations suggested by Moran (1982) and Szargut and Styrylska (1964) result in similar values. Example Table 4.11.2 lists the chemical exergy values used here.

The exergy destroyed within the system boundaries can be calculated by solving the unsteady exergy balance equation given as

$$\dot{ex}_{destroyed} = \sum_{in} \dot{m}_{in} ex_{in} - \sum_{out} \dot{m}_{out} ex_{out} - \dot{w} + \left(1 - \frac{T_0}{T_b}\right)\dot{q} + \dot{ex}_{light} - \left(\frac{d(mex)}{dt}\right)_{system}$$

Example Figure 4.11.4 shows the results of the energy and exergy balances.

Example 4.12: Heat Relase, Exergy Destruction, and Entropy Generation during Cultivation of *Neisseria meningitidis*

Neisseria meningitidis produces an antigen of the deadly disease meningitis (Değerli et al., 2015). Cultivation of this organism in a laboratory reactor is investigated in this example. The system, its boundaries, the inputs, and the outputs are described in Example Figure 4.12.1.

The molecular weight of *n*-acetylneuraminic acid is reported as 309.27 g/mol (Chemspider, 2013), and that of the serogroup C antigen is reported as 88,000 g/mol (Apicella and Robinson, 1970), implying that there are 285 repeating units of the *n*-acetylneuraminic acid monomers in the serogroup C antigen. Accordingly, the complex reaction chain that occurs in the reactor can be simplified as two overall reactions (Değerli et al., 2015) given in Example Figure 4.12.2.

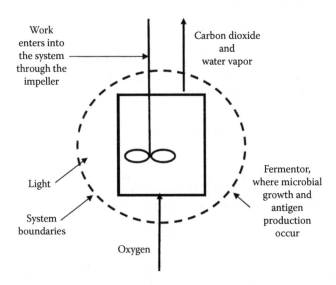

EXAMPLE FIGURE 4.12.1 Schematic description of the system and its boundaries.

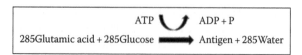

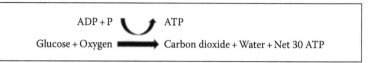

EXAMPLE FIGURE 4.12.2   Apparent kinetic reactions of antigen (top) and ATP production.

These reactions are comparative. Glucose is either consumed in the energy metabolism to produce ATP for cellular activities or converted into a longer-chain hydrocarbon (i.e., antigen) in the expense of ATP.

In this example, a kinetic model is developed to describe the continuous mathematical relations between the cell, antigen, and substrate concentrations at any moment of the process:

$$\frac{dx}{dt} = \mu x \left(1 - \frac{x}{x_{max}} \right)$$

$$\frac{dP}{dt} = \alpha x + \beta \frac{dx}{dt}$$

where

x_{max} is the maximum attainable biomass concentration
μ is the specific growth rate

The term αx of the antigen production model implies that it is formed in proportion with the size of the microbial population, and the next term, $\beta(dx/dt)$, represents the additional product formation rate in proportion with the growth rate. The glucose consumption rate is the sum of the rates of substrate consumption for growth, product formation, and maintenance:

$$-\frac{ds_s}{dt} = \frac{1}{Y_x} \frac{dx}{dt} + \frac{1}{Y_p} \frac{dp}{dt} + k_m x$$

where

Y_x is the growth yield (biomass/substrate consumed)
Y_p is the product formation yield (product formed/substrate consumed)
k_m is the maintenance coefficient (Bailey and Ollis, 1986)

Numerical values of the constants of the model are given in Example Table 4.12.1 (Değerli et al., 2015):

The kinetic parameters α, β, and Y change with the growth phase, subscript *exp* refers to the exponential growth phase and *st* refers to the stationary phase. The model equations are solved with the given kinetic constants via the MATLAB code E4.12.1 and compared to the experimentally measured data:

EXAMPLE TABLE 4.12.1 Numerical Values of the Constants of the Model When the Cultivation Is Achieved with the Dissolved Oxygen Control

μ (1/h)	x_{max} (g/L)	x_0 (g/L)	α_{exp}	β_{exp}	α_{st}	β_{st}	$Y_{x\,exp}$	$Y_{p\,exp}$	$Y_{x\,st}$	$Y_{p\,st}$	k_m (1/h)
0.56	2.85	0.048	5.0×10^{-3}	3.3×10^{-2}	3.0×10^{-3}	2.0×10^{-3}	4.0	0.18	1.0	0.07	0.06

MATLAB CODE E4.12.1

Command Line

```
clear all
close all
format compact
global mu xmax

% enter the constants of the kinetic models
mu=0.56; % specific growth rate
xmax=2.85; % maximum growth

% enter the experimental data (units are in g/L)
tData1=[0:2:16];
a=[0.07 0.300 0.410 0.840 2.225 2.499 2.510 2.800 2.800]; %
  growth data
ps = [0 0.005 0.019 0.043 0.085 0.125 0.150 0.175 0.185]; %
  polysaccharide antigen production data
s = [5 4.4 3.7 3.18 2.6 1.9 1.21 0.48 0]; % substrate
  consumption data

% microbial growth model
[t,x] = ode45('sode_O2', [0 16], [0.048 0 5]);
% [0 16] is time interval
% [0.048 0 5] are initial conditions of biomass, antigen and
  substrate, respectively.

% plot the microbial growth model
plot (t, x(:,1), 'k-.', 'LineWidth', 2.0);
hold on

% Plot the antigen production and the substrate utilization
[ax, h1, h2]=plotyy(t, x(:,3), t, x(:,2));

% plot of substrate consumption and antigen production models
set(ax(1),'ylim',[0 5],'ytick',[0:5],'ycolor','black');
set(ax(2),'ylim',[0 0.25],'ytick',[0:0.05:0.25],'ycolor','bl
  ack');
set(h1,'LineStyle',':','color','red','LineWidth',2);
set(h2,'LineStyle','-','color','blue','LineWidth',2);
legend ('biomass', 'substrate','product', 'Location',
  'NorthEast');
```

```
plot(tData1, a, 'k+')
[ax, h3, h4]=plotyy(tData1, s, tData1, ps);
set(ax(1),'ylim',[0 5],'ytick',[0:5],'ycolor','black');
set(ax(2),'ylim',[0 0.25],'ytick',[0:0.05:0.25],'ycolor',
  'black');
set(h3, 'LineStyle', '*', 'LineWidth', 2.0, 'Color', 'red');
set(h4, 'LineStyle', 'o', 'LineWidth', 2.0, 'Color', 'blue');
set(get(ax(1),'Ylabel'),'String','Substrate concentration
  (g/L)');
set(get(ax(2),'Ylabel'),'String','Biomass and antigen
  concentration (g/L)');
grid on
xlabel('t(days)')
```

M-File:

```
function dx= sode_O2(t,x);
% This function models the substrate consumption
mu=0.56; % specific growth rate
xmax=2.85; % maximum growth
km=0.06; % maintenance coefficient

if t<=12;
    alpha = 0.005; % growth associated coefficient
    beta = 0.033; % nongrowth associated coefficient
    Yx=4; % biomass yield
    Yp=0.18; % product yield
    dx1 = mu*x(1)*(1-(x(1)/xmax));
    dx2 = alpha*x(1) + beta*dx1;
    dx3 =-(((1/Yx)*dx1)+((1/Yp)*dx2)+(km*x(3)));
end

if t>12;
    alpha=0.003;
    beta=0.00002;
    Yx=1;
    Yp=0.07;
    dx1 =mu*x(1)*(1-(x(1)/xmax));
    dx2 = alpha*x(1)+beta*dx1;
    dx3 =-(((1/Yx)*dx1)+((1/Yp)*dx2)+(km*x(1)));
end

dx=[dx1; dx2;dx3];
```

Example Figure 4.12.3 will appear in the screen when we run the code.

Via the kinetic model, the rates of the reactions of Example Figure 4.12.2 are determined. Depending on these, the mass fluxes and the concentration of each chemical substance are calculated as a function of time.

In order to solve the energy and exergy balance equations, thermodynamic properties such as the enthalpy of formation, specific heat capacity, specific

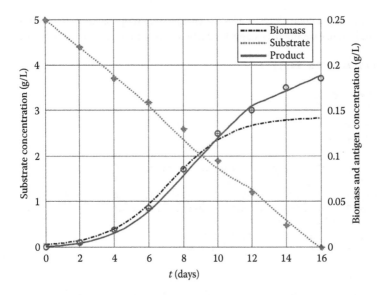

EXAMPLE FIGURE 4.12.3 Variation of the biomass, substrate, and the product, for example, antigen concentrations with time during cultivation of *Neisseria meningitides* when the cultivation is achieved with the dissolved oxygen control.

molar entropy, and the standard chemical exergy data are needed. In this laboratory scale reactor, chemicals are dissolved in a dilute aqueous solution at constant temperature (308 K) and pressure (1 atm). Therefore, their thermodynamic properties are estimated based on the *standard*-state conditions, rather than the *transformed* properties. The specific heat capacities of the chemicals are employed to calculate the properties at 308 K. Group contribution method (as described in Chapter 2) is employed to estimate the heat of formation and Gibbs free energy of *n*-acetylneuraminic acid and its polymer serogroup C polysaccharide antigen (Example Table 4.12.2). Their specific heats are calculated based on Kopp's rule. Thermodynamic properties of the biomass are determined by assuming a UCF of $C_{3.85}H_{6.69}O_{1.78}N$. Details of the calculations can be found in Değerli et al. (2015).

EXAMPLE TABLE 4.12.2 Molecular Weight and Thermodynamic Data of the Compounds

Chemical Compound	MW (g/mol)	$h^0_{f,298}$ (kJ/mol)	c_p (kJ/mol K)	s^0 (kJ/mol K)	ex^0 (kJ/mol)
Glucose	180	−1274.5	0.226	0.209	2955
Oxygen	32	0	2.94×10^{-2}	0.205	3.97
Glutamic acid	147.13	−1003.3	0.175	0.188	2393.2
Water	18	−285.8	7.53×10^{-2}	9.1×10^{-3}	0.9
Carbon dioxide	48	−394.0	3.71×10^{-2}	0.215	19.48
Antigen	8.8×10^4	-4.7×10^5	115	242	1.6×10^6
Biomass	95	−512.6	0.135	0.138	2037.1

Source: Adapted from Değerli, B. et al., *Int. J. Exergy*, 16, 1, 2015.

MATLAB code E4.12.2 solves the energy and exergy balances to determine the heat release and the exergy destruction during the course of the cultivation.

MATLAB CODE E4.12.2

Command Line

```
clear all
close all

% Input the molecular weights of glucose, glutamic acid,
  antigen (ps) and biomass
MW_glu=180; MW_ps=88000; MW_GA=147.13; MW_biomass=95;
RU=285; % repeating unit of n-acetylneuraminic acid is 285
  in antigen structure.
T=308; % reaction temperature
T0=298; % reference temperature

% Input the thermodynamic properties
% H, S, b of glucose, oxygen, GA, water, co2, ps, biomass
H_data=[-1272.2; 0.294; -1001.6; -285; -393.6;
  -467711;-511.3];
S_data=[0.209; 0.205;  0.188; 0.0091; 0.215; 242; 0.138];
b_data=[2955; 3.97; 2393.2; 0.9; 19.48 ; 1591953;2032.1];

% calculate biomass, antigen and substrate concentrations as
  a function of time
[t,x] = ode45('sode_O2', [0 17], [0.048 0 5]); % [0 17] is
  time interval and [0.048 0 5] are the initial conditions
  of biomass, antigen and substrate, respectively.

% ENERGY AND EXERGY BALANCE

% maintenance
for i=1:48;
    n_glu_m(i,:)=((x(i,3)-x(i+1,3))/MW_glu)-(((x(i+1,1)-
      x(i,1))*0.55)/MW_glu)-(((x(i+1,2)-x(i,2))*RU)/MW_ps);
    % mole of glucose used for maintenance

% subtract the amount of the glucose used for antigen and
  biomass from the total

% input the energy balance data
  Nh_glu_m=n_glu_m*H_data(1,:);
  Nh_oxygen=n_glu_m*6*H_data(2,:);
  Nh_CO2=n_glu_m*6*H_data(5,:);
  Nh_water_m=n_glu_m*6*H_data(4,:);

% input the exergy balance data for maintenance
  Nb_glu_m=n_glu_m*b_data(1,:);
  Nb_oxygen=n_glu_m*6*b_data(2,:);
```

```
    Nb_CO2=n_glu_m*6*b_data(5,:);
    Nb_water_m=n_glu_m*6*b_data(4,:);
end

% antigen production
for i=1:48;
    n_ps(i,:)=((x(i+1,2)-x(i,2))/MW_ps); % mole of
      polysaccharide antigen

    % energy balance terms for polysaccharide production
    Nh_glu_ps=n_ps*RU*H_data(1,:);
    Nh_GA_ps=n_ps*RU*H_data(3,:);
    Nh_water_ps=n_ps*RU*H_data(4,:);
    Nh_ps=n_ps*H_data(6,:);

    % exergy balance terms for polysaccharide production
    Nb_glu_ps=n_ps*RU*b_data(1,:);
    Nb_GA_ps=n_ps*RU*b_data(3,:);
    Nb_water_ps=n_ps*RU*b_data(4,:);
    Nb_ps=n_ps*b_data(6,:);
end

% substrate used for synthesizing the bacterial cell structure

for i=1:48;

    % the moles of biomass is found. Then, 55% of the
      biomass is glucose and
    % 15% of the biomass is nitrogen.
    m_biomass(i,:)=x(i+1,1)-x(i,1);
    n_biomass=m_biomass/MW_biomass;
    n_glu_biomass=(m_biomass*0.55)/MW_glu;
    n_GA_biomass=(m_biomass*0.15)/MW_GA;

    % input terms of the energy balance to calculate the
      substrate entering the cell structure
    Nh_glu_biomass=n_glu_biomass*H_data(1,:);
    Nh_GA_biomass=n_GA_biomass*H_data(3,:);
    Nh_biomass=n_biomass*H_data(7,:);

     % exergy balance terms for substrate entering bacteria
      structure
    Nb_glu_biomass=n_glu_biomass*b_data(1,:);
    Nb_GA_biomass=n_GA_biomass*b_data(3,:);
    Nb_biomass=n_biomass*b_data(7,:);
end

% energy balance Qout-Qin
Qin=Nh_glu_biomass+Nh_GA_biomass+Nh_glu_ps+Nh_GA_ps+Nh_glu_
  m+Nh_oxygen;
Qout=Nh_biomass+Nh_ps+Nh_water_ps+Nh_CO2+Nh_water_m;
Q_total=cumsum(Qout-Qin);
```

```
% exergy balance; exergy destruction is determined.
Nbin=Nb_glu_biomass+Nb_GA_biomass+Nb_glu_m+Nb_oxygen+Nb_glu_
  ps+Nb_GA_ps;
Nbout=Nb_biomass+Nb_CO2+Nb_water_m+Nb_ps+Nb_water_ps;
deltab=Nbin-Nbout;
Xdestroyed=cumsum(deltab)-Q_total*(1-(T0/T));

% antigen exergy efficiency calculation
Qin_ps=Nh_glu_ps+Nh_GA_ps;
Qout_ps=Nh_ps+Nh_water_ps;
Q_ps=cumsum(Qout_ps-Qin_ps);

disp('The percent exergetic efficieny of the antigen
  production is ')
eff=sum(n_ps*b_data(6,:))/sum(Xdestroyed)*100

disp('q_ratio = q_antigen/q_total')
q_ratio=sum(Q_ps)/sum(Q_total)

for i=1:48;
    time(i)=t(i+1);
end

[ax, h1, h2]=plotyy(time, Q_total, time, Xdestroyed);

set(ax,'xlim',[0 16],'xtick',[0:2:16]);
set(get(ax(1),'Ylabel'),'String','q (kJ/mole)')
set(get(ax(2),'Ylabel'),'String','ex_d_e_s_t_r_o_y_e_d (kJ/
  mole)')

xlabel ('time (days)')
set(ax(1),'ylim',[-100 0],'ytick',[-100:10:0],
  'ycolor','black');
set(ax(2),'ylim',[0 50],'ytick',[0:10:50],'ycolor','black');

set(h1,'LineStyle','-.','color','black','LineWidth',2);
set(h2,'LineStyle',':','color','black','LineWidth',2);
legend ('q', 'ex_d_e_s_t_r_o_y_e_d','Location', 'NorthEast');
legend ('q', 'ex_d_e_s_t_r_o_y_e_d','Location', 'NorthEast');

grid on
```

The following lines and Example Figure 4.12.4 will appear in the screen when we run the code:

```
The percent exergetic efficieny of the antigen production is
eff =
    0.5761
q_ratio = q_antigen/q_total
q_ratio = - 0.0036
```

MATLAB code E4.12.3 calculates the entropy generation:

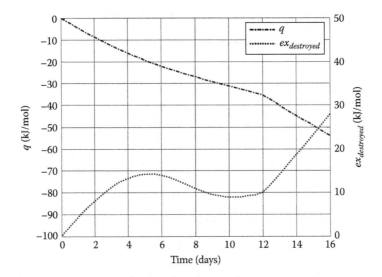

EXAMPLE FIGURE 4.12.4 Variation of the heat release and exergy destruction with time during cultivation of *Neisseria meningitides* with dissolved oxygen control.

MATLAB CODE E4.12.3

Command Line

```
clear all
close all

% this code calculates the entropy generation

% molecular weights of glucose, glutamic acid, antigen (ps)
  and biomass are
MW_glu=180;
MW_ps=88000;
MW_GA=147.13;
MW_biomass=95;
RU=285; % repeating unit of n-acetylneuraminic acid is 285
  in antigen structure.
T=308; % reaction temperature
T0=298; % reference temperature

% H, S, b of glucose, oxygen, GA, water, co2, ps, biomass
H_data=[-1272.2; 0.294; -1001.6; -285; -393.6; -467711;-511.3];
S_data=[0.209; 0.205; 0.188; 0.0091; 0.215; 242; 0.138];
b_data=[2955; 3.97; 2393.2; 0.9; 19.48 ; 1591953;2037.1];

[t,x] = ode45('sode_O2', [0 17], [0.048 0 5]);
% [0 17] is the time interval % days
```

```
% [0.048 0 5] are the initial conditions of biomass, antigen
  and substrate, respectively
% ENERGY & ENTROPY BALANCE for oxygen control

% maintenance
for i=1:48;

    % mole of glucose used for maintenance
    % the glucose used for antigen production and biomass is
      subtracted from the total glucose
    n_glu_m(i,:)=((x(i,3)-x(i+1,3))/MW_glu)-(((x(i+1,1)-
      x(i,1))*0.55)/MW_glu)-(((x(i+1,2)-x(i,2))*RU)/MW_ps);

    % energy balance terms for maintenance
    Nh_glu_m=n_glu_m*H_data(1,:);
    Nh_oxygen=n_glu_m*6*H_data(2,:);
    Nh_CO2=n_glu_m*6*H_data(5,:);
    Nh_water_m=n_glu_m*6*H_data(4,:);

    % entropy balance terms for maintenance
    NS_glu_m=n_glu_m*S_data(1,:);
    NS_oxygen=n_glu_m*6*S_data(2,:);
    NS_CO2=n_glu_m*6*S_data(5,:);
    NS_water_m=n_glu_m*6*S_data(4,:);
end

% ENERGY BALANCE FOR MAINTENANCE
Qin_m=Nh_glu_m+Nh_oxygen;
Qout_m=Nh_CO2+Nh_water_m;
Q_m=cumsum(Qout_m-Qin_m);

% ENTROPY BALANCE FOR MAINTENANCE
NSin_m=NS_glu_m+NS_oxygen;
NSout_m=NS_CO2+NS_water_m;
deltaS_m=cumsum(NSout_m-NSin_m);
Sgen_m=deltaS_m-(Q_m/T);

% polysaccharide production part
for i=1:48;

    n_ps(i,:)=((x(i+1,2)-x(i,2))/MW_ps); % mole of
      polysaccharide antigen

    % energy balance terms for polysaccharide production
    Nh_glu_ps=n_ps*RU*H_data(1,:);
    Nh_GA_ps=n_ps*RU*H_data(3,:);
    Nh_water_ps=n_ps*RU*H_data(4,:);
    Nh_ps=n_ps*H_data(6,:);

    % entropy balance terms for polysaccharide production
    NS_glu_ps=n_ps*RU*S_data(1,:);
```

```
    NS_GA_ps=n_ps*RU*S_data(3,:);
    NS_water_ps=n_ps*RU*S_data(4,:);
    NS_ps=n_ps*S_data(6,:);
end


% energy balance for polysaccharide production
Qin_ps=Nh_glu_ps+Nh_GA_ps;
Qout_ps=Nh_ps+Nh_water_ps;
Q_ps=cumsum(Qout_ps-Qin_ps);

% entropy balance for polysaccharide production
NSin_ps=NS_glu_ps+NS_GA_ps;
NSout_ps=NS_ps+NS_water_ps;
deltaS_ps=cumsum(NSout_ps-NSin_ps);
Sgen_ps=deltaS_ps-(Q_ps/T);

% substrate entering bacteria structure
for i=1:48;

    % the moles of biomass is found. Then, 55% of the biomass
      is glucose and
    % 15% of the biomass is nitrogen.
    m_biomass(i,:)=x(i+1,1)-x(i,1);
    n_biomass=m_biomass/MW_biomass;
    n_glu_biomass=(m_biomass*0.55)/MW_glu;
    n_GA_biomass=(m_biomass*0.15)/MW_GA;

    % energy balance terms for biomass
    Nh_glu_biomass=n_glu_biomass*H_data(1,:);
    Nh_GA_biomass=n_GA_biomass*H_data(3,:);
    Nh_biomass=n_biomass*H_data(7,:);

    % entropy balance terms for biomass
    NS_glu_biomass=n_glu_biomass*S_data(1,:);
    NS_GA_biomass=n_GA_biomass*S_data(3,:);
    NS_biomass=n_biomass*S_data(7,:);
end


% energy balance  for biomass
Qin_b=Nh_glu_biomass+Nh_GA_biomass;
Qout_b=Nh_biomass;
Q_b=cumsum(Qout_b-Qin_b);
  % entropy balance terms for biomass
NSin_b=NS_glu_biomass+NS_GA_biomass;
NSout_b=NS_biomass;
deltaS_b=cumsum(NSout_b-NSin_b);
Sgen_b=deltaS_b-(Q_b/T);


% TOTAL ENTROPY GENERATION
Sgen_total=Sgen_m+Sgen_ps+Sgen_b;
```

```
for i=1:48;
    time(i)=t(i+1);
end

hold all
plot (time,Sgen_total, 'k-', 'Linewidth',2)
plot (time,Sgen_m,'k:', 'Linewidth',2)
plot (time,Sgen_ps, 'k-.', 'Linewidth',2)
plot (time, Sgen_b, 'k--', 'Linewidth',2)

xlim([0 16])
ylim([-0.02 0.20])
xlabel('time (days)')
ylabel('s_g_e_n (kJ/K)')
legend('s_g_e_n total','s_g_e_n of maintenance','s_g_e_n
  of antigen production','s_g_e_n of biomass production',
  'location','Northwest')

grid on
```

When we run this code, Example Figure 4.12.5 will appear in the screen.

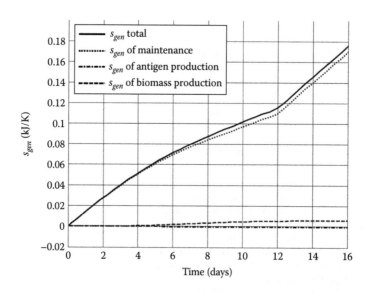

EXAMPLE FIGURE 4.12.5 Variation of the total entropy generation and its contributors with time during cultivation of *Neisseria meningitides* with dissolved oxygen control.

4.2 Eco-Exergy

Correlations suggested by Moran (1982) and Szargut et al. (1964) are originally derived to estimate the chemical exergy of hydrocarbons used as fuels. From this point of view, a fuel and a living cell would have the same chemical exergy, if their UCF are the same. However, a cell performs a number of biochemical reactions to stay alive. To do that, a living organism has to have the correct composition of enzymes and amino acids. The chemical exergy required for the synthesis of an organism may be expressed as

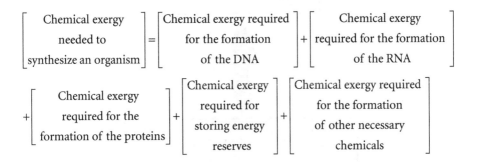

In this expression, the exergy of the stored energy reserves includes polymers of glucose, for example, cellulose, glycogen, and fats. The expression *chemical exergy required for the formation of other necessary chemicals* refers to all of the chemicals synthesized by the organism to stay alive and is not accounted for in the first four terms on the right-hand side. The turnover of the constituents, such as the genetic material and the enzymes, and development of the immune response are among the factors that are contributing to the exergy cost.

In an effort to estimate the exergy of living organisms, (Jorgensen and Nielsen, 2007; Jorgenson et al., 2007) defined *eco-exergy*. Eco-exergy is referred to as the work capacity possessed in the ecological network of organisms. Energy cost of synthesizing each organism is determined by the information embodied in their genome. As seen in Figure 4.9, the exergy of the ecosystem is calculated relative to a dead state at the thermodynamic equilibrium, where all the components are inorganic at their highest possible oxidation state. It is generally accepted that the ecosystems are able to utilize the influx of the solar radiation to move as far away from the thermodynamic equilibrium as possible (Jorgensen, 2015). Among the alternative combinations of the system components including various types of organisms, the combination of the components that move the system most away from the thermodynamic equilibrium becomes the prevailing ecosystem (Jorgensen et al., 2010). This statement is referred to as the thermodynamic version of *Darwin's survival of the fittest* hypothesis (Jorgensen, 2015).

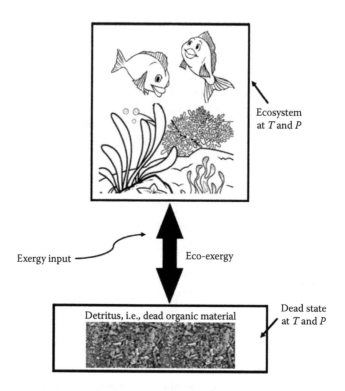

FIGURE 4.9 Both the ecosystem and the dead system are at the same temperature and pressure. The reference system is at the thermodynamic equilibrium. Additional exergy, named as the *eco-exergy*, is needed to produce a *living* ecosystem from the reference system.

The genome of an organism codes for the sequence and the amount of the amino acids of their proteins. Jorgensen et al. (2010) estimated the free energy of formation of a representative peptide of *Escherichia coli* from the following model chemical reaction:

$$C_{4.793}H_{7.669}O_{1.396}N_{1.370}S_{0.046} + 7.805O_2 \rightarrow 4.793CO_2$$

$$+ 3.1265H_2O + 3.1270NO_3^- + 0.046SO_4^{2-} + 1.416H^+$$

where $C_{4.793}H_{7.669}O_{1.396}N_{1.370}S_{0.046}$ is the model average peptide; this model chemical reaction represents its complete oxidation after burning with oxygen:

$$\Delta g_c = 4.793m\Delta g_{CO_2}^f + 3.1265\Delta g_{H_2O}^f + 3.1270\Delta g_{NO_3}^f + 0.046\Delta g_{SO_4}^f$$

$$+ 1.416\Delta g_{H^+}^f - \Delta g_{peptide}^f - 7.805\Delta g_{O_2}^f$$

TABLE 4.6 Gibbs Free Energy of Formation of Some Organisms

Organism	Number of Peptides	$\Delta g_{organism}^{f}$ (kJ/mol)
Virus	250	$(250)(-2723.26) = -6.80 \times 10^5$
Yeast	8,000,000	-2.18×10^{10}
Mosquito	80,250,000	-2.19×10^{11}
Homo sapiens	543,000,000	-1.48×10^{12}

The Gibbs free energy of combustion of this reaction is calculated by Jorgensen et al. (2010) as $\Delta g_c = -2723.26$ kJ/mol, which may be regarded as the exergy utilized to synthesize the peptide chain from its constituents at the dead state. After considering this piece of information as the standard yard stick, Jorgensen et al. (2010) calculated the Gibbs free energy of formation of other organisms (Table 4.6).

The total exergy of an ecological system may be expressed approximately as (Fonseca et al., 2002)

$$ex = RT \sum_{i=1}^{n} \beta_i c_i$$

where

$ex_i = RT\beta_i c_i$ is the exergy
β_i is the free energy
c_i is the concentration (in kg/m^3) of the ith species in the ecological system

Zhang et al.'s (2007) study regarding the recovery of ecological communities at the depths of the ocean near Chongming Island, near Shanghai City, with the data collected before and after an excavation work exemplifies the use of the previous equation with numerical data. Kernegger et al. (2008) used the same expression to estimate the variations of the exergetic value of the fisheries extracted from the oceans. Jorgensen et al. (2005) suggested the following expression for the estimation of the parameter β:

$$\beta = 1 + \frac{n(1-f)}{7.34 \times 10^5}$$

To demonstrate the eco-exergy concept, let us calculate the eco-exergy of the biomass of *C. reinhardtii*. For *C. reinhardtii* the exergy factor β can be taken as 20, which is the same exergy factor for an alga as reported by Jorgenson.

Jorgensen et al. listed the exergy of detritus, bacteria, algae, and human as 18.7, 50.5, 64.2, and 13,389 kJ/g, respectively. The exergy of living organisms are calculated based on the exergy of detritus. To compare these exergy values with our calculations, the difference in exergy of the organisms and detritus (which represents the dead state) has to be taken. The exergy content of the cell debris calculated according to Moran's correlation is 28.4 kJ/g. This is comparable to a bacteria cell. The eco-exergy difference between a bacteria and detritus is 31.8 kJ/g. The exergy values estimated based on LHV are about 10% less than the eco-exergy values reported for similar organisms.

4.3 Entropy as a Measure of Information in Biological Systems

In biological systems, entropy is used to measure irreversibilities of biochemical processes, to determine the information contained in a DNA sequence, and to predict life span of organisms. This chapter is devoted to explain how the information content of a living organism can be quantified by defining the information entropy.

There are two different types of information in the biological systems: the structural information and the functional information (Davies et al., 2013). The former is related to the neat and tight packing of various molecules into macromolecules and organelles to comprise the cell, which tends to improve the order and reduce the entropy. The functional information is related to the functioning of the cell and therefore related to the number and the rate of the chemical reactions taking place.

Proteins are synthesized in accordance with the information stored in the encoding gene; the DNA/RNA polymerase translocates along a DNA single strand, reads one nucleotide at a time, identifies a complementary nucleotide in the environment, and processes the information by catalyzing a phosphodiester bond in the emerging replicated strand as described in Figure 4.10.

Cells are equipped with mechanisms to correct the errors made during DNA expression to keep their structure well organized and achieve a level of the minimum structural entropy. The failure of those mechanisms disturbs the well-being of the organism. Macroscopic variables describing the chemical, morphological, structural, and physiological state of the cell can be used to define its entropy. It is hypothesized that entropy flow toward healthy cells causes diseases by destroying the order in the cell's chemical, morphological, structural, and physiological state. Luo (2009) argues that the cause of the spread of cancer is the entropy flow from the cancerous cells to healthy cells, which communicate the harmful information. Luo suggests that the reversal of the entropy flow can be a potential basis for an anticancer therapy development. Luo (2009) estimates the magnitude of the entropy production by the cancerous cells to be about twice as much as that of the healthy cells. The direction of the entropy flow may be reversed by using low-intensity electromagnetic field or ultrasound irradiation or by modifying the cellular pH. Similarly, West et al. (2012) described a cellular phenotype as a complex network of molecular interactions and demonstrated that the cancer cells are characterized by an increase in the network entropy. They argued that revealing the network properties that distinguish disease from the healthy cellular state is of critical importance for the development of cancer treatments.

In the DNA strain, three nucleotides make one codon. Since there are four nucleotides (A, G, C, and T in DNA or U in the RNA), it is theoretically possible to make 64 different combinations or codons ($4 \times 4 \times 4 = 64$); 3 codons (UAA, UAG, and UGA) specify a *stop* signal, indicating the termination of the polypeptide chain being synthesized on the ribosome. Each of the remaining sixty-one codons encodes an amino acid. The *start* signal is the codon AUG, which also codes the amino acid methionine. When all the nucleotides are available in abundance in a cell, we may estimate the probability of synthesizing a sequence of methionine–glutamic acid–alanine–arginine–lysine as

$$p_{synthesis} = (p_{methionine})(p_{glutamic\ acid})(p_{alanine})(p_{arginine})(p_{lysine})$$

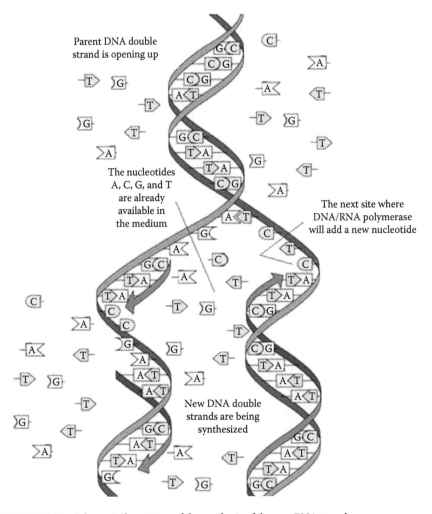

FIGURE 4.10 Schematic description of the synthesis of the new DNA strands.

There is only one codon coding for methionine, two codons are coding for glutamic acid, four codons are coding for alanine, and six for arginine and two for lysine. Therefore,

$$p_{synthesis} = \left(\frac{1}{64}\right)\left(\frac{2}{64}\right)\left(\frac{4}{64}\right)\left(\frac{6}{64}\right)\left(\frac{2}{64}\right) = 8.9 \times 10^{-8}$$

This calculation may be generalized after approximating p_k as (Sanchez, 2011)

$$p_k = \exp[-g_k n_k]$$

where

n_k is the number of the codons coding for the kth sequence ($n_k = 5$ in the previous example)

the parameter g is a positive number and referred to as the degeneracy

Then the probability of occurrence of the sequence with n_k codons is

$$p_N = \prod_{k=1}^{m} p_k = \exp\left[-\sum_{k=1}^{m} g_k n_k\right]$$

We may further rearrange the right-hand side of the equation as

$$p_N = \exp\left[-\left(\frac{\sum_{k=1}^{m} g_k n_k}{\sum_{k=1}^{m} n_k}\right)\sum_{k=1}^{m} n_k\right] = \exp\left[-\sum_{k=1}^{m} g_{average}\ N\right]$$

where

$\sum_{k=1}^{m} n_k = N$ is the total number of the coding codons in the genome

$g_{average}$ is the case-specific average degeneracy; it was calculated as $g_{average} = 3.24$ in the previous example (Sanchez, 2011)

If, in a genome with N codons, the maximum number of the ways the codons may be arranged is Ω, then the Gibbs entropy of the genome may be estimated as (Sanchez, 2011)

$$s = -k_B \sum p_i \ln(p_i) = -k_B \sum \frac{1}{\Omega} \ln\left(\frac{1}{\Omega}\right) = 3.06 k_B N$$

This equation implies that the larger the value of N is, the higher would be the magnitude of the genetic information and that of the entropy, which is crucial in determining the energy or exergy requirement for the synthesis of an organism. The longer the genetic strand is, the higher would the information needed to execute it be. Thus, it would contain an information entropy. It should be noted that all the codons on a strand do not code for an amino acid; the ratio of the codons that are coding for the amino acids is estimated to be about one-third of the total (Sanchez, 2011).

It was already stated in Chapter 1 that in statistical mechanics, a specific microscopic configuration of a system may occur with a certain probability in the course of its thermal fluctuations. If p_i is the probability of the occurrence of the thermodynamic state during these fluctuations, then the entropy of the system is

$$s = -k_B \sum p_i \ln(p_i)$$

where k_B is the Boltzmann constant. In a similar way, the information entropy is expressed as (Shannon, 1948)

$$s^{info} = -k \sum_{i=1}^{i=N} p_i \log_2(p_i)$$

where k is a constant. In information science, logarithms are usually expressed on the base of two, as in the case of $\log_2(p_i)$. Calculation of the information entropy for a gene strand is described with a MATLAB code in the following example.

Example 4.13: Information Entropy of a DNA Strand

Arias-Gonzalez (2012) devised an equilibrium pathway for the DNA replication. This example demonstrates the calculation of the information entropy of a DNA strand by referring to Shannon's theory. The DNA/RNA polymerase acts as a channel from the information point of view, since it passes the genetic information from a template strand to its copy. The DNA/RNA polymerase gets energy from deoxyribonucleotide triphosphate hydrolysis to move along the DNA in individual steps of 0.34 nM by developing 10–30 pN of force. A sequence in the template standard can be identified as a vector $y = (y_1,\dots y_i\dots y_n)$, where subscripts i and n are the index numbers referring to a position along the strand and the number of the nucleotides. The copied strand is represented as $x = (x_1\dots x_i\dots x_n)$. The probability of having n nucleotides in a sequence is $p = (p_1\dots p_i\dots p_n)$. Then the corresponding entropy is expressed as

$$s^{info} = -3.22 \sum_{i=1}^{n} p_i \log_{10}(p_i)$$

The probability of incorporating a nucleotide x in front of a nucleotide y is estimated from the experimental data as (Arias-Gonzalez, 2012)

$$p(x,y) = \frac{1}{z(y)} \exp\left(-\frac{\Delta g_y^x}{kT}\right)$$

where

$$z(y) = \sum_{x_1}^{x_4}\left(-\frac{\Delta g_y^x}{kT}\right)$$

The values of Δg_y^x are obtained from the experimental data as explained by SantaLucia and Hicks (2004). We will employ MATLAB code E4.13 to calculate $s$infoof a section of the strand of base pairs GC CG CG AT GC TA TA CG AT GC:

MATLAB CODE E4.13

```
Command Window

clear all
close all

% enter the dG0 values at 37 oC and 150 mM NaCl in
  alphabetic order
```

```
dG0=[1.01 1.7 0.34 -1.65; 1.5 1.78 -2.78 1.46; 0.03 -2.69
  -0.88 0.14;-1.64 1.39 -0.16 0.71] *4.184;
% data (in kT units) is taken from from SantaLucia and Hicks
  (2004), Appendix S1
% multiplication with 4.184 coverts kT units into kJ
kB=0.0083144621; % Boltzmann constant (kJ/mol)
T=37+273; % Body temperature (K)
z=exp(-dG0/(T*kB));
Z=sum(z); % Z(y) values for A, C, G, T respectively
Z4x4=[Z;Z;Z;Z]; %  Z(y) values as a 4x4 matrix to do element
  wise division
p=z./Z4x4;
disp('The p(x,y) values at  for A, C, G, T respectively at
  37C and 150 mM NaCl concentration')
disp(' ') % spacer
disp(p) % The p(x,y) values will appear on the screen, the
  columns will add up to 1
Sinfo_base=-3.22*(p.*log10(p));
disp('Information entropy values for each base pair
  combination in alphabetic order as a matrix')
disp(' ') % spacer
disp(Sinfo_base) % Sinfo values will appear on the screen


s= Sinfo_base
dna=input('Enter the DNA sequence','s');
dna=upper(dna); % make the sequence uppercase
dna(dna == ' ') = []; % remove spaces
disp(['The sequence is ' num2str(numel(dna)) ' bp long'])
nAA=numel(strfind(dna,'AA')); % number of AA/TT in the DNA
  sequence (This must be done for all 16 possible
  combinations)
nAC=numel(strfind(dna,'AC'));
nAG=numel(strfind(dna,'AG'));
nAT=numel(strfind(dna,'AT'));
nCA=numel(strfind(dna,'CA'));
nCC=numel(strfind(dna,'CC'));
nCG=numel(strfind(dna,'CG'));
nCT=numel(strfind(dna,'CT'));
nGA=numel(strfind(dna,'GA'));
nGC=numel(strfind(dna,'GC'));
nGG=numel(strfind(dna,'GG'));
nGT=numel(strfind(dna,'GT'));
nTA=numel(strfind(dna,'TA'));
nTC=numel(strfind(dna,'TC'));
nTG=numel(strfind(dna,'TG'));
nTT=numel(strfind(dna,'TT'));
% multiply the number of each combination with the
  corresponding entropy
Sinfo_DNA= nAA*s(1,1)+ nAC*s(1,2)+ nAG*s(1,3)+ nAT*s(1,4)+
  nCA*s(2,1)+  nCC*s(2,2)+ nCG*s(2,3)+ nCT*s(2,4)+
  nGA*s(3,1)+ nGC*s(3,2)+ nGG*s(3,3)+ nGT*s(3,4)+
  nTA*s(4,1)+ nTC*s(4,2)+ nTG*s(4,3)+ nTT*s(4,4);
```

```
disp(['The DNA with the given strand has an information
  entropy of ' num2str(Sinfo_DNA) ' kJ/(K mol).'])
% The length of the sequence and its Sinfo will be displayed
  on the screen
```

when we run the code, the following lines will appear on the screen:

```
The P(x,y) values at for A, C, G, T respectively at 37C and
150 mM NaCl concentration
```

```
   0.0125  0.0008  0.0059  0.9235
   0.0056  0.0007  0.9378  0.0059
   0.0612  0.9972  0.0429  0.0505
   0.9207  0.0013  0.0133  0.0200
```

```
Information entropy values for each base pair combination in
alphabetic order as a matrix
```

```
   0.0765  0.0080  0.0425  0.1028
   0.0408  0.0071  0.0842  0.0425
   0.2391  0.0040  0.1890  0.2109
   0.1064  0.0123  0.0805  0.1095
```

Then the following command will appear on the screen

```
Enter the DNA sequence
```

when we enter the code (Data, NCBI Reference Sequence: NM_001274316.1) to the command window, the following line will appear on the screen:

```
The sequence is 611 bp long
The DNA with the given strand has an information entropy of
  46.5998 kJ/(K mol).
Data (NCBI Reference Sequence: NM_001274316.1)
atcagtttctgaacactgcaaacatgaaggctttcatcgttctggttgccctggctct
  ggccgctcctgctcttggtcgcaccatggaccgttgctccctggcccgggagatgtcc
  aacctgggcgttcctcgtgaccaattggctcgttgggcctgcattgccgagcacgagt
  cctcctaccgcaccggagtggttggtcccgagaactacaacggctccaacgactacgg
  aatcttccagatcaacgactactactggtgcgctcctcccagcggtcgcttctcctac
  aatgagtgcgggttgagctgcaatgccctcttgaccgacgacatcacccactccgtcc
  gttgtgcccagaaggtcctcagccagcagggatggtccgcctggtccacctggcacta
  ctgcagcggatggttgccgtccatcgatgactgcttctaaaccgatttcgaccctgaa
  taaaatgattgcataacaccagcatctcctcttcgaattatagatctgttaatagctt
  acgttaaactggcatactagtttgtccaaattcgattcccttagtactagtaccagtg
  ctattagttttccacacttaactagaaaat
```

Example 4.14: Entropy Change in a Fertilized Egg

When we take an unfertilized egg out of a refrigerator at T_{cold} and leave it at a room temperature T_{room} until a thermal equilibrium is achieved, the total entropy change of the egg plus the air will be

$$\Delta s_{Total} = \left[\int_{T_{cold}}^{T_{room}} \frac{c_{egg} \, dT}{T}\right] + \left[\frac{q_{room}}{T_{room}}\right]$$

In this expression

Δs_{Total} is the total entropy change

$\left[\int_{T_{cold}}^{T_{room}} (c_{egg} \, dT/T)\right]$ is the entropy change of the unfertilized egg during its temperature change from T_{cold} to T_{room}

c_{egg} is the specific heat of the egg

q_{room} is the heat transferred from the air in the room to the egg at the room temperature T_{room}

q_{room}/T_{room} is the entropy change in the room

Δs_{Total} is positive because heat transfer through a finite temperature difference is an irreversible process.

When a fertilized egg goes through the same process, an extra term appears in the expression

$$\Delta s_{Total} = \left[\int_{T_{cold}}^{T_{room}} \frac{c_{egg} \, dT}{T}\right] + \left[\frac{q_{room}}{T_{room}}\right] + \Delta s_{reaction}$$

where $\Delta s_{reaction}$ is the entropy change, which occurs because of the biochemical reactions occurring in the fertilized egg. In the fertilized egg, genes are expressed to develop an embryo in the egg (Sanchez, 2011). The first two terms of the right-hand side of this equation tend to increase the entropy of the universe as suggested by the second law of thermodynamics, whereas the last term $\Delta s_{reaction}$ is negative, since it is related to a process that increases the order (Dinçer and Çengel, 2001; Sanchez, 2011). The longer is the genetic code, the more will be the energy needed to carry out the reactions associated with the entropy change term $\Delta s_{reaction}$.

The minimum work that the biosystem *fertilized egg* has to perform is $w_{min} = T_0 \Delta s_{reaction}$. This work has to be performed by ATP hydrolysis (Example Figure 4.14.1).

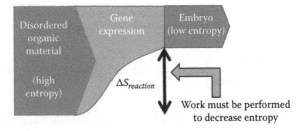

EXAMPLE FIGURE 4.14.1 Description of the entropy change during the growth of the embryo in the fertilized egg.

Example 4.15: Estimation of the Energy Cost of mRNA and Amino Acid Turnover in *Saccharomyces cerevisiae*

The energy cost of synthesizing the mRNA is the sum of the energy costs of the nucleotides of both the unexpressed part of the mRNA and those of the sequence coding for the amino acids and the energy cost of the polymerization, for example, that of connecting the nucleotides along the polymer chain. The elongation rate of an mRNA strand may be expressed as (Wagner, 2005)

$$\frac{dn_{mRNA}}{dt} = r_s - k_d n_{mRNA}$$

where
  n_{mRNA} is the number of the nucleotides in an mRNA strand
  r_s is the synthesis rate of the mRNA
  k_d is the dissociation constant of the mRNA

Under the steady-state conditions, when $dn_{mRNA}/dt = 0$, the energy cost of the turnover may be calculated as $r_s C_s = (k_d n_{mRNA})C_s$, where C_s is the energy cost of synthesizing the specific mRNA. After collecting data from the literature, Wagner (2005) states that the mean length of a yeast RNA is 1474 nucleotides and the mean cost of precursor synthesis for per nucleotide, as derived from the base composition of yeast-coding regions, is 49.3 high-energy phosphate bonds. There are 1.2 mol of mRNA/cell, and the decay constant $k_{d,mRNA} = 5.6 \times 10^{-4}$ s^{-1}; then the energy cost of the mRNA synthesis is calculated as

$$\begin{bmatrix} \text{Energy cost of the} \\ \text{mRNA synthesis} \end{bmatrix} = [5.6 \times 10^{-4} \text{ s}^{-1}][1.2][1474][49.3 \text{ P}] = 48.8 \text{ P/s}$$

where P stands for the energy of a high-energy phosphate bond.
  Wagner (2005) argues that the median length of a yeast protein is 385 amino acids, the polymerization cost is 30.3 high-energy phosphate bonds/amino acids,

there are 2460 protein molecules in a cell, and the decay constant is $k_{d,protein} =$ 1.92 × 10^{-5} s^{-1} and then estimates the energy cost of the protein turnover as

$$\begin{bmatrix} \text{Energy cost} \\ \text{of protein synthesis} \end{bmatrix} = [1.92 \times 10^{-5} \, s^{-1}][385][2460][30.3 \, P] = 551 \, P/s$$

When the ATP to ADP conversion reaction provides the high-energy phosphate bond with the reaction ATP + H_2O →ADP + P, we may estimate the energy cost as 30,514 kJ/mol high-energy phosphate bond; therefore, the energy cost of mRNA synthesis is estimated as (48.8 mol high-energy phosphate bond/s) × (30,514 J/mol high-energy phosphate bond) = 1.5 MJ/s and that of the protein turnover as (551 mol high-energy phosphate bond/s) × (30,514 J/mol high-energy phosphate bond) = 16.8 MJ/s.

4.4 Behavioral Homeostasis Theory

According to the *behavioral homeostasis theory* in the case of external changes, an organism rapidly rearranges itself to cope with the new stimuli and minimize unnecessary energy expenditure (Eisenstein et al., 2012). The variations in the protein structure are typical examples to the behavioral homeostasis theory at the molecular level (Figure 4.11). At low environmental temperature, the primary structure of a peptide chain folds to make the secondary structure, α-helix and β-sheets, to achieve a lower energy level. Hydrogen bonds are made in this process to lower the energy level of the system. The α-helixes and the β-sheets fold further together to achieve an energy level lower than what they achieved by making the secondary structure. The protein chains that fold to make the tertiary structure still search for a lower energy level and make aggregates, which is referred to as the quaternary structure (Schulz and Schirmer, 1979).

When the environmental temperature increases, the quaternary structure dissociates. If this protein is an enzyme, it loses its activity upon dissociation. After going back to a low-temperature state, it may fold back to the quaternary structure and regain its activity, which is a problem in food processing. Upon thermal treatment, if the primary structure of the protein is destroyed, no enzyme activity may be reestablished after returning back to the low-temperature state (Machado and Saraiva, 2005).

There are a vast number of examples for the minimization of the energy expenditure in biological systems: a well-known example is the loss of the muscles of hospitalized people (Pavy-Le Traon et al., 2007). If a patient spends an elongated period of time in bed, then the body starts depleting the unused muscles. Similarly, microorganisms produce some metabolites, such as amino acids, only in required quantities. In order to achieve industrial overproducer species, the control mechanism is eliminated via mutations (Özilgen, 1988). Another example can be given considering the energy metabolism of yeast species. Some yeast species employ only the glycolytic pathway of the

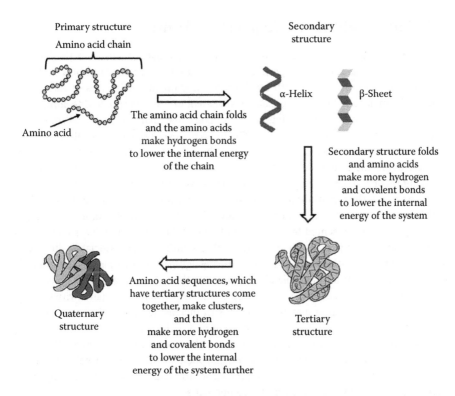

Primary structure
Amino acid chain

Secondary structure

Amino acid

The amino acid chain folds and the amino acids make hydrogen bonds to lower the internal energy of the chain

α-Helix

β-Sheet

Secondary structure folds and amino acids make more hydrogen and covalent bonds to lower the internal energy of the system

Quaternary structure

Amino acid sequences, which have tertiary structures come together, make clusters, and then make more hydrogen and covalent bonds to lower the internal energy of the system further

Tertiary structure

FIGURE 4.11 Processes occurring during the establishment of the protein structure at low temperature. The process reverses itself and the quaternary structures dissociate themselves back into amino acid chains in a high-temperature environment.

energy metabolism. Under unfavorable environmental conditions, such species minimize their energy expenditure, via minimizing the synthesis of the enzymes, which would not be needed for the survival (Kitagaki and Takagi, 2014).

Biological processes, such as protein synthesis carries substantial costs beyond those of the chemical exergy or internal energy of formation of the amino acids themselves, such as the energy costs of charging of the tRNAs with the amino acids, initiation of the translation at the ribosome, translocation of the ribosome along the mRNA, and the termination of the process. In a yeast cell, under the respiratory conditions, the median energy cost of the synthesis of an amino acid of a protein was estimated to be 30.3 high-energy phosphate bonds (Wagner, 2005). We need to carry out a lot of research to obtain further information regarding the other costs that may still remain unaccounted.

Living beings have their own energy budget; when they experience an increased demand for thermoregulation or parental care, they may reduce the energy allocated for costly maintenance processes such as immunological defense or DNA repair and reallocate it for the emerging necessity (Nilsson, 2002).

4.5 Thermodynamic Aspects of Aging

Aging research follows two major paths. One of these paths focuses on the biological nature of aging and uses experimental observations to understand how and why the cells age. Another major path employs the fundamental principles of thermodynamics (Hamalainen, 2005). With the advancement of the experimental techniques, it became possible to trace the aging-related microscopic changes in the cell morphology (Toussaint et al., 1995; Nyström, 2007). The gene-level aging occurs with distortion of the DNA strands (Giese et al., 2002), and the cells attempt to reverse this process via destroying the mutated strands (Davies et al., 2013). The telomere-induced cellular aging theory suggests the presence of a gene-level built-in program, which triggers death to prevent the mutations to propagate to the forthcoming generations (Kosmadaki and Gilhrest, 2004). A comprehensive review of the biological phenomena occurring in aging muscle is presented by Carmeli et al. (2002). Metabolic activity is reported to be decreasing with aging (Zotin and Zotin, 1997). Significant age-associated alterations are observed in oxidative phosphorylation during aging of the cytoplasmic hybrid cells carrying mitochondria from skeletal muscle and brain cells (Li et al., 2013). Toledo and Goodpaster (2013) argue that the lower mitochondrial capacity associated with obesity, type 2 diabetes, and aging may be reversed with increasing the physical activity level, which may improve the mitochondrial content and perhaps functioning of individual mitochondria. While describing the aging of a fibroblast stem cell system, Toussaint et al. (1995) described the evolution of seven morphologically distinguishable groups of strains within single species and related the structural changes with the entropy accumulating in the DNA and other cellular components.

If we analyze aging from a thermodynamic point of view, then entropy can be used as a measure for age, in accordance with Schrödinger's statement, *living organisms must maintain a state of high organization* (1944). The entropy balance equation for an open biological system is written as

$$\underbrace{\sum_{in} \dot{m}s - \sum_{out} \dot{m}s}_{\text{Entropy transfer via mass transfer}} + \underbrace{\sum_{j} \frac{\dot{Q}_j}{T_j}}_{\substack{\text{Entropy transfer} \\ \text{via heat transfer}}} + \underbrace{\dot{m}s_{gen}}_{\substack{\text{Entropy} \\ \text{generation}}} = \underbrace{\frac{d(ms)_{system}}{dt}}_{\text{Entropy accumulation}}$$

In order to understand the effect of entropy on aging, let us evaluate the physiological meaning of each term in this equation one by one.

- *Entropy transfer via mass transfer*: We have already differentiated between biomass-consuming and biomass-generating processes. In biomass-producing processes, reactants have a high entropy level, and molecular bonds between these are formed to store energy. This process increases order, so that the entropies of the products are low. Then, the net entropy transfer via mass transfer is positive. Work (either extracted from solar energy in photosynthetic species or extracted from the energy metabolism) must be performed to enable this process.

- All biological systems must perform biomass-consuming processes (like the energy metabolism in a cell) to generate work for intracellular activities. In the energy metabolism, nutrients are oxidized, molecular bonds are broken, and work is generated from the released Gibbs free energy. Thus, the net value of the term $\sum_{in} \dot{m}s - \sum_{out} \dot{m}s$ is negative for biomass-consuming processes.
- *Entropy transfer via heat transfer*: In the energy metabolism, nutrients that have a high enthalpy are oxidized, and low-enthalpy products are extracted out of the biological system. For a resting biological system, the difference in the inflow and outflow enthalpy has to be equal to the heat released out of the system, that is, $\dot{q} = \Delta h_{metabolic\ reactions} < 0$. So, for biomass-consuming processes, heat transfer is always negative as shown in Figure 4.12. As the metabolic heat is transferred out of the system, entropy flows out of the system, too. The term entropy transfer via heat transfer can be written as $\sum_j \dot{q}_j / T_j = (\dot{q}/T_{surface}) < 0$ for a system that is only in contact with one thermal reservoir (in Figure 4.12, this thermal reservoir is the surrounding air). Heat flows through a finite temperature difference; therefore, the entropy flowing into the outer layer of the body is $\dot{q}/T_{inner\ core}$ and the entropy flowing out of the outer layer of the body is $\dot{q}/T_{surface}$. Since $T_{innercore} > T_{surface}$, entropy is generated in the outer layer of the body.
- *Entropy generation*: Internal irreversibilities in a process generate entropy. In an ideal system, every process occurs reversibly; thus, $s_{gen} = 0$. In actual systems $s_{gen} > 0$. Consider a neuron cell, which needs to transfer K^+ from the extracellular region in order to return to its resting condition. If the concentration of K^+, c_{K^+}, in the extracellular region is larger than c_{K^+} in the cell, then K^+ can flow in the direction of the concentration gradient, like in the case of a passive diffusion. This is

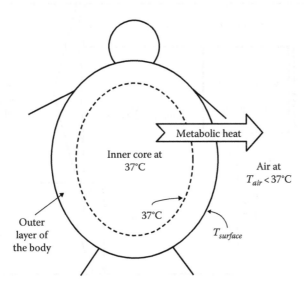

FIGURE 4.12 Schematic description of the entropy generation with the metabolic heat release.

an irreversible process and generates entropy. But, if the c_{K^+} in the extracellular region is low, then the cell has to perform work to take the necessary K^+ into the cell. That would require a minimum work of

$$w_{min} = T_0 s_{gen} = RT_0 \ln\left(\frac{c_{K^+, \, cell}}{c_{K^+, \, extracellular \, region}}\right).$$

- *Entropy accumulation*: Biological systems are rarely at steady state; therefore, we may observe an entropy change in the system with respect to time. Our previous discussion shows that both of the entropy transfer terms are negative for biomass-consuming processes, and entropy generation term is always positive for actual processes. Hence, the accumulation term may be positive, negative, or zero, depending on the numerical values of the transfer and generation terms (Figure 4.13). For an ideal system, where $s_{gen} = 0$, accumulation would be negative. As the irreversibilities in the system increases, accumulation term increases and may become zero or positive.

There are numerous theories that aim to relate entropy with aging. Ilya Prigogine was awarded the Nobel Prize in Chemistry in 1977 for his pioneering work on the dissipative structures (Prigogine, 1977; Macklem, 2008). The dissipative structure theory states that the open systems are capable of continuously importing energy from the environment while exporting entropy (Prigogine, 1977; Demirel, 2010). The concept of dissipative

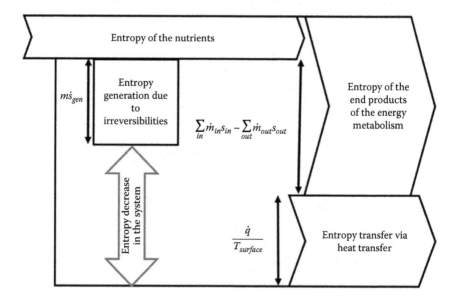

FIGURE 4.13 Schematic description of how entropy decreases in the biological systems due to biomass-consuming processes.

structures when considered together with the other topics of nonequilibrium thermodynamics is proved to be quite useful in chemical, biological, and mechanical engineering applications (Demirel and Sandler, 2004). To sustain their growth, the dissipative systems must not only increase their entropy exporting potential but also eliminate the entropy accumulation over time. Otherwise, the accumulation of positive entropy in the system will eventually bring it to thermodynamic equilibrium, a state in which the system cannot maintain its order and organization. It has already been more than three decades since Prigogine had received the Nobel Prize; the recent version of this theory is the entropy and the exergy balances pertinent to the living systems.

Hershey (1974) and Hershey and Wang (1980) calculated the human life span entropy in terms of the metabolic heat generation divided by the body temperature. Their work led to the definition of the concepts like entropic age, expected life span, and senile death. Silva and Annamalai (2008, 2009) related the concept of the life span entropy and related it with the level of the physical activity of the people and the composition of their diet. Demetrius et al. (2009) draw attention to the point that body size is a fundamental property of an organism, which regulates the rate at which an organism transforms the energy of resources into biological work, hence the metabolic rate, and showed that the life span entropy is related to the life span and the body size. Prigogine and Wiame's (1946) theory has been interpreted by Balmer (1982) that the *organisms have evolved over the years to minimize entropy generation*, for example, $d(ms)_{gen}/dt$. Within the context of this statement, biological systems may be considered as open systems living under near-steady-state conditions and evolving very slowly in the direction of reducing the entropy generation, which reaches to a minimum when the organisms' adaptation to their environment has concluded (Hollinger and Zenzen, 1982; del Castillo and Vera-Cruz, 2011).

Biological systems have limited capacity of accumulating entropy. When the total entropy accumulation in a living being reaches a certain limit, which is referred to as the *life span entropy*, it dies. In the previous studies, life span entropy was related to the nutrient consumption (Silva and Annamalai, 2009) and heat loss from the body (Aoki, 1989, 1994; Silva and Annamalai, 2008). Increase of the cellular entropy is related to aging according to the literature (Hayflick, 2007; Salminen and Kaarniranta, 2010), which occurs because the changed energy states of biomolecules render them inactive or malfunction, while increasing the cellular entropy. Genetically induced repair and replacement processes are capable of maintaining the balance in favor of the functioning molecules. With aging, the balance between these processes tends toward inactivation and malfunctioning (Hayflick, 2007). The events that lead to the increase of the cellular entropy during aging is hypothesized (Chinnery et al., 2002), and its indications are evidenced in the literature (Polyak et al., 1998; Elson et al., 2001), but the attempts for quantifying the life span entropy is yet very limited (Silva and Annamalai, 2008, 2009). Entropy generation in the human body is also associated with cancer development, which changes the way of energy allocation in the body and leads to higher energy dissipation (Garland, 2013).

After stating that the genes linked in cancer induction are extensive, Garland (2013) draws attention to the point that all the cancer cells behave in identical and highly predictable fashion. This behavior involves changes in motility, invasion, replication, nuclear and chromosomal fragmentation, structure degradation, and phenotypic fluidity. Garland (2013) also argues that cancer is a diversion of the energy needed for structural organization into maximum entropy dissipation. The channels that are distributing the entropy generation are organized along fractal networks. The oncogenic mutations and molecular alterations in the cell redirect these channels. Oncogenic alterations create cumulative effects by permanently stabilizing parts of the fractal network. Garland (2013) draws attention to the need for multidisciplinary research achieving for further understanding of the phenomena; in such a case, bridging the gap between the thermodynamic modeling studies and the proposals of the medical researchers would probably be among the best attempts to be done.

Example 4.16: Balmer's Model for Aging

Balmer (1982) presented some data showing that $(\dot{q}/mT)$, where m is the body mass, which decreases with age, and $[d(ms)/dt]_{system}$ remains negative as long as the system is alive and becomes positive upon the death of the system. Having a $[d(ms)/dt]_{system}<0$ value implies that the entropy of the living system decreases throughout its life span, with $(ms)_{system}$ being larger at the earlier stages of the life and smaller at the later stages. Balmer (1982) argues that the crude chemical analysis of the living systems support this view, where proteins, chemicals with higher disorderly molecular structure, are replaced with fat, molecules with more orderly structure, and soluble collagens are cross-linked to make insoluble polymers, leading to cataract and wrinkling of the skin. Recent multiscale entropy analysis studies, which may be regarded as entropy accumulation in the system, have revealed that the aging process reduces complexity and functional responsiveness of electroencephalographic measurements in response to photic stimulation (Takahashi et al., 2009). Quantitative measurement of aging using image texture entropy, which may be regarded as entropy production, revealed that aging of transparent round worm *Caenorhabditis elegans* has several distinct stages (Shamir et al., 2009). In a 12-day-long study, the texture entropy increases as the worm ages, with sharper increase measured between day 2 and day 4, and between day 8 and day 10. It should be noticed that the predictions of the Balmer's model, for example, decreasing entropy accumulation rate in the system and positive entropy production rate, are valid when we consider the entire system. Zotin and Zotin (1997) draw attention to the point that although Balmer's (1982) principles are valid for the integrity of the system, when we consider a subsystem of the body only, like a healing, regenerating, or a tissue being malignant, a reverse of the Balmer's principles may also be observed.

Stewart et al.'s (2005) observations with doubling *E. coli* strengthen the entropy accumulation hypothesis of aging. *E. coli* grows to form elongated cells like a rod, before division. Upon the beginning of the division, a new pole sites start to appear in the middle of the elongated cell, and then the cells are separated into two new cells (Stewart et al., 2005). The DNA and the lipids are distributed in the mother and the daughter cells (Lin et al., 1971). The structure of the old pole survives in the mother cell (de Pedro et al., 1997), and some of the cellular constituents cannot diffuse fast enough to the daughter cell and accumulate in the mother cell; although the mother and the daughter cells look the same morphologically, they are different physiologically (Stewart et al., 2005; Nyström, 2007). During the process of series of cell divisions, the cells keep producing entropy. The older the cell is, the higher would be the entropy it produced. The inheritance of the old pole is regarded as an aging sign, related to cumulatively slowed growth, less offspring biomass production, and increased probability of death (Stewart et al., 2005).

The second law of thermodynamics state that all the real processes are irreversible, which therefore causes an increase in the entropy of the universe. All the living beings have limited entropy generation capacity during their lifetimes, which is referred to as the *life span entropy*; when they reach to that limit, they die (Aoki, 1994; Boregowda et al., 2009; Annamalai and Silva, 2012).

Example 4.17: Estimation of the Life Span of the Masseter Muscle

Çatak et al. (2015) developed a model to estimate the life span of the masseter muscle based on the entropy generated by this muscle group. Çatak et al. (2015) divided the overall system into three subsystems and performed mass, energy, and exergy analyses for the oral cavity (system I), ipsilateral condylar muscle (system II), and contralateral condylar muscle (system III) subsystems, which are shown in Example Figure 4.17.1:

Material balance

$$m_{almond} + m_{saliva} + m_{solution} = 0$$

Energy balance

$$q_{OC} + w_{OC} + (mh)_{almond} + (mh)_{saliva} + (mh)_{solution} = 0$$

Exergy balance

$$ex_{destroyed, OC} = q_{OC}\left(1 - \frac{T_0}{T_{OC}}\right) - w_{OC} + (mex)_{almond} + (mex)_{saliva} + (mex)_{solution} = 0$$

where
T_{OC} is the temperature of the boundary of the oral cavity
T_0 is the reference temperature

The specific exergy of the species are calculated based on their chemical composition and thermophysical state ($ex = ex_{ch} + h - T_0 s - \sum x_i \mu_i^0$). In this analysis, reference temperature is set as equal to the oral cavity ($T_0 = T_{OC} = 37°C$). Assuming that the solution heading to the esophagus is ideal, the enthalpy

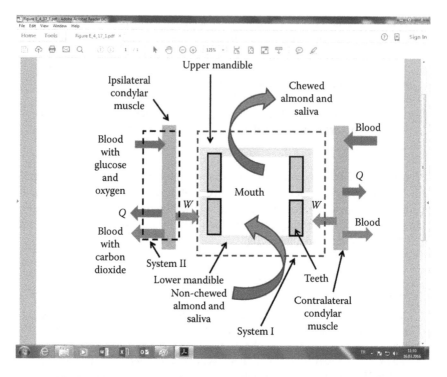

EXAMPLE FIGURE 4.17.1 The schematic description of the systems involved in the chewing process. Within the context of this study, the ipsilateral muscle is on the right side of the jaw of the subject while the contralateral muscle is on the left. Almond is the source of the glucose.

difference in the energy balance is zero. Then, the heat released can be calculated by simplifying this equation as

$$q_{OC} + w_{OC} = 0$$

Then the entropy generated is calculated as

$$s_{gen, OC} = \frac{ex_{destroyed, OC}}{T_0}$$

Çatak et al. (2015) calculated the work done by the masseter muscles to move the lower jaw in one loop with the expression

$$W_{muscle} = \oint_{il} F dl + \oint_{d} F dl$$

by using the force versus elongation data of Hannam et al. (2008). Throughout one chewing cycle, the ipsilateral muscles create forces as large as 40 N, whereas

EXAMPLE TABLE 4.17.1 Numerical Values of the Work Performed, Exergy Destroyed, and Entropy Generated during the Chewing Process

	Contralateral and Ipsilateral Masseter Muscles (η_{II} = 0.3)	
	Masseter Muscle	Oral Cavity
Glucose consumed (mmol/min)	5.95×10^{-3}	—
Work (J)	2.30	1.59
Exergy destroyed (25 cycle) (J/K)	5.77	28.5
Entropy generation (25 cycle) (J/K)	1.94×10^{-2}	9.55×10^{-2}

the maximum force created by the contralateral muscles reaches only up to 25 N. However, contraction of the ipsilateral muscles stops during high-force period. Therefore, the total work done by the contralateral muscles is about 55 times larger than the ipsilateral muscles.

Example Table 4.17.1 provides the numerical values of the work performed, exergy destroyed, and entropy generated when η_{II} = 0.3 with the uptake of 5.95×10^{-3} mmol glucose during 25 cycles of the chewing process.

MATLAB code E4.17 computes the work done by the contralateral and ipsilateral muscles and the cumulative entropy generation by the masseter muscles during the life span:

MATLAB CODE E4.17

Command Window

```
clear all
close all

% enter the data
saliva_production=5; % 5 grams per minute
length_of_a_cycle = 20; % seconds
number_of_cycles=25;      % total cycle

% SYSTEM: ORAL CAVITY INPUT DATA
almond.lipid.mass_in=0.4942;
almond.lipid.enthalpy_in=-2.717;
%almond.lipid.entropy_in=0.0027615;
almond.lipid.exergy_in=39.54;
almond.protein.mass_in=0.2122;
almond.protein.enthalpy_in=-6.805;
almond.protein.entropy_in=0.188;
almond.protein.exergy_in=16.27;
almond.carbonhydrate.mass_in=0.0389;
almond.carbonhydrate.enthalpy_in=-6.474;
almond.carbonhydrate.entropy_in=0.392;
almond.carbonhydrate.exergy_in=17.52;
```

```
% almond.starch.mass_in=0.122;
% almond.starch.enthalpy_in=5.2524;
% almond.starch.entropy_in=0.016982;
% almond.starch.exergy_in=-0.0119;
almond.fiber.mass_in=0.122;
almond.fiber.enthalpy_in=-5.654;
almond.fiber.entropy_in=0.0011718;
almond.fiber.exergy_in=18.86;
almond.mineral.mass_in=0.0173;
almond.mineral.enthalpy_in=0;
almond.mineral.entropy_in=0.0016424;
almond.mineral.exergy_in=9.38;
almond.water.mass_in=0.047;
almond.water.enthalpy_in=-15.819;
almond.water.entropy_in=0.0005053;
almond.water.exergy_in=0.05;


%%%%%%%% Concentration data %%%%%
almond.lipid.mol_in=almond.lipid.mass_in/282.46;
almond.protein.mol_in=almond.protein.mass_in/147.13;
almond.carbonhydrate.mol_in=almond.carbonhydrate.
  mass_in/342.29;
almond.fiber.mol_in=almond.fiber.mass_in/162.14;
almond.mineral.mol_in=almond.mineral.mass_in/39.09;
almond.water.mol_in=almond.water.mass_in/18.01;
sum_in=almond.lipid.mol_in+almond.protein.mol_in+almond.
  carbonhydrate.mol_in+almond.fiber.mol_in+almond.mineral.
  mol_in+almond.water.mol_in;
% Almond_fields=fieldnames(almond);
% % for i=1:numel(Almond_fields)
% %      sum_in=sum(almond.(Almond_fields{i}).mol_in);
% % end
almond.lipid.fraction_in=almond.lipid.mol_in/sum_in;
almond.protein.fraction_in=almond.protein.mol_in/sum_in;
almond.carbonhydrate.fraction_in=almond.carbonhydrate.
  mol_in/sum_in;
almond.fiber.fraction_in=almond.fiber.mol_in/sum_in;
almond.mineral.fraction_in=almond.mineral.mol_in/sum_in;
almond.water.fraction_in=almond.water.mol_in/sum_in;


% Saliva
saliva.mass_in=saliva_production* length_of_a_cycle/60;
saliva.enthalpy_in=-15.819;
saliva.entropy_in=0.0005053;
saliva.exergy_in=0.05;
saliva.water.mol_in=saliva.mass_in/18.01;
saliva.water.fraction_in=1;


% ORAL CAVITY SYSTEM OUTPUT DATA
almond.lipid.mass_out=0.4942;
almond.lipid.enthalpy_out=-2.717;
almond.lipid.entropy_out=0.0027615;
```

```
almond.lipid.exergy_out=39.54;
almond.protein.mass_out=0.2122;
almond.protein.enthalpy_out=-6.805;
almond.protein.entropy_out=0.188;
almond.protein.exergy_out=16.27;
almond.carbonhydrate.mass_out=0.0389;
almond.carbonhydrate.enthalpy_out=-6.474;
almond.carbonhydrate.entropy_out=0.392;
almond.carbonhydrate.exergy_out=17.52;
% almond.starch.mass_in=0.122;
% almond.starch.enthalpy_in=5.2524;
% almond.starch.entropy_in=0.016982;
% almond.starch.exergy_in=-0.0119;
almond.fiber.mass_out=0.122;
almond.fiber.enthalpy_out=-5.654;
almond.fiber.entropy_out=0.0011718;
almond.fiber.exergy_out=18.86;
almond.mineral.mass_out=0.0173;
almond.mineral.enthalpy_out=0;
almond.mineral.entropy_out=0.0016424;
almond.mineral.exergy_out=9.38;
almond.water.mass_out=0.047+saliva.mass_in;
almond.water.enthalpy_out=-15.819;
almond.water.entropy_out=0.0005053;
almond.water.exergy_out=0.05;
%%%%%%%% Concentration data %%%%%
almond.lipid.mol_out=almond.lipid.mass_out/282.46;
almond.protein.mol_out=almond.protein.mass_out/147.13;
almond.carbonhydrate.mol_out=almond.carbonhydrate.
  mass_out/342.29;
almond.fiber.mol_out=almond.fiber.mass_out/162.14;
almond.mineral.mol_out=almond.mineral.mass_out/39.09;
almond.water.mol_out=almond.water.mass_out/18.01;
sum_out=almond.lipid.mol_in+almond.protein.mol_out+almond.
  carbonhydrate.mol_out+almond.fiber.mol_out+almond.mineral.
  mol_out+almond.water.mol_out;
% Almond_fields=fieldnames(almond);
% for i=1:numel(Almond_fields)
%      sum_out=sum(almond.(Almond_fields{i}).mol_out);
% end

almond.lipid.fraction_out=almond.lipid.mol_out/sum_out;
almond.protein.fraction_out=almond.protein.mol_out/sum_out;
almond.carbonhydrate.fraction_out=almond.carbonhydrate.
  mol_out/sum_out;
almond.fiber.fraction_out=almond.fiber.mol_out/sum_out;
almond.mineral.fraction_out=almond.mineral.mol_out/sum_out;
almond.water.fraction_out=almond.water.mol_out/sum_out;
%%%%%%%%%% Saliva%%%%%%%%%%%%%%%%
saliva.mass_out=saliva_production*length_of_a_cycle/60;
 saliva.enthalpy_out=-15.819;
 saliva.entropy_out=0.0005053;
 saliva.exergy_out=0.05;
```

```
% BLOOD CYCLE INPUT DATA
% blood.glucose.mass_in ==> will be calculated in bloodcycle.m
blood.glucose.enthalpy_in=-7.035;
blood.glucose.entropy_in=0.0011601;
blood.glucose.exergy_in=11.47;
% blood.oxygen.mass_in ==> will be calculated in bloodcycle.m
blood.oxygen.enthalpy_in=-0.366;
blood.oxygen.entropy_in=0.0064083;
blood.oxygen.exergy_in=0.27;


% BLOOD CYCLE OUTPUT DATA
% blood.co2.mass_in ==> will be calculated in bloodcycle.m
blood.co2.enthalpy_out=-15.923;
blood.co2.entropy_out=0.0048864;
blood.co2.exergy_out=-3.11;
% blood.water.mass_in ==> will be calculated in bloodcycle.m
blood.water.enthalpy_out=-15.919;
blood.water.entropy_out=0.0005053;
blood.water.exergy_out=-9.24;


%% CONTRALATERAL WORK
%%%%%%%%%%%%%%%%%%%%%%%%%%%%%%%%%%%%%%%%%%%%%%%%%%%%%%%%%
% enter the data
%disp('Calculating Contralateral Work...')
Lc =[0 0.2 0.4 0.6 0.75 0.8 0.9 0.95 1.0 1.05 1.2 1.3 1.35
    1.4 1.5 1.6...
        1.65 1.75 1.8 1.9 2.1 2.3 2.5 2.85 3.0 3.15 3.3 3.5
    3.5 3.5 3.5...
        3.5 3.5 3.5 3.5 3.5 3.5 3.4 3.3 3.25 3.2 2.9 2.6 1.75
    0.0 0 0 0 ...
        0 0 0 0 0 0 0]; % muscle length (mm)
Fc= [1 0.8 0.6 0.8 1.1 1.5 1.7 2 2.4 2.8 3 3.25 3.5 3.8 4.2
    4.35 4.5 ...
        4.65 4.8 4.9 5 5.2 5.4 5.3 5.2 5.1 5 4.9 4.8 4.7 4.6
    4.55 4.5 ...
        4.25 4 4.75 5 6 7 10 14 17.5 20 23 24 23 22 22.5 20 15
    13.5 ...
        10.5 6 4 3 1]; % Force (N)


% calculate the muscle work
for i=2:length(Fc)
    deltaW1(i)=(Fc(i)*(Lc(i)-Lc(i-1))); % incremental work
end

W1 = sum(deltaW1);

%% IPSILATERAL WORK
%%%%%%%%%%%%%%%%%%%%%%%%%%%%%%%%%%%%%%%%%%%%%%%%%%%%%%%%%%%
% enter the data
%disp('Calculating Ipsilateral Work...')

Li =[0 0.5 1 1.3 1.4 1.5 1.6 1.7 1.8 1.9 2 2.2 2.4 2.4 2.1 ...
        1.8 1.3 0.8 0.8 0 0]; % muscle length (mm)
```

```
Fi= [1 1 1.5 2.0 2.3 2.6 3.0 3.5 4 4.5 5 5 4.5 4 3.5 3.3 ...
    2.6 2.3 2.3 3.6 1]; % Force (N)

% calculate the muscle work
for i=2:length(Fi)
    deltaW2(i)=(Fi(i)*(Li(i)-Li(i-1))); % incremental work
end

W2 = sum(deltaW2);

W=(W1+W2)/0.69;

%% ORAL ENERGY BALANCE
%%%%%%%%%%%%%%%%%%%%%%%%%%%%%%%%%%%%%%%%%%%%%%%%%%%%%%%%%%%%
%disp('Calculating the Oral Energy Balance...')
Almond_fields=fieldnames(almond);

mh_almond_in=0;
for i=1:numel(Almond_fields)
    mh_almond_in = mh_almond_in+...
        almond.(Almond_fields{i}).mass_in *
almond.(Almond_fields{i}).enthalpy_in;
end

mh_saliva_in=saliva.mass_in * saliva.enthalpy_in;

mh_almond_out=0;
for i=1:numel(Almond_fields)   %water mass_out yok kendi
  kontrol edecek
    mh_almond_out = mh_almond_out+...
almond.(Almond_fields{i}).mass_out*almond.(Almond_
  fields{i}).enthalpy_out;
end
%mh_saliva_out=saliva.mass_out * saliva.enthalpy_out;
W_total=((W/1000)* number_of_cycles)/1000;
W_net=W_total*0.69;
mh_in = mh_almond_in + mh_saliva_in;
mh_out = mh_almond_out;

Q_oral = mh_out - mh_in - W_net;

%% BLOOD-CYCLE ANALYSIS
% disp('Performing Blood-Cycle Analysis...')
ii=1;
for ii =1:26;
    efficiency=0.17:0.01:0.42;
total_chemical_energy = abs(W_total) / efficiency(ii); %in kJ
one_mol_ATP_energy = 3868; % ATP produced for 1 mol glucose
mol_glucose = total_chemical_energy / one_mol_ATP_energy ; %
  in mol

blood.glucose.mass_in=mol_glucose*180; %1 mol glucose = 180 gr
blood.oxygen.mass_in=mol_glucose*6*32; %1 mol O2 = 32 gr
```

```
blood.co2.mass_out=mol_glucose*6*44;    %1 mol CO2 = 44 gr
blood.water.mass_out=mol_glucose*6*18; %1 mol H2O = 18 gr

reaction_time = 20 ; % in seconds
glucose_reaction_rate = (mol_glucose/reaction_time)*...
                        (60*1000); % in milimol per minute

 flow_diameter = 1.9; % in milimeters
 flow_velocity = 254; % milimeters per second
 v_blood = (pi*(flow_diameter/2)^2)*flow_velocity*(60/1e6);
   %litres/mm

 glucose_absorbed = glucose_reaction_rate / v_blood; % in
   mmol/lt

%% BLOOD - ENERGY CYCLE
 %%%%%%%%%%%%%%%%%%%%%%%%%%%%%%%%%%%%%%%%%%%%%%%%
 %disp('Calculating the Energy Balance for Blood...')
 Blood_fields=fieldnames(blood);

mh_blood_in=0;
for i=1:2 %glukose & oxygene
    mh_blood_in = mh_blood_in+...

blood.(Blood_fields{i}).mass_in*blood.(Blood_fields{i}).
  enthalpy_in;
end

mh_blood_out=0;
for i=3:4 %co2 & water
    mh_blood_out = mh_blood_out+...

blood.(Blood_fields{i}).mass_out*blood.(Blood_fields{i}).
  enthalpy_out;
end

Q_blood = mh_blood_out - mh_blood_in + abs(W_total);

%% BLOOD - EXERGY BALANCE
 %%%%%%%%%%%%%%%%%%%%%%%%%%%%%%%%%%%%%%%%%%%%%%%%%%%%
 % disp('Calculating the Exergy Destruction for Blood...')
  T_out=298;
  T_in=310;
 Blood_fields=fieldnames(blood);

mb_blood_in=0;
for i=1:2
    mb_blood_in = mb_blood_in+...
        blood.(Blood_fields{i}).mass_in *
blood.(Blood_fields{i}).exergy_in;
end
```

```
mb_blood_out=0;
for i=3:4
    mb_blood_out = mb_blood_out+...
        blood.(Blood_fields{i}).mass_out *
blood.(Blood_fields{i}).exergy_out;
end

x_destroyed_blood = mb_blood_in - mb_blood_out - abs(W_
  total) - ...
            Q_blood*(1-(T_out/T_in)); % in kJ

Sgen_blood(ii)=x_destroyed_blood/298;
%ii=ii+1;
end
%% ORAL - EXERGY BALANCE
%%%%%%%%%%%%%%%%%%%%%%%%%%%%%%%%%%%%%%%%%%%%%%%%%%%%%%
% disp('Calculating the Exergy Destruction for Blood...')
Almond_fields=fieldnames(almond);

mb_almond_in=0;
for i=1:numel(Almond_fields)
    mb_almond_in = mb_almond_in+...
        almond.(Almond_fields{i}).mass_in *
  almond.(Almond_fields{i}).exergy_in;
end

mb_almond_out=0;
for i=1:numel(Almond_fields)-1
    mb_almond_out = mb_almond_out+...
        almond.(Almond_fields{i}).mass_out *
almond.(Almond_fields{i}).exergy_out;
end
mb_almond_concentration_in=0;
for i=1:numel(Almond_fields)
    mb_almond_concentration_in =
mb_almond_concentration_in+...
        almond.(Almond_fields{i}).mol_in *8.314*310*log(
  almond.(Almond_fields{i}).fraction_in);
end
mb_saliva_concentration_in=saliva.water.mol_
  in*8.314*310*log(saliva.water.fraction_in);
mb_almond_concentration_out=0;
for i=1:numel(Almond_fields)-1
    mb_almond_concentration_out =
mb_almond_concentration_out+...
        almond.(Almond_fields{i}).mol_out
*8.314*310*log(almond.(Almond_fields{i}).fraction_out);
end
x_destroyed_oralc = mb_almond_in - mb_almond_out + abs(W_
  total) - ...
            Q_oral*(1-(T_out/T_in))+(mb_almond_
concentration_in+mb_saliva_concentration_in-mb_almond_
concentration_out)/1000; % in kJ
```

```
Sgen_oralc=x_destroyed_oralc/298;
%% LIFE EXPECTANCY
%%%%%%%%%%%%%%%%%%%%%%%%%%%%%%%%%%%%%%%%%%%%%%
% disp('Calculating Expected Life...')
age = 20:1:100;
age = age' ;
reference_age = 25;

% Survival Probability
St = [ 98.3 98.2 98.1 98.0 97.9 97.8 97.6 97.5 97.4 97.2 ...
       97.1 96.9 96.8 96.6 96.4 96.2 96.1 95.8 95.6 95.4 ...
       95.2 94.9 94.6 94.4 94.1 93.7 93.4 93.0 92.6 92.2 ...
       91.8 91.3 90.8 90.2 89.6 89.0 88.3 87.5 86.6 85.7 ...
       84.6 83.5 82.3 81.0 79.6 78.2 76.6 75.0 73.3 71.5 ...
       69.6 67.6 65.5 63.3 60.9 58.5 55.9 53.2 50.5 47.6 ...
       44.6 41.4 38.2 34.9 31.4 28.0 24.7 21.6 18.7 16.0 ...
       13.5 11.3 9.3 7.5 6 4.7 3.6 2.7 1.9 1.4 1];

for i = 1:numel(St)-1
    Lt(i) = (St(i)+St(i+1))/2;
end

% Person-years lived at and above age 25 (T25)
T25 = sum(Lt(find(age==reference_age):end));

S25 = St(find(age==reference_age));

% Life Expectancy
E25 =    T25 / S25 ;

life = round(reference_age + E25);

%% ENTROPY GENERATION FOR NORMAL AND OBESE PERSON
%%%%%%%%%%%%%%%%%%%%%%%%%%%%%%%%%%%%%%%%%%%%%%%%%%%%%
% Calorie of 1 gram almond
%disp('Calculating Entropy Generation for Normal and Obese
  Person...')
age = 1:life; age=age';
%age_2=1:90; age_2=age_2';
almond_cal = 5.85 * 4.18; % kcal
%%%%%%%%%%%%NORMAL%%%%%%%%%%%%%%%%%%
%%%%%% 10% OFF %%%%%%%%%%%%%%
%%%%%%% EFF=0.17%%%%%%%%%
epd_normal_10_1 =
[100;120;120;130;130;130;130;130;180;180;...
            180;180;180;220;220;220;220;220;250;250;...
            250;250;250;250;250;250;250;250;250;250;...
            230;230;230;230;230;230;230;230;230;230;...
            230;230;230;230;230;230;230;230;230;230;...
            210;210;210;210;210;210;210;210;210;210;...
            210;210;210;210;210;210;210;210;210;210;...
            210;210;210;210;210;210] * 4.18;
```

```
entpd_normal_10_1 = (epd_normal_10_1 / almond_cal) *
Sgen_blood(1) ;
epy_normal_10_1=365*cumsum(epd_normal_10_1);
entpy_normal_10_1=365*cumsum(entpd_normal_10_1);
epd_fit_normal_10_1=fit(age,epd_normal_10_1,'fourier2');
epd_fit_normal_10_1=epd_fit_normal_10_1(age);
entpd_fit_normal_10_1=fit(age,entpd_normal_10_1,'fourier2');
entpd_fit_normal_10_1=entpd_fit_normal_10_1(age);
%%%%%%% EFF=0.3%%%%%%%%%%
epd_normal_10_2 =
[100;120;120;130;130;130;130;130;180;180;...
             180;180;180;220;220;220;220;220;250;250;...
             250;250;250;250;250;250;250;250;250;250;...
             230;230;230;230;230;230;230;230;230;230;...
             230;230;230;230;230;230;230;230;230;230;...
             210;210;210;210;210;210;210;210;210;210;...
             210;210;210;210;210;210;210;210;210;...
             210;210;210;210;210;210] * 4.18;

entpd_normal_10_2 = (epd_normal_10_2 / almond_cal) *
Sgen_blood(14) ;
epy_normal_10_2=365*cumsum(epd_normal_10_2);
entpy_normal_10_2=365*cumsum(entpd_normal_10_2);
epd_fit_normal_10_2=fit(age,epd_normal_10_2,'fourier2');
epd_fit_normal_10_2=epd_fit_normal_10_2(age);
entpd_fit_normal_10_2=fit(age,entpd_normal_10_2,'fourier2');
entpd_fit_normal_10_2=entpd_fit_normal_10_2(age);
%%%%%%% EFF=0.42%%%%%%%%%%
epd_normal_10_3 =
[100;120;120;130;130;130;130;130;180;180;...
             180;180;180;220;220;220;220;220;250;250;...
             250;250;250;250;250;250;250;250;250;250;...
             230;230;230;230;230;230;230;230;230;230;...
             230;230;230;230;230;230;230;230;230;230;...
             210;210;210;210;210;210;210;210;210;210;...
             210;210;210;210;210;210;210;210;210;210;...
             210;210;210;210;210;210] * 4.18;

entpd_normal_10_3 = (epd_normal_10_3 / almond_cal) *
Sgen_blood(26) ;
epy_normal_10_3=365*cumsum(epd_normal_10_3);
entpy_normal_10_3=365*cumsum(entpd_normal_10_3);
epd_fit_normal_10_3=fit(age,epd_normal_10_3,'fourier2');
epd_fit_normal_10_3=epd_fit_normal_10_3(age);
entpd_fit_normal_10_3=fit(age,entpd_normal_10_3,'fourier2');
entpd_fit_normal_10_3=entpd_fit_normal_10_3(age);
%%%%% 50% OFF %%%%%%%%%%%%%%
%%%%%% EFF=0.17%%%%%%%%%%
epd_normal_50_1 =
[500;600;600;650;650;650;650;650;900;900;...
900;900;900;1100;1100;1100;1100;1100;1100;1250;...
1250;1250;1250;1250;1250;1250;1250;1250;1250;...
```

```
1150;1150;1150;1150;1150;1150;1150;1150;1150;1150;...
1150;1150;1150;1150;1150;1150;1150;1150;1150;1150;...
1050;1050;1050;1050;1050;1050;1050;1050;1050;1050;...
1050;1050;1050;1050;1050;1050;1050;1050;1050;1050;...
            1050;1050;1050;1050;1050;1050] * 4.18;

entpd_normal_50_1 = (epd_normal_50_1 / almond_cal)
* Sgen_blood(1) ;
epy_normal_50_1=365*cumsum(epd_normal_50_1);
entpy_normal_50_1=365*cumsum(entpd_normal_50_1);
epd_fit_normal_50_1=fit(age,epd_normal_50_1,'fourier2');
epd_fit_normal_50_1=epd_fit_normal_50_1(age);
entpd_fit_normal_50_1=fit(age,entpd_normal_50_1,'fourier2');
entpd_fit_normal_50_1=entpd_fit_normal_50_1(age);
%%%%%% EFF=0.3%%%%%%%%%%
epd_normal_50_2 =
[500;600;600;650;650;650;650;650;900;900;...
900;900;900;1100;1100;1100;1100;1100;1100;1250;...
1250;1250;1250;1250;1250;1250;1250;1250;1250;1250;...
1150;1150;1150;1150;1150;1150;1150;1150;1150;1150;...
1150;1150;1150;1150;1150;1150;1150;1150;1150;1150;...
1050;1050;1050;1050;1050;1050;1050;1050;1050;1050;...
1050;1050;1050;1050;1050;1050;1050;1050;1050;1050;...
            1050;1050;1050;1050;1050;1050] * 4.18;

entpd_normal_50_2 = (epd_normal_50_2 / almond_cal) *
Sgen_blood(14) ;
epy_normal_50_2=365*cumsum(epd_normal_50_2);
entpy_normal_50_2=365*cumsum(entpd_normal_50_2);
epd_fit_normal_50_2=fit(age,epd_normal_50_2,'fourier2');
epd_fit_normal_50_2=epd_fit_normal_50_2(age);
entpd_fit_normal_50_2=fit(age,entpd_normal_50_2,'fourier2');
entpd_fit_normal_50_2=entpd_fit_normal_50_2(age);
%%%%%% EFF=0.42%%%%%%%%%%
epd_normal_50_3 =
[500;600;600;650;650;650;650;650;900;900;...
900;900;900;1100;1100;1100;1100;1100;1100;1250;...
1250;1250;1250;1250;1250;1250;1250;1250;1250;1250;...
1150;1150;1150;1150;1150;1150;1150;1150;1150;1150;...
1150;1150;1150;1150;1150;1150;1150;1150;1150;1150;...
1050;1050;1050;1050;1050;1050;1050;1050;1050;1050;...
1050;1050;1050;1050;1050;1050;1050;1050;1050;1050;...
            1050;1050;1050;1050;1050;1050] * 4.18;

entpd_normal_50_3 = (epd_normal_50_3 / almond_cal) *
Sgen_blood(26) ;
epy_normal_50_3=365*cumsum(epd_normal_50_3);
entpy_normal_50_3=365*cumsum(entpd_normal_50_3);
```

```
epd_fit_normal_50_3=fit(age,epd_normal_50_3,'fourier2');
epd_fit_normal_50_3=epd_fit_normal_50_3(age);
entpd_fit_normal_50_3=fit(age,entpd_normal_50_3,'fourier2');
entpd_fit_normal_50_3=entpd_fit_normal_50_3(age);

%%%%% 80% OFF %%%%%%%%%%%%%
%%%%% EFF=0.17 %%%%%%%%%
epd_normal_80_1 =
[800;960;960;1040;1040;1040;1040;1040;1440;1440;...
1440;1440;1440;1760;1760;1760;1760;1760;2000;2000;...
2000;2000;2000;2000;2000;2000;2000;2000;2000;2000;...
1840;1840;1840;1840;1840;1840;1840;1840;1840;1840;...
1840;1840;1840;1840;1840;1840;1840;1840;1840;1840;...
1680;1680;1680;1680;1680;1680;1680;1680;1680;1680;...
1680;1680;1680;1680;1680;1680;1680;1680;1680;...
            1680;1680;1680;1680;1680;1680] * 4.18;

entpd_normal_80_1 = (epd_normal_80_1 / almond_cal) *
Sgen_blood(1) ;
epy_normal_80_1=365*cumsum(epd_normal_80_1);
entpy_normal_80_1=365*cumsum(entpd_normal_80_1);
epd_fit_normal_80_1=fit(age,epd_normal_80_1,'fourier2');
epd_fit_normal_80_1=epd_fit_normal_80_1(age);
entpd_fit_normal_80_1=fit(age,entpd_normal_80_1,'fourier2');
entpd_fit_normal_80_1=entpd_fit_normal_80_1(age);
%%%%% EFF=0.3 %%%%%%%%%
epd_normal_80_2 =
[800;960;960;1040;1040;1040;1040;1040;1440;1440;...
1440;1440;1440;1760;1760;1760;1760;1760;2000;2000;...
2000;2000;2000;2000;2000;2000;2000;2000;2000;2000;...
1840;1840;1840;1840;1840;1840;1840;1840;1840;1840;...
1840;1840;1840;1840;1840;1840;1840;1840;1840;1840;...
1680;1680;1680;1680;1680;1680;1680;1680;1680;1680;...
1680;1680;1680;1680;1680;1680;1680;1680;1680;...
            1680;1680;1680;1680;1680;1680] * 4.18;

entpd_normal_80_2 = (epd_normal_80_2 / almond_cal) *
Sgen_blood(14) ;
epy_normal_80_2=365*cumsum(epd_normal_80_2);
entpy_normal_80_2=365*cumsum(entpd_normal_80_2);
epd_fit_normal_80_2=fit(age,epd_normal_80_2,'fourier2');
epd_fit_normal_80_2=epd_fit_normal_80_2(age);
entpd_fit_normal_80_2=fit(age,entpd_normal_80_2,'fourier2');
entpd_fit_normal_80_2=entpd_fit_normal_80_2(age);
%%%%% EFF=0.42 %%%%%%%%%5
epd_normal_80_3 =
[800;960;960;1040;1040;1040;1040;1040;1440;1440;...
1440;1440;1440;1760;1760;1760;1760;1760;2000;2000;...
2000;2000;2000;2000;2000;2000;2000;2000;2000;2000;...
```

```
1840;1840;1840;1840;1840;1840;1840;1840;1840;1840;...
1840;1840;1840;1840;1840;1840;1840;1840;1840;1840;...
1680;1680;1680;1680;1680;1680;1680;1680;1680;1680;...
1680;1680;1680;1680;1680;1680;1680;1680;1680;1680;...
                1680;1680;1680;1680;1680;1680] * 4.18;


entpd_normal_80_3 = (epd_normal_80_3 / almond_cal) *
Sgen_blood(26);
epy_normal_80_3=365*cumsum(epd_normal_80_3);
entpy_normal_80_3=365*cumsum(entpd_normal_80_3);
epd_fit_normal_80_3=fit(age,epd_normal_80_3,'fourier2');
epd_fit_normal_80_3=epd_fit_normal_80_3(age);
entpd_fit_normal_80_3=fit(age,entpd_normal_80_3,'fourier2');
entpd_fit_normal_80_3=entpd_fit_normal_80_3(age);
%%%%%%%%%%%%%%%%%%%%%%%%%%%%%%%%%%%%%OBESE%%%%%%%%%%%%%%%%%%%%%%%%%
  %%%%%%%%
%%%%%%%%%%%%%%%%%%% 10% OFF %%%%%%%%%%%%%%%%
%%%%%% EFF=0.17 %%%%%%%
epd_obese_10_1 = [
110;132;132;143;143;143;143;143;198;198;...
                198;198;198;242;242;242;242;242;275;275;...
                275;275;275;275;275;275;275;275;275;275;...
                253;253;253;253;253;253;253;253;253;253;...
                253;253;253;253;253;253;253;253;253;253;...
                231;231;231;231;231;231;231;231;231;231;...
                231;231;231;231;231;231;231;231;231;231;...
                231;231;231;231;231;231] * 4.18;


epy_obese_10_1=365*cumsum(epd_obese_10_1);
entpd_obese_10_1 = (epd_obese_10_1 / almond_cal) *
Sgen_blood(1) ;
entpy_obese_10_1=365*cumsum(entpd_obese_10_1);
epd_fit_obese_10_1=fit(age,epd_obese_10_1,'fourier2');
epd_fit_obese_10_1=epd_fit_obese_10_1(age);
entpd_fit_obese_10_1=fit(age,entpd_obese_10_1,'fourier2');
entpd_fit_obese_10_1=entpd_fit_obese_10_1(age);
%%%%%% EFF=0.3 %%%%%%%
epd_obese_10_2 = [
110;132;132;143;143;143;143;143;198;198;...
                198;198;198;242;242;242;242;242;275;275;...
                275;275;275;275;275;275;275;275;275;275;...
                253;253;253;253;253;253;253;253;253;253;...
                253;253;253;253;253;253;253;253;253;253;...
                231;231;231;231;231;231;231;231;231;231;...
                231;231;231;231;231;231;231;231;231;231;...
                231;231;231;231;231;231] * 4.18;


epy_obese_10_2=365*cumsum(epd_obese_10_2);
entpd_obese_10_2 = (epd_obese_10_2 / almond_cal) *
Sgen_blood(14) ;
entpy_obese_10_2=365*cumsum(entpd_obese_10_2);
```

```
epd_fit_obese_10_2=fit(age,epd_obese_10_2,'fourier2');
epd_fit_obese_10_2=epd_fit_obese_10_2(age);
entpd_fit_obese_10_2=fit(age,entpd_obese_10_2,'fourier2');
entpd_fit_obese_10_2=entpd_fit_obese_10_2(age);
%%%%%% EFF=0.42 %%%%%%%
epd_obese_10_3 = [
110;132;132;143;143;143;143;198;198;...
                198;198;198;242;242;242;242;242;275;275;...
                275;275;275;275;275;275;275;275;275;275;...
                253;253;253;253;253;253;253;253;253;253;...
                253;253;253;253;253;253;253;253;253;253;...
                231;231;231;231;231;231;231;231;231;231;...
                231;231;231;231;231;231;231;231;231;231;...
                231;231;231;231;231;231] * 4.18;

epy_obese_10_3=365*cumsum(epd_obese_10_3);
entpd_obese_10_3 = (epd_obese_10_3 / almond_cal) *
Sgen_blood(26) ;
entpy_obese_10_3=365*cumsum(entpd_obese_10_3);
epd_fit_obese_10_3=fit(age,epd_obese_10_3,'fourier2');
epd_fit_obese_10_3=epd_fit_obese_10_3(age);
entpd_fit_obese_10_3=fit(age,entpd_obese_10_3,'fourier2');
entpd_fit_obese_10_3=entpd_fit_obese_10_3(age);
%%%%%%%%%%%%%%%%%% 50% OFF %%%%%%%%%%%%%%%%%
%%%%%% EFF=0.17 %%%%%%
epd_obese_50_1 = [
550;660;660;715;715;715;715;715;990;990;...

990;990;990;1210;1210;1210;1210;1210;1375;1375;...

1375;1375;1375;1375;1375;1375;1375;1375;1375;1375;...

1265;1265;1265;1265;1265;1265;1265;1265;1265;1265;...

1265;1265;1265;1265;1265;1265;1265;1265;1265;1265;...

1155;1155;1155;1155;1155;1155;1155;1155;1155;1155;...

1155;1155;1155;1155;1155;1155;1155;1155;1155;1155;...
                1155;1155;1155;1155;1155;1155] * 4.18;

epy_obese_50_1=365*cumsum(epd_obese_50_1);
entpd_obese_50_1 = (epd_obese_50_1 / almond_cal) *
Sgen_blood(1) ;
entpy_obese_50_1=365*cumsum(entpd_obese_50_1);
epd_fit_obese_50_1=fit(age,epd_obese_50_1,'fourier2');
epd_fit_obese_50_1=epd_fit_obese_50_1(age);
entpd_fit_obese_50_1=fit(age,entpd_obese_50_1,'fourier2');
entpd_fit_obese_50_1=entpd_fit_obese_50_1(age);
%%%%%% EFF=0.3 %%%%%%%
epd_obese_50_2 = [
550;660;660;715;715;715;715;715;990;990;...

990;990;990;1210;1210;1210;1210;1210;1375;1375;...

1375;1375;1375;1375;1375;1375;1375;1375;1375;1375;...

1265;1265;1265;1265;1265;1265;1265;1265;1265;1265;...

1265;1265;1265;1265;1265;1265;1265;1265;1265;1265;...
```

```
1155;1155;1155;1155;1155;1155;1155;1155;1155;1155;...
1155;1155;1155;1155;1155;1155;1155;1155;1155;1155;...
             1155;1155;1155;1155;1155;1155] * 4.18;

epy_obese_50_2=365*cumsum(epd_obese_50_2);
entpd_obese_50_2 = (epd_obese_50_2 / almond_cal) *
Sgen_blood(14) ;
entpy_obese_50_2=365*cumsum(entpd_obese_50_2);
epd_fit_obese_50_2=fit(age,epd_obese_50_2,'fourier2');
epd_fit_obese_50_2=epd_fit_obese_50_2(age);
entpd_fit_obese_50_2=fit(age,entpd_obese_50_2,'fourier2');
entpd_fit_obese_50_2=entpd_fit_obese_50_2(age);
%%%%%% EFF=0.42 %%%%%%%
epd_obese_50_3 = [
550;660;660;715;715;715;715;715;990;990;...
990;990;990;1210;1210;1210;1210;1210;1375;1375;...
1375;1375;1375;1375;1375;1375;1375;1375;1375;1375;...
1265;1265;1265;1265;1265;1265;1265;1265;1265;1265;...
1265;1265;1265;1265;1265;1265;1265;1265;1265;1265;...
1155;1155;1155;1155;1155;1155;1155;1155;1155;1155;...
1155;1155;1155;1155;1155;1155;1155;1155;1155;...
             1155;1155;1155;1155;1155;1155] * 4.18;

epy_obese_50_3=365*cumsum(epd_obese_50_3);
entpd_obese_50_3 = (epd_obese_50_3 / almond_cal) *
Sgen_blood(26) ;
entpy_obese_50_3=365*cumsum(entpd_obese_50_3);
epd_fit_obese_50_3=fit(age,epd_obese_50_3,'fourier2');
epd_fit_obese_50_3=epd_fit_obese_50_3(age);
entpd_fit_obese_50_3=fit(age,entpd_obese_50_3,'fourier2');
entpd_fit_obese_50_3=entpd_fit_obese_50_3(age);
%%%%%%%%%%%%%%%%%% 80% OFF %%%%%%%%%%%%%%%%
%%%%%% EFF=0.17 %%%%%%%
epd_obese_80_1 = [
880;1056;1056;1144;1144;1144;1144;1144;1584;1584;...
1584;1584;1584;1936;1936;1936;1936;1936;2200;2200;...
2200;2200;2200;2200;2200;2200;2200;2200;2200;2200;...
2024;2024;2024;2024;2024;2024;2024;2024;2024;2024;...
2024;2024;2024;2024;2024;2024;2024;2024;2024;2024;...
1848;1848;1848;1848;1848;1848;1848;1848;1848;1848;...
1848;1848;1848;1848;1848;1848;1848;1848;1848;1848;...
             1848;1848;1848;1848;1848;1848] * 4.18;

epy_obese_80_1=365*cumsum(epd_obese_80_1);
entpd_obese_80_1 = (epd_obese_80_1 / almond_cal) *
Sgen_blood(1) ;
```

```
% calculate the cumulative entropy generation by the obese
  person who uptakes 80 % of his food in liquid form

entpy_obese_80_1=365*cumsum(entpd_obese_80_1);

epd_fit_obese_80_1=fit(age,epd_obese_80_1,'fourier2');
epd_fit_obese_80_1=epd_fit_obese_80_1(age);
entpd_fit_obese_80_1=fit(age,entpd_obese_80_1,'fourier2');
entpd_fit_obese_80_1=entpd_fit_obese_80_1(age);

%%%%%% EFF=0.3 %%%%%%%
epd_obese_80_2 = [
880;1056;1056;1144;1144;1144;1144;1144;1584;1584;...
1584;1584;1584;1936;1936;1936;1936;1936;2200;2200;...
2200;2200;2200;2200;2200;2200;2200;2200;2200;2200;...
2024;2024;2024;2024;2024;2024;2024;2024;2024;2024;...
2024;2024;2024;2024;2024;2024;2024;2024;2024;2024;...
1848;1848;1848;1848;1848;1848;1848;1848;1848;1848;...
1848;1848;1848;1848;1848;1848;1848;1848;1848;1848;...
              1848;1848;1848;1848;1848;1848] * 4.18;

epy_obese_80_2=365*cumsum(epd_obese_80_2);
entpd_obese_80_2 = (epd_obese_80_2 / almond_cal) *
Sgen_blood(14) ;
entpy_obese_80_2=365*cumsum(entpd_obese_80_2);
epd_fit_obese_80_2=fit(age,epd_obese_80_2,'fourier2');
epd_fit_obese_80_2=epd_fit_obese_80_2(age);
entpd_fit_obese_80_2=fit(age,entpd_obese_50_2,'fourier2');
entpd_fit_obese_80_2=entpd_fit_obese_80_2(age);
%%%%%% EFF=0.42 %%%%%%%
epd_obese_80_3 = [
880;1056;1056;1144;1144;1144;1144;1144;1584;1584;...
1584;1584;1584;1936;1936;1936;1936;1936;2200;2200;...
2200;2200;2200;2200;2200;2200;2200;2200;2200;2200;...
2024;2024;2024;2024;2024;2024;2024;2024;2024;2024;...
2024;2024;2024;2024;2024;2024;2024;2024;2024;2024;...
1848;1848;1848;1848;1848;1848;1848;1848;1848;1848;...
1848;1848;1848;1848;1848;1848;1848;1848;1848;1848;...
              1848;1848;1848;1848;1848;1848] * 4.18;

epy_obese_80_3=365*cumsum(epd_obese_80_3);
entpd_obese_80_3 = (epd_obese_80_3 / almond_cal) *
Sgen_blood(26) ;
entpy_obese_80_3=365*cumsum(entpd_obese_80_3);
epd_fit_obese_80_3=fit(age,epd_obese_80_3,'fourier2');
epd_fit_obese_80_3=epd_fit_obese_80_3(age);
```

```
entpd_fit_obese_80_3=fit(age,entpd_obese_80_3,'fourier2');
entpd_fit_obese_80_3=entpd_fit_obese_80_3(age);
%% RESULTS
%%%%%%%%%%%%%%%%%%%%%%%%%%%%%%%%%%%%%%%%%%%%%%%%%%%%%%%
disp('Contralateral Analysis Results:')

disp(['Total work for ' num2str(number_of_cycles)...
      ' cycles is ' num2str(abs(W_total)) ' kJ.'])

disp(['Glucose absorbed to the cell is ' ...
    num2str(glucose_absorbed) ' mmol/lt.'])

disp(['Q for blood is ' num2str(Q_blood) ' kJ'])

disp(['Q for oral cavity is ' num2str(Q_oral) ' kJ'])

disp(['X_destroyed for blood is '
num2str(x_destroyed_blood) ' kJ'])

disp(['X_destroyed for oral cavity is '
num2str(x_destroyed_oralc) ' kJ'])

disp(['Expected life is ' num2str(life) ' years'])


%plot entrpy generation versus 2ND LAW Efficiency %
figure;
plot(efficiency,Sgen_blood,'LineStyle','-.','Marker','x','co
  lor','blue','LineWidth',2)
xlabel('2nd law efficiency')
ylabel('Sgen,muscle')
title('Entropy Generation versus 2nd law Efficiency')
ylim([0 5e-5]); % limit of the range of the y axis
grid on
%%%%%%%Work loop%%%%%%%%%%%%%
figure;
plot(Lc,Fc,'k-', 'LineWidth',2.85)
hold on
plot(Li,Fi,'r-','LineWidth',2.85)
xlabel('L (mm)')
ylabel('F (N)')
title('Work Loop')
legend('Contralateral','Ipsilateral','Location','NorthEast');
ylim([0 25]); % limit of the range of the y axis
grid on
% hold on
% plot(L,F,'k-', 'LineWidth',2.85)
% xlabel('L (mm)')
% ylabel('F (N)')
% title('Workloop (Ipsilateral)')
% ylim([-20 50]); % limit of the range of the y axis
```

```
% plot the cumulative Sgen & energy expenditure for the case
  that the person consumes 80 % of its food in liquid form
  and the second law efficiency is =.42

figure
[x,y1,y2]=plotyy(age,epd_fit_normal_80_3,age,entpy_
  normal_80_3);
set(x,'xlim',[0 80],'xtick',[0 1:5:80]);
set(x(1),'ylim',[3e3 1.4e4],'ytick',[3e3:1e3:1.4e4],'ycolor'
  ,'black');
set(x(2),'ylim',[0 550],'ytick',[0:50:550],'ycolor','black');
set(y1,'LineStyle','-','Marker''','o','color','black','LineW
  idth',2);
set(y2,'LineStyle','--','Marker','*','color','red','LineWi
  dth',2);
hold on
[x,y3,y4]=plotyy(age,epd_fit_obese_80_3,age,entpy_
  obese_80_3);
grid on
set(x,'xlim',[0 80],'xtick',[0 1:5:80]);
set(x(1),'ylim',[3e3 1.4e4],'ytick',[3e3:1e3:1.4e4],'ycolor'
  ,'black');
set(x(2),'ylim',[0 550],'ytick',[0:50:550],'ycolor','black');
set(y3,'LineStyle','-.','Marker','x','color','blue','LineWi
  dth',2);
set(y4,'LineStyle',':','Marker','+','color','green','LineWi
  dth',2);
title(' Energy utilization and entropy generation by normal
  and obese people ');
xlabel('Age (years)');
ylabel('Energy (J/day)');
set(get(x(2),'ylabel'), 'string', ' Cumulative Entropy
  (kJ/K)');

legend('Entropy obese','Energy normal','Energy
  obese','Location','West');
legend(y2,'Entropy normal','Location','East');
legend(y2,'Entropy normal','Location','East');
```

When we run the code, Example Figures 4.17.2 through 4.17.4 will appear on the screen:

```
Contralateral Analysis Results:
Total work for 25 cycles is 0.0023027 kJ.
Glucose absorbed to the cell is 0.098411 mmol/lt.
Q for blood is -0.0041982 kJ
Q for oral cavity is 0.0015889 kJ
X_destroyed for blood is 0.003438 kJ
X_destroyed for oral cavity is 0.028474 kJ
Expected life is 76 years
```

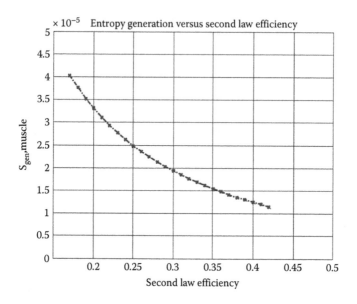

EXAMPLE FIGURE 4.17.2   Entropy generation as a function of the second law efficiency during the chewing process.

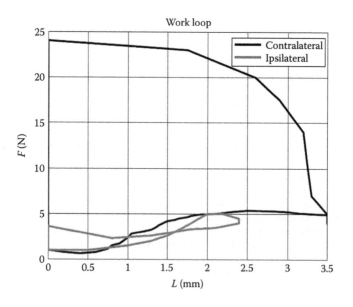

EXAMPLE FIGURE 4.17.3   Work loops of the chewing process.

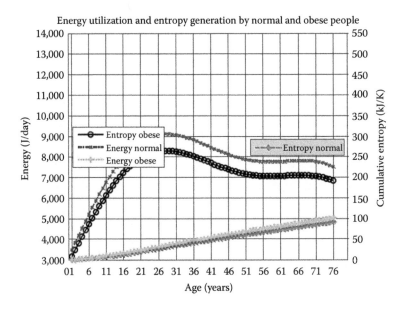

EXAMPLE FIGURE 4.17.4   Energy utilization and entropy generation by normal and obese people.

5

Thermodynamic Assessment of the Industrial Bioprocesses

Today, scarcity of energy resources and carbon dioxide concentration in the atmosphere are considered the two major concerns of humanity. Since the industrial processes are run irreversibly, they cause entropy accumulation in the environment (Lewins, 2011). Utilization of fossil fuels in the industry causes carbon dioxide and greenhouse gas emissions. Since the beginning of the industrial revolution in the 1850s, energy demand is rising, resources are diminishing, and carbon dioxide concentration in the atmosphere is increasing. Consequences are publicly blamed for numerous environmental and climatic adverse observations. Increasing thermodynamic efficiency and decreasing CO_2 emission are essential.

The Kyoto Protocol was accepted in 1997 and entered into force in 2005; Turkish Parliament approved ratification in 2009. Up until July 2010, 191 states had signed or ratified the protocol. Turkey is among the Annex I countries, which agree to reduce their collective greenhouse gas (carbon dioxide, methane, nitrous oxide, sulfur hexafluoride) emissions by 5.2% from the 1990 level in the period of 2008–2012. In most countries, progress toward clean environment is achieved only with the support of the public, which comes only after informing the people. One way of determining the carbon footprint of production processes is to estimate the cumulative CO_2 emission, which is defined as the sum of the total CO_2 emission in each unit operation to produce a product.

To assess the impact of the industrial bioprocesses on the environment thermodynamically, we make use of the mass, energy, and exergy balance equations. Mass balance allows us to estimate the flow rates of chemical species, including the CO_2 emission, thus enabling us to determine the CO_2 footprint of each process. Energy balance allows us to determine, among others, the first law efficiency, heat and work transfer, and the related fuel consumption. Entropy and exergy analyses allow us to pinpoint the entropy generation or exergy loss sources, compare actual systems with the ideal one, and determine the theoretical limits. In industrial production processes, it is also useful to evaluate the cumulative energy consumption (CEnC), cumulative exergy consumption (CExC), and cumulative carbon dioxide emission (CCO_2E).

In the following, several examples of industrial bioprocesses are described to demonstrate the use of the laws of thermodynamics to assess the overall impact of a process to environment.

5.1 CEnC, CExC, and CCO_2E

So far, we have demonstrated the use of the first law of thermodynamics in various processes. The first law analysis states that the total energy change in a system is equal to the sum of all the energy transfers, via heat, work, and mass. Industrial processes can be evaluated via the first law of thermodynamics, by comparing the energy flows to the system via various sources. Another method for the energetic evaluation of industrial production processes is to calculate the CEnC. The total energy used to produce one product is defined as the cumulative energy consumption.

CEnC of a product summarizes the energetic impact of the production method of this product on the environment. CEnC does take into account not only the direct energy consumption but also indirect energy consumptions such as (Szargut et al., 1988) extracting nonrenewable primary energy and raw materials; producing materials and semiproducts; processing primary energy into final energy carriers; transporting raw materials, semiproducts, materials, and energy carriers; and constructing production plants and installations in which the given product (energy carrier) is manufactured.

Szargut et al. (1988) proposed the concept of cumulative exergy consumption, which is the sum of exergy of all resources consumed in all the steps of a production process. *CExC* represents the available work consumed to obtain a product. It involves the exergy cost of raw materials, transportation, work, and heat transfer. Thus, it is a function of the pathway that the process follows and quantifies the total consumption of the exergy, including those of the raw materials, transportation, work, and heat transfer for production. CExC of various fuels and industrial products has been calculated during the last decade (Szargut et al., 1988; Sorgüven and Özilgen, 2010). Some studies started to appear in the literature focusing on the exergy consumption of a single unit (Gungor et al., 2011a,b; Murr et al., 2011) or the entire process (Pellegrini and de Oliveira, 2011) employed in food plants, but the CExC calculations focusing on production of foods from farm to the market are rare in the food industry (Özilgen and Sorgüven, 2011).

The cumulative degree of perfection (CDP) is the ratio of the chemical exergy of the product to the sum of the exergies of all the raw materials and the fuel consumed during production (Szargut et al., 1988):

$$CDP = \frac{(mex)_{product}}{\sum (mex)_{raw\ materials} + \sum (mex)_{fuels}} \tag{5.1}$$

Example 5.1: Calculation of the CDP of Paper

Chemical exergy of paper is 16.5 MJ/kg. We will calculate the CDP of paper when (i) paper is produced from standing timber and (ii) when paper is produced from wastepaper.

To calculate the CDP, we need the CExC values of all the consumed materials and chemical exergy for paper. An extensive list of CExC can be found in Szargut (1988).

Mainly timber and fuel are used to produce paper. A simplified flowchart given in Example Figure 5.1.1 can be prepared to represent the exergy content of the consumed resources.

Accordingly, CDP is 0.19. If production is performed in an integrated plant, where waste products are used as fuels, CDP rises to 0.28, since CExC of the consumed raw materials decreases to 19.0 MJ/kg and CExC of the consumed fuel becomes 40.9 MJ/kg.

If wastepaper is used as raw material, then the exergy cost of the raw materials decreases dramatically as described in Example Figure 5.1.2.

When we use the wastepaper, CDP rises to 0.74. As this example illustrates, low CDP values indicate the need for improved production techniques.

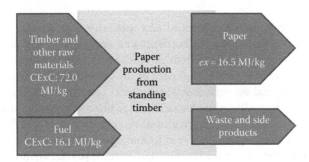

EXAMPLE FIGURE 5.1.1 Exergy chart representing the exergy content of the consumed resources when paper is produced from timber.

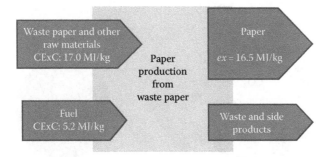

EXAMPLE FIGURE 5.1.2 Exergy chart representing the exergy content of the consumed resources when paper is produced from wastepaper.

5.2 Agriculture

Plants uptake CO_2 and convert it into organic compounds using the solar energy. The environmental effect of photosynthesis is quite different from industrial processes, since a vast portion of the input energy comes from a renewable source. Nonetheless, some nonrenewable resources like fertilizers and microelements are used up during photosynthesis (Figure 5.1). Besides, fossil fuels and electricity are consumed to run agricultural machines or to transport agricultural goods.

Photosynthesis involves a series of reactions, where ATP and $NADPH_2$ are produced and consumed to generate cell components. A summary of this complex set of reactions is given in Example 4.9 and a detailed thermodynamic analysis is provided by Petela (2008). Photosynthesis can be summarized with the following equation:

$$CO_2 + H_2O + \text{Inorganics} \xrightarrow{\text{Solar energy}} \underbrace{\begin{array}{c} \text{Cell debris} + \\ \text{Carbohydrates} + \\ \text{Lipids} + \\ \text{Proteins} \end{array}}_{\text{Biomass}} + O_2 \qquad (5.2)$$

Agriculture involves both biochemical and mechanical processes. The vegetable performs photosynthesis; takes in nutrients and energy from the soil, water, and sun; and grows (Figure 5.2). Hundreds of biochemical reactions occur simultaneously while the vegetable grows. The speed of the growth, size of the ripe vegetable, and its quality depend on various environmental conditions, such as light intensity, temperature, humidity, and chemical composition of the soil. To speed up the growth and enhance the quality, agricultural machinery is employed. Depending on the geographical location and the use of technological aids, quality and quantity of the agricultural goods may vary extensively. Thermodynamic analyses may help to gain insight, pinpoint the least efficient steps, and also assess the overall environmental impact.

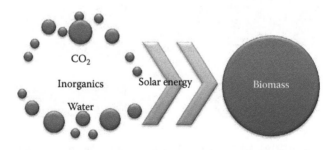

FIGURE 5.1 A simple schematic description of photosynthesis.

FIGURE 5.2 A simple schematic description of the process of agriculture.

The first step of any thermodynamic analysis is, as usual, to define the system and its boundaries. Agriculture occurs in farms, where a continuous interaction between the vegetable and the surroundings occurs. Since we want to determine the overall impact on the environment, we include the hydrosphere, lithosphere, and atmosphere within the system boundaries. All the chemical ingredients, for example, the minerals, which are absorbed from water and soil and integrated into the biomass of the vegetable, flow out of the system boundaries as a part of the vegetable's biomass. Therefore, the energy and exergy of these chemicals are accounted for as nonrenewable resources depleted from the atmosphere, hydrosphere, and lithosphere. These nonrenewable chemicals have to be fed into the system to maintain a steady (or at least quasi-steady) chemical state in the system.

During agriculture, soil is enriched with both chemical fertilizers and manure. To protect the vegetable, agrochemicals (herbicides and insecticides) are used. The fertilizers and pesticides used in agriculture are nonrenewable chemicals and the environmental cost for these raw materials has to be accounted for, just like the minerals. Furthermore, agricultural machinery and, depending on the environmental conditions, a water irrigation system may be employed. Thus, other nonrenewable resources consumed during agriculture include electricity and fuel (Figure 5.2). These can be generated from fossil fuels or renewable resources, and depending on that, CEnC, CExC, and CCO_2E of these nonrenewables will vary dramatically.

5.2.1 Inputs of Agriculture

5.2.1.1 Fertilizers

Chemical fertilizer production is an energy-intensive process. In 1998, 1.2% of the world's energy demand and approximately 1.2% of the total greenhouse gas emissions were associated with fertilizer production; fortunately, energy consumption

TABLE 5.1 Values of Cumulative Energy Consumption and Cumulative Exergy Consumption for the Nitrogenous, Potassium, and Phosphorous Fertilizers and Lime

	Δh_f^0 (MJ/kg)	ex_{ch}^0 (MJ/kg)	CEnC (MJ/kg)	CExC (MJ/kg)	CCO_2E
Nitrogenous fertilizer	66.1 (Houshyar et al., 2010)	3.68 (Szargut et al., 1988)	78.20 (Helsel, 1992); 66.14 (Banaeian et al., 2011)	32.70 (Szargut et al., 1988)	0.09 kg/MJ (Kongshaug, 1998)
Potassium fertilizer	11.2 (Houshyar et al., 2010)	4.40 (Woods and Garrels, 1987)	13.8 (Helsel, 1992); 11.15 (Banaeian et al., 2011)	4.56 (Pimentel, 1991)	0.51 kg/MJ (Kongshaug, 1998)
Phosphorous fertilizer	12.4 (Houshyar et al., 2010)	2.70 (Woods, 1987)	17.5 (Helsel, 1992); 12.44 (Banaeian et al., 2011)	7.52 (Wittmus et al., 1975)	0.15 kg/MJ (Kongshaug, 1998)
Lime		1.96 (Patzek, 2004)	1.33 (Pimentel and Patzek, 2005)	10 (Patzek, 2004)	0.79 kg/kg (Patzek, 2004)

for chemical fertilizer production is decreasing over the years and approaching to the theoretical minimum in modern factories, for example, 40 MJ/kg of nitrogenous fertilizer (Kongshaug, 1998; Anundskas, 2000). The CEnC and CExC values for the nitrogenous, potassium, and phosphorous fertilizers (including packaging, transportation, and application) employed in this thermodynamic analysis are listed in Table 5.1. If the fertilizers were produced in technologically improved factories, then these values would drop.

Organic manure is also employed in agriculture. Ptasinski et al. (2007) give the lower heating value of manure as 7.506 MJ/kg and its chemical exergy as 8.427 MJ/kg.

Example 5.2: Calculation of the Values of CEnC, CExC, and CCO_2E of Organic Manure

Fadere et al. (2010) states that 4,221 kg cow droppings and 12,663 kg market refuse are processed to produce 9,000 kg of organic manure. During the manure production, a total of 0.35 MJ/kg of energy is consumed:

$$CEnC = 0.35 \text{ MJ/kg}$$

Of the total energy consumed, 0.33 MJ/kg is electrical work and the rest is human labor. Szargut calculated the CExC to generate 1 MJ of electricity from fossil fuel as 4.17 MJ/MJ:

$$CExC = Ex_{raw\ materials} + Ex_{fuel} = 8.427 \frac{4221 \text{ kg}}{9000 \text{ kg}} + 0.33 \times \underbrace{4.17 \text{ MJ/MJ}}_{CExC \text{ of electricity}} = 5.33 \text{ MJ/kg}$$

As a result, the CExC of the organic manure is calculated to be 5.3 MJ/kg. Note that the exergy of the market refuse is not taken into account during the CExC calculation, since it is a waste.

The CO_2 emission factor of electricity is 0.14 kg CO_2/MJ. So, the CCO_2E of manure is

$$CCO_2E = 0.33 \text{ MJ/kg} \times 0.14 \text{ kg} CO_2/\text{MJ} = 0.0462 \text{ kg} CO_2/\text{kg}$$

5.2.1.2 Pesticides

If we know the exact chemical formula of the pesticide, then thermodynamic properties such as enthalpy of formation, Gibbs free energy, chemical exergy, or specific heat can be estimated.

Example 5.3: Calculation of the Enthalpy of Formation, Gibbs Free Energy, and the Chemical Exergy of the Insecticide Methomyl and the Fungicide Thiram

This example demonstrates the calculation of the thermodynamic properties of two agrochemicals, that is, the insecticide methomyl and the fungicide thiram.

EXAMPLE TABLE 5.3.1 Atomic Group Contributions Employed for Estimating Δh^0 and Δg^0 of Methomyl and Thiram

Atomic Group	Occurrence	Δh_f^0 (kJ/mol)	Δg_f^0 (kJ/mol)
Methomyl			
$-CH_3$	3	−76.45	−43.96
$-S-$	1	41.87	33.12
$-C=$	1	83.99	92.36
$=N-$	1	23.61	—
$-O-$	1	−132.22	−105.00
$-C=O$	1	−133.22	−120.50
$-NH-$	1	53.47	89.39
		$\Delta h_f^0 = -223.56$ kJ/mol	$\Delta g_f^0 = -88.63$ kJ/mol
Thiram			
$-CH_3$	4	−76.45	−43.96
$-N-$	2	123.34	163.16
$-C=$	2	83.99	92.36
$-S-$	2	41.87	33.12
$=S$	2	−17.33	−22.99
		$\Delta h_f^0 = -226.23$ kJ/mol	$\Delta g_f^0 = 409.34$ kJ/mol

Source: Green, D.W. and Perry, R.H., *Perry's Chemical Engineers' Handbook*, 7th ed., McGraw-Hill, New York, 1997.

EXAMPLE TABLE 5.3.2 Increments to Be Added
for the Estimation of the Standard Chemical Exergy

Molecule	$ex_{ch, element}$/Molecule (kJ/mol)
C	410.25
H_2	236.10
O_2	3.97
N_2	0.72
S	607.05

EXAMPLE TABLE 5.3.3 Estimates of the Standard Chemical Exergies of
Methomyl and Thiram

	$\sum n_i ex_{ch, element}$ (kJ/mol)	Chemical Exergy ex_{ch}^0 (kJ/mol)
Methomyl	3844.0	3755.4
Thiram	6307.0	6716.3

The method of atomic group contributions given in Example Table 5.3.1 are employed to estimate the Δh^0 and Δg^0.

The chemical exergy of the contributing elements and those of methomyl and thiram is given in Example Tables 5.3.2 and 5.3.3:

$$ex_{ch}^0 = \Delta g^0 + \sum_{i=1}^{n} n_i \left(ex_{ch,i}^0 \right)$$

The state-dependent thermodynamic properties can be estimated as demonstrated in the previous example for any chemical species with a known chemical structure. However, CEnC, CExC, and CCO$_2$E are path-dependent properties and therefore cannot be estimated solely based on the chemical structure. Details of the production process have to be known so that all of the mass, heat, and work transfers can be accounted for. For example, Pimentel (1980) calculated the CEnC for a specific herbicide Treflan as the sum of the energy consumed during production (147.2 MJ/kg) and the energy required for packaging and transportation (24.2 MJ/kg), that is, 171.4 MJ/kg. We can find some examples like these, where a specific pesticide production process is analyzed and the CEnC, CExC, and CCO$_2$E values are calculated. There are also some studies that give ranges for CEnC, CExC, and CCO$_2$E for several pesticides with different technological methods. Example Table 5.3.4 provides a summary of the literature survey on the thermodynamic properties and cumulative energy, exergy, and CO$_2$ values of pesticides.

EXAMPLE TABLE 5.3.4 Summary of the Literature Survey on the Thermodynamic Properties and Cumulative Energy, Exergy, and CO_2 Values of Pesticides

	Δh_f^0	ex_{ch}	CEnC (MJ/kg)	CExC (MJ/kg)	CCO_2E
Herbicides					
General		131 MJ/L ton (Hovelius and Hansson, 1999)	198.8 MJ/kg (Banaeian et al., 2011); 418.2 MJ/kg (Polychchronaki et al., 2007)	32.7 MJ/kg (Szargut et al., 1988); 172–564 MJ/kg (Brehmer, 2008)	6.3 ± 2.7 kg/kg (Lal, 2004)
Dicamba	2.3 MJ/kg (calculated with the group contribution method)	18.4 MJ/kg (calculated with the group contribution method)			
Treflan			171.4 MJ/kg (Pimentel, 1980)		
Insecticides					
General		51.4 MJ/L (Hovelius, 1999)	198.8 MJ/kg (Banaeian et al., 2011); 363.8 MJ/kg (Polychchronaki et al., 2007)	7.52 MJ/kg (Wittmus et al., 1975); 21–667 MJ/kg (Brehmer, 2008)	5.1 ± 3.0 kg/kg (Lal, 2004)
Methomyl	1.4 MJ/kg (calculated with the group contribution method)	23.2 MJ/kg (calculated with the group contribution method)			
Fungicides					
General		91.2 MJ/L ton (Hovelius et al., 1999)	198.8 MJ/kg (Banaeian et al., 2011)	4.56 MJ/kg (Pimentel, 1980); 38–474 MJ/kg (Brehmer, 2008)	3.9 ± 2.2 kg/kg (Lal, 2004)
Thiram	0.9 MJ/kg (calculated with the group contribution method)	27.9 MJ/kg (calculated with the group contribution method)			

TABLE 5.2   Values of Cumulative Exergy Consumption and Cumulative Carbon Dioxide Emission for Irrigation

	ex_{ch} (MJ/kg)	CEnC (MJ/kg)	CExC (MJ/kg)	CCO_2E (kg/kg)
Water irrigation	0.0428 (Woods, 1987)	0.00102 (Banaeian et al., 2011)	0.00425	0.000595

5.2.1.3 Water

Electricity is used for water irrigation. Banaeian (2011) states that 0.00102 MJ of electricity is consumed for the irrigation of 1 kg water. Taking the exergy cost of electricity as 4.17 MJ/MJ and CCO_2E as 0.14 kg/MJ, then CExC and CCO_2E of water irrigation are calculated as listed in Table 5.2.

5.2.1.4 Solar Radiation

The plants convert the energy from the sun into chemical energy with photosynthesis, and then the chemical energy is used to fuel the biological activities of the plants, including their growth. Only the light that is in the wavelength range of 400–700 nm (which constitutes about 45% of the total solar energy) may be used in photosynthesis. About 25% of the referred 45% may be absorbed by the plants; the rest is lost by various reasons including reflection. Receiving nonoptimal levels of radiation reduces the efficiency of photosynthesis further; therefore, only 3%–6% of total solar radiation may be used in photosynthesis (Todd, 1982; Miyamato, 1997). Ways to increase the photosynthetic efficiency of the plants are being actively researched to improve their yields, including those of the grain crops (Long et al., 2006). The pollutants such as sulfur dioxide and herbicides (weed killers) inhibit the photosynthesis (Gunsolus, 1999; Appleton et al., 2000). The rate of photosynthesis is affected by the temperature; the optimum is reported as 25°C for the winter wheat (Snyder et al., 2009). Soils deficient in nitrate, magnesium, or iron give rise to chlorophyll-deficient plants and reduce the rate of photosynthesis (Hopkins and Hüner, 2009).

Petala (2010) showed that the energy, entropy, and exergy of photon can be calculated with the following equations:

$$u_{solar} = aT^4$$

$$s_{solar} = \varepsilon \frac{4}{3} aT^3$$

$$ex_{solar} = \frac{a}{3}\left(3T^4 + T_0^4 - 4T_0T^3\right)$$

where
$a = 7.564 \cdot 10^{-16}$ J/(m³K⁴) is universal constant
ε is the emissivity of surface
T is the absolute temperature
T_0 is the environmental temperature (usually $T_0 = 300$ K)

The exergy of the direct solar radiation that passes through the earth's atmosphere is calculated as 1.2835 kW/m² by Petela (2008). Only a fraction of this total exergy is available in the photosynthetically active radiation, which is 0.5155 kW/m².

Example 5.4: Calculation of the Exergy Loss during the Photosynthesis of Microalgae

Microalgae in the form of slurry with 50% lipid content will be extracted from an open pond with 20 g of microalgae per liter at a rate of 60 g/m² day. The cell debris of the algae is described with the following empirical formula:

$$C_{4.17}H_8O_{1.25}N_1P_{0.10}S_{0.03}K_{0.03}Na_{0.04}Ca_{0.01}Mg_{0.02}C_{10.01}Fe_{0.004}$$

Algal lipid contains only docosahexaenoic acid ($C_{69}H_{98}O_6$). First, we need to make a mass balance. During this photosynthesis, CO_2 from the atmosphere and inorganic materials from the water will be extracted and cellular debris and lipid will be formed. Let us assume the inorganics that will make up the cell debris are available in water in the form of $Ca(OH)_2$, KCl, $MgSO_4$, Na_2SO_4, $FeCl_2$, NH_4OH, and H_3PO_4. The third column of Example Table 5.4.1 (mass flow rate) presents the results of the mass balance.

EXAMPLE TABLE 5.4.1 Details of the Exergy Analysis

Input	MW (kg/ kmol)	Mass Flow Rate g/(m² day)	ex_{ch} kJ/(m² day)	$RT_0 \ln (y)$ kJ/(m² day)	$ex = ex_{ch} + RT_0\ln(y)$kJ/ (m² day)	ex/(Total Exergy Input) (%)
CO_2	44.0	24.03	0.00	0.00	0.00	
H_2O	18.0	486.63	20.80	−0.15	20.65	
$Ca(OH)_2$	74.1	0.04	0.05	−0.01	0.04	
KCl	74.6	0.11	0.02	−0.04	−0.02	
$MgSO_4$	120.4	0.12	0.03	−0.03	0.01	0.08
Na_2SO_4	142.0	0.14	0.02	−0.03	−0.01	
$FeCl_2$	126.8	0.03	0.03	−0.01	0.03	
NH_4OH	35.0	1.76	16.48	−0.78	15.70	
H_3PO_4	98.0	0.49	0.40	−0.11	0.30	
Sunlight			44,539.20		44,539.20	99.92
Total mass input		*513.35*	*Total exergy input*		*44,575.90*	
Output	kg/kmol	g/(m² day)	kJ/(m² day)	kJ/(m² day)	kJ/(m² day)	
Debris	99.8	5.00	142.09	−0.78	141.31	0.77
Lipid	1023.5	5.00	201.86	−0.10	201.75	
O_2	32.0	22.16	0.00	0.00	0.00	
HCl	36.5	0.05	0.07	−0.03	0.03	0.05
H_2SO_4	98.1	0.05	0.08	−0.01	0.07	
H_2O	18.0	481.09	20.56	−0.14	20.42	
Total mass output		*513.35*	*Total exergy output*		*363.58*	

Chemical exergy of the inorganics are found from literature (exergoecology), and the chemical exergy of the debris and lipid is calculated based on the group additivity method (Sorgüven and Özilgen, 2010). The last four columns of Example Table 5.4.1 show the results of the exergy analysis.

Most of the exergy flowing into the system comes from solar energy (i.e., 99.92%). Exergy flowing in due to inorganics and water is negligible. Converted into biomass (i.e., cell debris + lipid) is 0.77% of the total inlet exergy. A small part is flowing out of the system as water solution (0.05%). The rest of the total inlet exergy is lost.

This example demonstrates the role of the solar exergy in photosynthesis. Since the exergy content of the solar energy is very large, taking the solar exergy into account makes the contribution of all the other ingredients appear negligibly small. Therefore, a common practice is to disregard the solar energy contribution to be able to compare the energetic and exergetic effect of the non-renewables adequately.

5.2.1.5 Electricity and Fuel

Electricity and fuel consumed by the agricultural machinery should be taken into energy and exergy analyses as well. Traditionally both electricity and fuel used in farming are obtained from fossil resources, which have a large CExC. In recent years, the number of farms employing renewable energy sources increases. Producing electricity from wind turbines or solar panels reduces its exergy cost dramatically. Similarly, exergy cost of fuels can be reduced by producing fuel from vegetable or algal oil or via pyrolysis of wastes.

Example 5.5: Energy and Exergy Utilization in the Agriculture of 1000 Ton of Olives, Soybeans, and Sunflower Seeds

Example Figure 5.5.1 shows a simple system to investigate the agricultural process. Chemical fertilizers, water (either from natural sources like rain or from an electrically driven irrigation system), chemical pesticides, and diesel (including utilization for machine work) are the inputs of the system. And the vegetable is the only output of the system.

Note that as we have discussed in Example 5.4, the solar energy will not be included in the energy and exergy balance, because it is a renewable resource and its value is too large that makes the contribution of the other inputs seem negligibly small.

Data on the agriculture of olives (Polychchronaki et al., 2007), sunflowers (Kallivroussis et al., 2002), and soybeans (De et al., 2001) are found in the literature. The extracted data include the amounts of fertilizers and pesticides consumed, the amount of the agricultural product harvested, whether water irrigation is employed or not, and the amount of electricity and diesel oil used. The amount of the harvest per unit area of the allocated land varies drastically. For example, only 500 kg of olive can be harvested from 1 ha land, whereas up

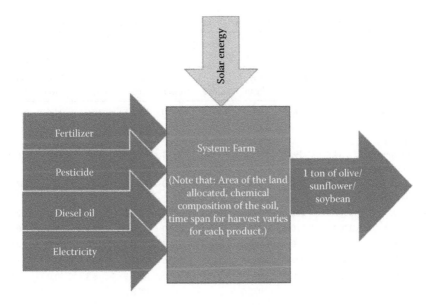

EXAMPLE FIGURE 5.5.1 Simplified flow chart of the olive, sunflower, and soybean agriculture.

to 2700 kg of soybean can be harvested from the same land area. We have normalized all the data per 1000 kg of harvest. Results of this literature survey are summarized in Example Table 5.5.1.

There are differences in the agricultural techniques. For example, water irrigation is only employed for olives. Another example is that farmyard manure is used in the production of soybeans. Example Table 5.5.2 summarizes results of the energy, exergy, and CO_2 emission analyses. Comparison of CEnC, CExC, and CCO_2E of olive, sunflower, and soybean agriculture is presented in Example Figure 5.5.2.

EXAMPLE TABLE 5.5.1 Amounts of Fertilizers, Pesticides, Diesel Oil, and Electricity Consumed to Produce 1000 kg of Olives, Sunflowers, and Soybeans

	Olive	Sunflower	Soybean
Nitrogenous fertilizer (NH_4NO_3) (kg)	2.8	33.3	11.4
Phosphorous fertilizer (P_2O_5) (kg)	0.8	16.7	0.0
Potassium fertilizer (K_2O) (kg)	0.8	0.0	64.4
Herbicides (kg)	4.8	0.6	0.4
Insecticides (kg)	3.6	0.0	0.0
Fungicides (kg)	0.0	0.0	0.0
Diesel oil (kg)	85.4	35.6	36.5
Electricity (MJ)	60.4		356.1

EXAMPLE TABLE 5.5.2   Results of the Energy, Exergy, and CO_2 Emission Analyses

	Energy Utilization (MJ)			Exergy Consumption (MJ)			CO_2 Emission (kg)		
	Olive	Sun-flower	Soy-bean	Olive	Sun-flower	Soy-bean	Olive	Sun-flower	Soy-bean
Nitrogenous fertilizer (NH_4NO_3)	222.1	2606.7	888.7	92.9	1090.0	371.6	20.0	234.6	80.0
Phosphorous fertilizer (P_2O_5)	14.4	291.7	0.0	6.2	125.3	0.0	2.2	43.8	0.0
Potassium fertilizer (K_2O)	10.5	0.0	888.7	3.5	0.0	293.7	5.3	0.0	453.2
Herbicides	2007.4	238.1	150.2	1766.4	318.8	132.2	30.2	5.6	2.3
Insecticides	1309.7	0.0	0.0	1238.4	0.0	0.0	18.4	0.0	0.0
Fungicides	0.0	0.0	0.0	0.0	0.0	0.0	0.0	0.0	0.0
Diesel oil	4908.8	2042.7	2096.6	4545.5	1891.6	1941.4	80.3	33.4	34.3
Electricity	60.4	0.0	356.1	251.9	0.0	1484.9	8.5	0.0	49.9
Seed	—	—	972.6	—	—	1358.1	—	—	36.4
Manure	—	—	719.8	—	—	511.0	—	—	100.8
Total	8533.2	5179.2	6072.7	7904.7	3425.7	6092.9	164.9	317.4	756.8

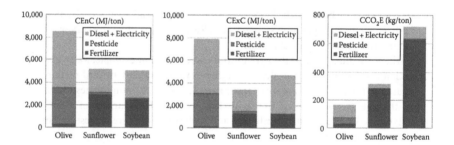

EXAMPLE FIGURE 5.5.2   Comparison of cumulative energy consumption, cumulative exergy consumption, and cumulative carbon dioxide emission of olive, sunflower, and soybean agriculture.

Among the three investigated vegetables, olive is the most energy- and exergy-intensive agricultural product. But surprisingly, CO_2 emitted during its production is very low.

If we compare the amounts of the consumed fertilizer, then we see that nearly the same amount of fertilizers are consumed for sunflower and soybean. However, in the agriculture of soybean, a large portion of the fertilizers is potassium based, which has the highest CCO_2E. Thus, in this comparison, the overall CCO_2E depends mostly on the amount of the potassium fertilizer consumed.

5.2.2 Effect of Different Farming Methods on the Overall Energy and Exergy Consumption

Chemical fertilizer use in agriculture is generally not based on soil analysis and is much more than the actual need (Esengun et al., 2007). Introducing some microorganisms to the soil may stimulate nitrogen fixation, solubilize the insoluble minerals, and reduce the need to produce and transport large amounts of chemical fertilizers (Aslantas et al., 2007). Fertilizer management practices may reduce energy utilization up to 72%; consequently, herbicide utilization and pollution also decrease (Clements et al., 1995; Hülsbergen et al., 2001; Snyder et al., 2009; Ren et al., 2010).

Example 5.6: How do the CEnC, CExC, and CCO₂E Values Change Based on Different Agricultural Techniques?

In order to demonstrate how different agricultural techniques change the results, CEnC, CExC, and CCO₂E for soybean production are calculated based on four different data (Pimentel, 1991; Sheehan et al., 1998a; De et al., 2001; Fore et al., 2011). Energy and exergy consumption and carbon dioxide emission associated with seeds are omitted in this comparison. Results are summarized in Example Figures 5.6.1 through 5.6.3 and Example Table 5.6.1.

The total amount of fertilizers consumed is reported as 17.4 kg/ton by Fore et al. (2011), 43.4 kg/ton by Sheehan et al. (1998b), and 75.8 kg/ton by De et al. (2001). Pimentel and Patzek (2005) reported that 4800 kg/ha of lime is used in addition to the nitrogenous, phosphorous, and potassium fertilizer utilization, which is 3.7, 37.8, and 14.8 kg/ha, respectively. According to their data, nearly half of the total consumed energy (15,658 MJ/ha) comes from lime (5,639 MJ/ha). In spite of this rather large consumption of fertilizers, the total

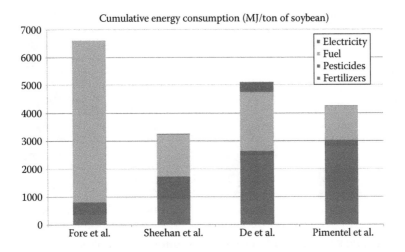

EXAMPLE FIGURE 5.6.1 Comparison of cumulative energy consumptions in soybean production.

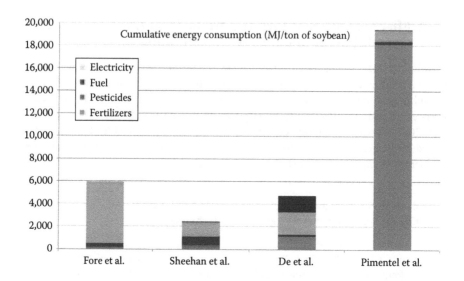

EXAMPLE FIGURE 5.6.2 Comparison of cumulative exergy consumptions in soybean production.

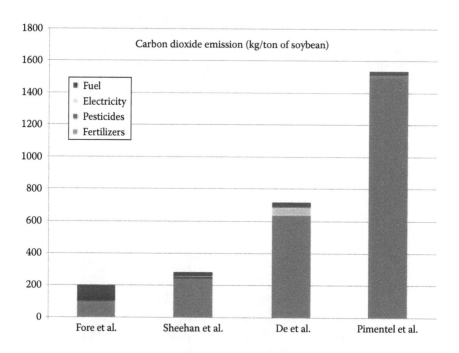

EXAMPLE FIGURE 5.6.3 Comparison of carbon dioxide emission in soybean production.

EXAMPLE TABLE 5.6.1 Raw Materials and Fuels Consumed to Produce 1 Ton of Soybean

	Fore et al. (2011)	Sheehan et al. (1998)	De et al. (2001)	Pimentel and Patzek (2005)
Nitrogenous fertilizer (kg)	1.79	4.58	11.36	1.39
Phosphorous fertilizer (kg)	5.85	14.37	0.00	14.17
Potassium fertilizer (kg)	9.79	24.46	64.40	5.55
Lime (kg)	0.00	0.00	0.00	1799.10
Herbicides (kg)	1.06	1.88	0.36	0.49
Diesel (kg)	100.81	18.62	36.49	13.19
Gasoline (kg)	0.00	10.00	0.00	10.52
Electricity (MJ)	0.00	16.91	356.10	13.49

energy consumption is comparable to other data sets, since diesel and electricity consumptions are comparatively less than the rest of the published data sets. However, exergy consumption and carbon dioxide emission values are effected dramatically. Lime production is an energy-intensive process, where 10 MJ of exergy is consumed and 0.79 kg of CO_2 is emitted per kg of lime (Pimentel and Patzek, 2005). The CExC value is calculated based on Pimentel and Patzek (2005) to be 19,427 MJ/ton of soybean, where 93.5% of the total exergy consumption is due to fertilizers. The CO_2 emission per ton of soybeans calculated based on Pimentel and Patzek (2005) is 1534 kg, where 98.3% of this is due to fertilizer consumption. Example Table 5.6.1 and Example Figures 5.6.1 through 5.6.3 draw attention to the point about excessive fertilizer utilization as criticized by Esengun et al. (2007).

This example shows that the CEnC, CExC, and CCO_2E values may vary in a quite large range depending on the agricultural practices. For soybeans, CEnC values vary between 3,261 and 6,612 MJ/ton, CExC values between 2,428 and 19,427 MJ/ton, and CCO_2E values between 198 and 1,534 kg/ton.

Example 5.7: Calculation of CEnC, CExC, and CCO_2E for the Agriculture of Strawberries

Values of CEnC, CExC, and CCO_2E will be calculated for the agriculture of strawberries based on the data given in Example Table 5.7.1; we will ignore the energy, exergy, and CO_2 emission related to the seed production.

Strawberry production can be simplified as shown in the flowchart (Example Figure 5.7.1). Raw strawberries, which are harvested very early in the morning, are put into the reusable plastic crates and transported with no cooling. We assumed that heavy-duty trucks with a 10 ton capacity are employed to transport raw strawberries from the farm to a cleaning facility, which is 30 km away. The strawberry cases are unloaded from the trucks manually and stems are removed on the sorting table. Strawberry cleaning machine (Model JY X700, Zhengzhou Jun Yu Trade Co., Ltd., Zhengzhou,

EXAMPLE TABLE 5.7.1 Inputs of the Strawberry Agriculture When the Growth Yield Is 64.2 Ton/(ha Year)

Fertilizer			3369.2 kg/(ha Year)
	Nitrogenous (NH_4NO_3)	18%	
	Phosphorous (P_2O_5)	55%	
	Potassium(K_2O)	27%	
Microelements			51.0 kg/(ha Year)
	Fe	92%	
	B	2%	
	Mn	4%	
	Zn	2%	
Pesticides			42.9 kg/(ha Year)
	Herbicides	33%	
	Insecticides	33%	
	Fungicides	33%	
Diesel oil consumed during agriculture			10,976.1 kg/(ha year)
Electricity consumed during agriculture			11,076.4 MJ/(ha year)

Source: Banaeian, N. et al., *Energy Convers. Manage.*, 5, 1020, 2011.

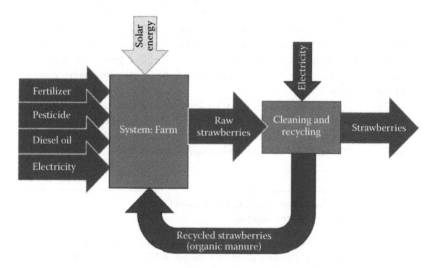

EXAMPLE FIGURE 5.7.1 Flowchart of strawberry agriculture.

Henan, China; capacity = 500 kg/h) utilizes 5.4 MJ of electricity per ton of strawberries. The sorting machine (Jiadi, China, Model JD-JG-5; capacity = 5 ton/h) utilizes 1.1 MJ of electricity per ton of strawberries. About 5% of the strawberries, which are inappropriate for processing, are ground in an industrial waste grinder (Tsingtao Donghao Plastics Machinery Co., Ltd., Qingdao, Shandong, China; capacity = 500–2000 kg/h; energy utilization = 324 MJ/h) and recycled to the fields in 50 g polylactic acid (PLA) bags as organic manure.

EXAMPLE TABLE 5.7.2 Energy and Exergy Consumed and CO_2 Emitted during Production of 1 Ton of Strawberry

Activity	CEnC (MJ/ton)	CExC (MJ/ton)	CCO_2E (kg/ton)
Consumption of chemical fertilizers	1,540.0	598.7	516.6
Agrochemicals, such as pesticides, herbicides and fungicides	132.8	215.4	3.4
Diesel consumption	9,822.4	9,095.5	136.4
Electric power consumption	172.5	719.5	24.2
Transportation of the strawberries to the factory	41.2	38.1	0.6
Cleaning machinery	5.4	22.5	0.8
Sorting machinery	1.1	4.6	0.2
Recycling of the strawberries	23.0	88.3	3.2
Total	11,738.4	10,782.6	685.4

Energy and exergy consumed and CO_2 emitted during the agriculture of strawberry are listed in Example Table 5.7.2.

About 86% of the total energy and 92% of the total exergy consumed during the strawberry production originate from fossil fuels; the rest is contributed by nonrenewable chemicals (fertilizers, pesticides, and micronutrients).

Example 5.8: Calculation of the CDP of Strawberries

Chemical composition and the chemical exergy of each constituent of the strawberries are described in Example Table 5.8.1. The CExC of 1 ton of strawberries is estimated as 10,782.6 MJ.

EXAMPLE TABLE 5.8.1 Chemical Composition and the Chemical Exergy of Each Constituent of the Strawberries

Chemical Composition of Strawberry (%)		Estimation Method	ex_{ch} (MJ/kg)	$y_l R T_0 \ln (y_l)$
Carbohydrate	8.30	In strawberries, carbohydrate is found as a mixture of glucose, sucrose, and fructose. The chemical exergy value used here is the chemical exergy of sugar as listed in Szargut (1988).	25.95	−0.0005
Protein	0.80	Assume all proteins are polymers of alanine, and calculate based on group additivity method.	25.35	−0.0001
Fat	0.50	Assume fat is based on oleic acid, and calculate based on group additivity method.	37.14	−0.0001
Ash	0.50	There are no available data on the chemical exergy of ash in literature. Let us neglect the contribution of ash, since it makes up only a very small portion of strawberries.	0.0	−0.0001
Water	89.90	Woods (1987).	0.043	−0.0002

The chemical exergy of strawberries is calculated with the given data:

$$(\text{Chemical composition of strawberry})\,(ex_{ch}) = 2.58 \text{ MJ/kg}$$

The CDP of strawberries is then

$$\text{CDP} = \frac{2.58 \text{ MJ/kg}}{10.78 \text{ MJ/kg}} = 0.24.$$

Example 5.9: Calculation of the CEnC, CExC, and CCO₂E Values of the Sugar Beet Agriculture

Mass flow rates of the nonrenewable inputs for sugar beet production are listed in Example Table 5.9.1 and the flowchart is shown in Example Figure 5.9.1. Assuming that 5% of the sugar beet is found inappropriate for processing, it is ground and recycled to the field as fertilizer. Sugar beets are transported to the factory at an average distance of 60 km with heavy-duty trucks (Roy et al., 2008). Since the trucks are empty, or carrying the recycled beets only, while going back to the fields, calculations are performed for the two-way trip. Results are summarized in Example Table 5.9.2.

The diesel and the chemical fertilizers have large shares of energy and exergy utilization in the agriculture of both strawberry and sugar beet (Example Table 5.9.2). The diesel utilization may be reduced by using energy-efficient trucks

EXAMPLE TABLE 5.9.1 Inputs to Produce Sugar Beet with a Yield of 60.8 Ton/(ha Year)

Fertilizer			480.3 kg/(ha Year)
	Nitrogenous (NH_4NO_3)	53%	
	Phosphorous (P_2O_5)	33%	
	Potassium (K_2O)	13%	
Organic Manure			9100.0 kg/(ha Year)
	Hemicellulose	21%	
	Cellulose	25%	
	Lignin	13%	
	Protein	12%	
	Ash	9%	
Pesticides			0.60 kg/(ha Year)
	Herbicides	18%	
	Insecticides	25%	
	Fungicides	57%	
Diesel oil consumed during agriculture			141.9 kg/(ha year)
Electricity consumed during agriculture			1884.8 MJ/(ha year)

Source: Adapted from Erdal, G. et al., *Energy*, 32, 35, 2007.

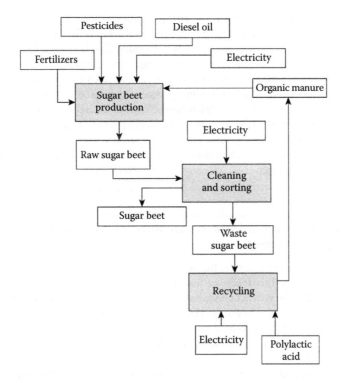

EXAMPLE FIGURE 5.9.1 Flowchart of sugar beet agriculture and sugar production.

EXAMPLE TABLE 5.9.2 Inputs Energy and Exergy Consumed and CO_2 Emitted to Produce 1 Ton of Sugar Beet

Activity	CEnC per Ton of Sugar Beet (MJ/Ton)	CExC per Ton of Sugar Beet (MJ/Ton)	CCO_2E per Ton of Sugar Beet (kg/Ton)
Consumption of chemical fertilizers	434.1	2317.8	61.5
Consumption of chemicals (pesticides and manure)	2.0	3.0	0.0
Diesel consumption	134.1	124.2	1.9
Electric power consumption	31.0	129.3	4.3
Transportation of the sugar beet to the sugar factory	164.6	152.4	2.4
Cleaning, sorting, and recycling of the sugar beet	34.3	135.5	4.7
Total	800.1	2862.2	74.8

and optimizing delivery plans. The energy consumption decreased in modern chemical fertilizer factories over the years and approached to the theoretical minimum (Kongshaug, 1998; Anundskas, 2000). Providing fertilizers from the energy-efficient plants may contribute to energy savings. The chemical fertilizer use in agriculture is generally not based on soil analysis and much more than the actual need (Esengun et al., 2007). Introducing some microorganisms to the soil may stimulate nitrogen fixation, solubilize insoluble minerals, and reduce the need to produce and transport large amounts of chemical fertilizers (Aslantas et al., 2007). Successful fertilizer management practices may reduce energy utilization up to 72%; consequently, herbicide utilization and pollution also decrease (Clements et al., 1995; Hülsbergen et al., 2001; Snyder et al., 2009; Ren et al., 2010).

Example 5.10: Comparison of the Agriculture of Wheat and Rye in Turkey and in Germany

Similar to previous analyses, the system boundaries involve the atmosphere, hydrosphere, and lithosphere so that the soil in which the wheat or rye seeds grow is within the system boundaries. All the nonrenewables (including chemicals and fuels) are fed to the system, and the amount of the consumed nonrenewables are determined based on a literature survey. Example Table 5.10.1 lists the amounts of the nonrenewables consumed to produce wheat or rye in 1 ha of land area.

Even though there are only small differences between the agricultural inputs in Turkey and in Germany, the yield per 1 ha of allocated land area differs immensely (Oren and Ozturk, 2006; Toprak Mahsulleri Ofisi, 2014; United States Department of Agriculture, 2014). As seen in Example Figure 5.10.1, both grain and straw yields in Turkey are much lower than in Germany. In Germany, 3.3 times more wheat grain and 2.3 times more rye grain are harvested than in Turkey.

Example Table 5.10.2 lists the energy and exergy flowing into and out of the system boundaries. The values listed are calculated by multiplying the enthalpy or chemical exergy values of the species with the mass flow rate. Comparing the input values for Turkey and Germany shows only small differences. But since the yield in Germany is considerably larger, the energy and exergy content of the product in Germany is also much larger than the product obtained in Turkey. The large difference in the CDP values demonstrates this fact.

Example Table 5.10.3 shows the results of the cumulative energy, exergy, and CO_2 emission analyses.

In Example 6.6, we have compared soybean production at different sites, where the amount of the yield per land area varies between 1977 and 2737 kg/ha. So the variation in the soybean yield was less than 40%. The main difference in the CEnC, CExC, and CCO_2E values came about because of the large variation in the fertilizer consumption. Here, we see that even though similar amounts of fertilizers and pesticides are used and similar agricultural techniques are employed, amount of the yield may change drastically. The answer to the question remains to be unanswered: "Why wheat and the rye production in Germany is more efficient?" It may possibly be due to the genetic characteristics

EXAMPLE TABLE 5.10.1   Nonrenewable Inputs of the Wheat and Rye Agriculture in Turkey and Germany

	Wheat		Rye	
	Turkey	Germany	Turkey	Germany
Diesel oil (L/ha)	165.6 (Tipi et al., 2009)	165.6 (Tipi et al., 2009)	124.3 (Houshyar et al., 2010)	124.3 (Houshyar et al., 2010)
Nitrogen fertilizer (kg/ha)	101.88 (Oren and Ozturk, 2006)	145 (Phillips and Norton, 2012)	40 (Erzurum Valiligi, 2013)	98 (Magid et al., 2001)
Phosphorus fertilizer (kg/ha)	72.2 (Oren and Ozturk, 2006)	37 (Phillips and Norton, 2012)	50 (Erzurum Valiligi, 2013)	24 (Magid et al., 2001)
Potassium fertilizer (kg/ha)	—	41 (Phillips and Norton, 2012)	—	44 (Magid et al., 2001)
Herbicide (kg/ha)	1.12 (Everman, 2015)	1.12 (Everman, 2015)	1.12 (Everman, 2015)	1.12 (Everman, 2015)
Insecticide (kg/ha)	0.56 (Penn State College, 2014)	0.56 (Penn State College, 2014)	0.56 (Penn State College, 2014)	0.56 (Penn State College, 2014)
Fungicide (kg/ha)	1.69 (Clemson Cooperative Extension, 2014)	1.69 (Clemson Cooperative Extension, 2014)	1.69 (Clemson Cooperative Extension, 2014)	1.69 (Clemson Cooperative Extension, 2014)
Seed (kg/ha)	227.7 (Oren and Ozturk, 2006)	227.7 (Oren and Ozturk, 2006)	200 (Oren and Ozturk, 2006)	200 (Oren and Ozturk, 2006)
Irrigation water (kg/ha)	1195 (Hellevang, 1995)	6350 (Hellevang, 1995)	1982 (Hellevang, 1995)	4731 (Hellevang, 1995)
Transportation (L diesel/km)	0.287 (Roy et al., 2008)	0.287 (Roy et al., 2008)	0.287 (Roy et al., 2008)	0.287 (Singh, 2002)

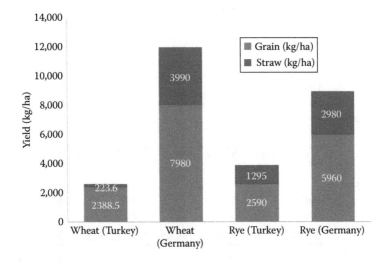

EXAMPLE FIGURE 5.10.1   Wheat and rye yields in Turkey and Germany.

EXAMPLE TABLE 5.10.2   Total Energy and Exergy Input and Output during the Agriculture of the Wheat Grain and Rye Grain in Turkey and Germany

	Turkey		Germany	
	Energy Inflow (MJ/ha)	Exergy Inflow (MJ/ha)	Energy Inflow (MJ/ha)	Exergy Inflow (MJ/ha)
Wheat grain inputs				
Diesel oil	5,842	5,675	5,842	5,675
Nitrogen fertilizer	2,694	150	3,834	213
Phosphate fertilizer	269	227.6	89	117
Potassium fertilizer	—	—	46	1
Herbicide	2.60	20.6	2.60	20.6
Insecticide	0.78	12.8	0.78	12.8
Fungicide	1.01	34.5	1.01	34.5
Seed	3,575	4,508	3,575	4,508
Irrigation	120	60	635	318
Electricity	543	543	1,573	1,573
Total	$\sum (mh)_{in}$: 13,047	$\sum (mex)_{in}$: 11,232	$\sum (mh)_{in}$: 15,598	$\sum (mex)_{in}$: 12,473
Wheat grain outputs				
Grain	35,037	41,950	117,306	140,448
Straw	2,795	474	49,875	8,379
Total	$\sum (mh)_{out}$: 37,832	$\sum (mex)_{out}$: 42,424	$\sum (mh)_{out}$: 167,181	$\sum (mex)_{out}$: 148,827
CDP_{grain}		3.73		11.26
Rye grain inputs				
Diesel oil	4,378	4,255	4,378	4,255
Nitrogen fertilizer	1,058	58.9	2,593	144.3
Phosphate fertilizer	186	157.6	89	75.7
Potassium fertilizer	—	—	51	1.14
Herbicide	2.60	20.6	2.60	20.6
Insecticide	0.78	12.8	0.78	12.8
Fungicide	1.01	34.5	1.01	34.5
Seed	3,140	3,960	3,140	3,960
Irrigation	198	99	473	237
Transportation	6	3.65	6	3.65
Electricity	591	591	1,286	1,286
Total	$\sum (mh)_{in}$: 9,561	$\sum (mex)_{in}$: 9,193	$\sum (mh)_{in}$: 12,020	$\sum (mex)_{in}$: 10,031
Rye grain outputs				
Grain	38,073	45,584	87,612	104,896
Straw	16,188	2,745	37,250	6,258
Total	$\sum (mh)_{out}$: 54,261	$\sum (mex)_{out}$: 48,329	$\sum (mh)_{out}$: 12,862	$\sum (mex)_{out}$: 111,154
CDP_{grain}		4.96		10.46

EXAMPLE TABLE 5.10.3 Values of the Cumulative Energy Consumption, Cumulative Exergy Consumption, and Cumulative Carbon Dioxide Emission as Calculated for Wheat and Rye Grain Agriculture in Turkey and Germany

	Turkey			Germany		
	CEnC (MJ/kg)	CExC (MJ/kg)	CCO$_2$E (kg/kg)	CEnC (MJ/kg)	CExC (MJ/kg)	CCO$_2$E (MJ/kg)
Wheat grain	5.4	6.7	0.3	1.9	2.7	0.2
Rye grain	3.5	5.4	0.2	1.8	2.4	0.2

of the seed varieties and that one type of seed is better adapted to the environmental conditions (temperature, humidity, chemical composition of the soil) and therefore performs photosynthesis at a higher efficiency. Nevertheless, thermodynamic analysis allows us to pinpoint the root of the inefficiencies in the agricultural process.

5.2.3 Use of Plants as Fuel Source

Traditionally, vegetable were consumed only as food. But today, because of the fossil fuel crisis, some vegetable oils are also used as biodiesel (Peiro et al., 2010, a,b). Usually biodiesels are called *renewable* energy resources. But renewability has many aspects. Renewability can be thought in terms of thermodynamic, social, economic, and environmental aspects, which include—among other criteria—net carbon dioxide emission, competition with the food supply, and waste management (Cramer et al., 2006; Lobo, 2007; Worldwatch Institute, 2007; Goldemberg et al., 2008). Some energy sources are found to be energetically, socially, and economically renewable, although the exergy required to restore the environmental damages emerged during the production of the energy source is larger than the useful work produced by the energy source. A good example for this is bioethanol. Production of ethanol from corn gives a positive energy balance, emits no net carbon dioxide to the atmosphere, and creates new workplaces. Hence, it is considered as a renewable energy source. However, the exergy analysis of the use of ethanol from corn as a biofuel shows that the process is not renewable, since exergy loss is due to the consumption of nonrenewable materials; resource processing and waste treatment are larger than the exergy produced. Exergy consumed during the process is nearly five times larger than the available work produced by ethanol (Berthiaume et al., 2001; Yang et al., 2009).

5.2.4 Renewability Assessment

A renewable energy source should not destroy exergy. This means that the total work used to produce the energy source plus the total work used to restore environment to its initial state has to be smaller or equal to the work produced by this energy source. Fossil fuels, for example, are nonrenewable, since the work required for the environmental restoration is much larger than the produced work. During the production of biofuels, however, nonrenewable inputs are limited. Thus, these are closer to full renewability. In the following, the methodology to quantify renewability in terms of exergy is explained.

The hydrosphere, lithosphere, and atmosphere act as water and carbon reservoirs. Nutrient-rich water is consumed during agriculture. Vegetables extract the necessary minerals and nutrients from water and integrate those into their own biomass. Therefore, the chemical composition of the outlet water is different. To keep system conditions unchanged, the degraded nonrenewable chemicals have to be recycled. Similarly, the water employed in all stages of the overall process needs to be fully recycled, too. Carbon dioxide used during photosynthesis is released back to the atmosphere during the work production step, which may be, for example, the combustion process in a diesel engine. Hence, there is no mass transfer through the system boundaries, and the system remains at constant chemical composition (Figure 5.3).

Berthiaume et al. (2001) extended Szargut's definition of CExC by defining the cumulative net exergy consumption (*CNEx*) and the restoration work (w_r). CNEx is the difference between the CExC and the exergy content of the main product (ex_p):

$$CNEx = CExC - ex_p$$

The restoration work (w_r) represents the net exergy cost to produce an energy source, without causing any environmental damage, like water and air pollution. It is defined as the sum of the CNEx during the production process and waste treatment:

$$w_r = CNEx_{production} + CNEx_{waste\ treatment}$$

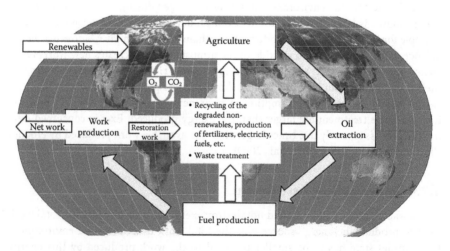

FIGURE 5.3 Schematic representation of the phenomena occurring in the hydrosphere, lithosphere, and atmosphere.

Accordingly, renewability indicator, I, is defined as

$$I_r = \frac{(w_p - w_r)}{w_p}$$

where w_p is the useful work obtained by the product. If the maximum work potential of the product is extracted via reversible processes, then w_p equals to ex_p.

Example 5.11: Calculation of the Cumulative Net Exergy Consumed to Produce Gasoline in a Petroleum Refinery

The chemical exergy of gasoline (ex_p) is 35.6 MJ/kg. If the total amount of the exergy consumed (including the chemical exergy of raw materials, electricity, and fuels used during the process) to produce gasoline in a petroleum refinery (*CExC*) is 42.4 MJ/kg, then *CNEx* is 6.8 MJ/kg (Berthiaume et al., 2001). In other words, 42.4 MJ of available energy is consumed to obtain a product, which has 35.6 MJ work potential.

Example 5.12: Is Biodiesel Produced from Microalgae a Renewable Energy?

The system has to include the atmosphere, lithosphere, and hydrosphere as described in the previous section. As shown in Example Figure 5.12.1, only a few nonrenewable chemicals and solar energy are entering the system boundaries. All the chemicals that can be recycled (here carbon dioxide, water, ethanol, etc.) are recycled and fed into the unit processes. The work required for this recycling is taken from the work produced during the combustion of the biodiesel in the diesel engine.

Mass, energy, and exergy analyses are performed for each unit process. Algae are grown in a glass pond via photosynthesis. The cellular inorganic chemicals needed for the photosynthesis are provided by adding the following salts (per 100 kg of debris produced) into the pond: makeup water, 72 kg; Ca(OH)$_2$, 2 kg; KCl, 6 kg; MgSO$_4$, 1 kg; NaH$_2$PO$_4$, 4.7 kg; FeCl$_2$, 1 kg; NH$_4$Cl, 16 kg. Photosynthesis is simplified as a one-step reaction:

$$4.17CO_2 + 0.04NH_4^+ + 4H_2O \rightarrow C_{4.17}H_8O_{1.25}N_{0.04} + \text{Cellular Inorganics}$$

Algae grow in the pond and mature cells flocculate, settle in continuously, and are fed into an expeller press. In the expeller press, algal oil is separated from cell debris. Algal oil is pumped to a reactor.

The dominant chemical reaction occurring in the reactor is

$$C_{69}H_{98}O_6 + 6C_2H_5OH \rightarrow 3C_{24}H_{36}O_2 + C_3H_8O_3 + 3C_2H_5OH$$

The rest of the processes in the biodiesel production is separation and recycling processes. Example Table 5.12.1 summarizes the consumed exergy to produce algal biodiesel.

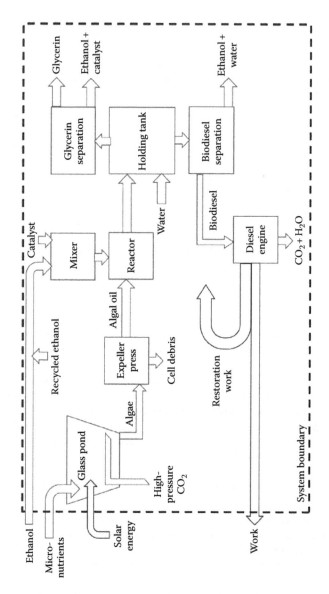

EXAMPLE FIGURE 5.12.1   Sketch of the overall system.

EXAMPLE TABLE 5.12.1 Summary of the Exergy Consumption for Algal Biodiesel Production

Input		CNEx
$Ca(OH)_2$	7.4 kg	1,395 kJ/kg
KCl	22.2 kg	190 kJ/kg
$MgSO_4$	3.7 kg	271 kJ/kg
NaH_2PO_4	17.4 kg	351 kJ/kg
$FeCl_2$	3.7 kg	1,259 kJ/kg
NH_4Cl	59.2 kg	6,059 kJ/kg
Compressor work	25,685 kJ of electricity	4.17 kJ/kJ
Water recycling	4,006 kJ of electricity	4.17 kJ/kJ
Press work	170,600 kJ of electricity	4.17 kJ/kJ
Transportation	32.5 kg of diesel fuel	53,200 kJ/kg
Ethanol	129 kg	26,400 kJ/kg
Exergy destroyed	402,519 kJ of electricity	4.17 kJ/kJ
Water recycling	153,258 kJ of electricity	4.17 kJ/kJ
Ethanol recycling	30,527 kJ of electricity	4.17 kJ/kJ
Catalyst recycling	349 kJ of electricity	4.17 kJ/kJ
Total exergy consumed to produce 1,000 kg of algal biodiesel		8,802,864 kJ

Biodiesel produced from algal oil is combusted in a diesel engine to produce work. The first law efficiency of a diesel engine cycle is a function of the compression ratio, r; cutoff ratio, r_c; and specific heat ratio, k (Çengel and Boles, 2007):

$$\eta_{th} = 1 - \frac{1}{r^{k-1}}\left(\frac{r_c^k - 1}{k(r_c - 1)}\right)$$

The exergy balance for the diesel cycle is (Sorgüven and Özilgen, 2010)

$$ex_{destroyed} = T_0(s_{out} - s_{in}) - \oint \frac{\delta q_{in}}{T_{in}} + \oint \frac{\delta q_{out}}{T_{out}}$$

where

q_{in} and q_{out} are the inlet and outlet heat transfers, respectively
T_{in} and T_{out} are the corresponding boundary temperatures

In a cyclic process, the first term in the right-hand side is zero, since thermodynamic properties at the inlet and outlet are the same. During the heat release, exhaust gas is ejected to the surroundings. In the following calculations, the boundary temperature is assumed to remain constant at the environmental temperature during the heat rejection process. Heat is added into the system due to the combustion of the biodiesel. Through the combustion process, system temperature increases. If heat is assumed as a linear function of the temperature, then δq_{in} can be represented as

$$\delta q_{in}(T) = \int_{T-dT}^{T+dT} a\, dT$$

where parameter a is the proportionality constant. The entropy increases due to the heat addition process, which occurs across a finite temperature difference, which becomes

$$\int_{T_2}^{T_3} \frac{\delta q_{in}}{T_{in}} = 2a \ln\left(\frac{T_3}{T_2}\right) = 2a \ln(r_c)$$

Finally, the destroyed exergy in the cycle can be represented as a function of the heat input, compression ratio, cutoff ratio, and specific heat ratio:

$$ex_{destroyed} = \left(\frac{q_{out}}{T_{out}} - 2a \ln(r_c)\right) = q_{in} \frac{1}{kr^{k-1}(r_c-1)}\left(r_c^k - k \ln r_c - 1\right)$$

In the operating conditions for a regular diesel engine, values of the compression ratio, cutoff ratio, and specific heat ratio can be approximated as $r = 15$, $r_c = 5$, and $k = 1.4$. Thus, the first and the second law analyses result in (Sorgüven and Özilgen, 2010)

$$\eta_{th} = 0.56$$

and

$$ex_{destroyed} = 0.38 q_{in}$$

To produce 1000 kg of algal biodiesel, 8,802,864 kJ of exergy is consumed, and when the produced biodiesel is burned in a diesel engine, then a work of 12,042,584 kJ can be produced (Sorgüven and Özilgen, 2010). These results show that the algae–biodiesel–carbon dioxide cycle provides a positive amount of useful work, which is (12,042,584−8,802,864) kJ/1,000 kg = 3,239 kJ/kg. The renewability indicator is 0.27. This means that nearly three-fourth of the work potential of algal biodiesel is used for its production and to restore the environment. Thus, the net available work gain from this process is about one-fourth of the work produced by the algal biodiesel. The renewability indicator can be increased, if electricity from renewable sources (e.g., hydropower) is used, or the lipid content of the algae is enhanced. Genetic engineering techniques may be helpful to improve this efficiency drastically.

Example 5.13: Which Processes Are Limiting the Overall Efficiency in Algal Biodiesel Production? How Can We Improve the Efficiency?

In the previous example, the overall system is analyzed, and the renewability is estimated based on the inputs and outputs of the overall system. To analyze the process in detail and to determine the efficiency-limiting steps, each unit operation should be analyzed separately. This example demonstrates how to estimate the exergetic efficiency of each unit operation.

The exergy balance equation for a steady process with several inlets and outlets can be formulated as

$$ex_{destroyed} = \sum \left(\sum_{i=1}^{N} n_i ex_i \right)_{in} - \sum \left(\sum_{i=1}^{N} n_i ex_i \right)_{out} - q\left(1 - \frac{T_0}{T}\right) - w$$

where

n is the number of species in an inlet or outlet stream
ex_i is the exergy of the ith species in a mixture

$$ex_i = \underbrace{ex_i^{ch}}_{\substack{chemical \\ exergy}} + \underbrace{h_{f,i}^0 + c_{p,i}(T - T_0)}_{enthalpy} - T_0 \underbrace{\left[s_i^0 + c_{p,i} \ln\frac{T}{T_0} - R_u \ln x_i \right]}_{entropy}$$

Note that in this equation the exergy values of inlet and outlet streams are employed, rather than the CExC.

The exergy balance equation shows us that exergy can be destroyed due to

1. Mass transfer

$$ex_{destr, mass} = \sum \left(\sum_{i=1}^{N} n_i ex_i \right)_{in} - \sum \left(\sum_{i=1}^{N} n_i ex_i \right)_{out}$$

2. Heat transfer

$$ex_{destr, Q} = -\sum q\left(1 - \frac{T_0}{T}\right)$$

3. Work transfer

$$ex_{destr, Q} = -w$$

Mixing is an irreversible process. The minimum work required for separation equals to $ex_{destr, mass}$. If we assume a second law efficiency of 70% for each separation process, then the results can be summarized as shown in Example Table 5.13.1.

These results show that most of the exergy destruction (67%) occurs in the reactor. In the reactor, irreversibilities arise due to the uncontrolled and fast chemical reactions. The second largest exergy destruction source is the holding tank,

EXAMPLE TABLE 5.13.1   Exergy Destruction in Each Unit Operation

	$ex_{destr,\,mass}$ (kJ)	$ex_{destr,\,Q}$ (kJ)	$ex_{destr,\,W}$ (kJ)	$ex_{destroyed}$ (kJ)	Contribution of the Component in the Overall Exergy Destruction (%)
Mixer	1,426	0	0	1,426	0.75
Reactor	−305,063	52,311	380,357	12,7605	67.49
Holding tank	52,541	853	0	53,394	28.24
Biodiesel separation	−14,378	0	20,541	6,162	3.26
Glycerin separation	−1,134	0	1,621	486	0.26
Total exergy destroyed				189,074	

where the biodiesel and glycerin are separated. Here, exergy is mainly destroyed due to the change in the composition of the inlet and outlet streams. Variations in the inlet water temperature, operating temperature, and the compositions of the outlet streams may decrease the total exergy loss in this process and, hence, increase the second law efficiency of the unit. The mixer and the separation units make a small contribution to the total exergy loss.

Note that this example demonstrates the basic principles of exergy analysis and the calculations are performed for a simplified theoretical plant. Only the main processes, which are the dominant sources of exergy destruction, are considered. Auxiliary utilities such as pumps, pipelines, and heat exchangers are not taken into account. Furthermore, in an actual plant, there will be deviations from the ideal operation conditions, which will cause incomplete reactions, side products, incomplete separation, heat losses, etc. If an actual plant is to be analyzed, then temperature, pressure, and chemical composition of each stream should be measured and $\sum_{i=1}^{N} n_i ex_i$ for inlet and outlet streams should be calculated based on these measurements.

5.2.5 Improvements Based on Exergy Analysis

It is worth to mention the effects of the algal type and its lipid content on the efficiency of the algal biodiesel cycle. Algae consist of nucleic acid, proteins, carbohydrates, and lipids in varying proportions. Lipids are extracted from the harvested algae and converted into algal oil. Algal oil contains unsaturated fatty acids, like arachidonic acid, eicosapentaenoic acid, docosahexaenoic acid, and linoleic acid. The chemical exergy of lipids containing these fatty acids is listed in Table 5.3. Proportional to the molecular size of a fatty acid, its chemical exergy increases. Lipid content of algae varies in a large range. For instance, the lipid content of *Scenedesmus quadricauda* is only 1.9% of its dry weight, whereas *Scenedesmus dimorphus* can have a lipid content of up to 40%. It is possible to genetically modify algae to increase their lipid content. In order to demonstrate the increase in the total chemical exergy of algal oil with respect to algae's lipid content, the total chemical exergy per 1 kg of algae is calculated as a function of lipid content (Table 5.4). The linear relationship between the lipid content and the total exergy shows

TABLE 5.3 Exergy Content of Lipids Found in Different Algal Species

Lipids with Three Branches of	Molecular Formula of the Fatty Acid	Molecular Weight of the Lipid (kg/kmol)	e_{ch} (kJ/kg)
Linoleic acid	$C_{18}H_{32}O_2$	879	39,930
Arachidonic acid	$C_{20}H_{32}O_2$	921	40,150
Docosahexaenoic acid	$C_{22}H_{32}O_2$	1023	40,352

TABLE 5.4 Exergy Content of Algal Oil per Mass of Algae

Lipid Content of Algae (on Dry Matter Basis) (%)	Total Chemical Exergy of Algal Oil (kJ/kg Algae)
10	4,035
20	8,070
40	16,141
80	32,281

that if the lipid content of algae could be doubled by genetic modifications, the amount of algal oil required to produce the same amount of work would be halved.

5.3 Livestock

Thermodynamic sustainability of livestock farms is considerably variable depending on the technology employed. Livestock farming techniques may have a large impact on the ecosystems, biodiversity, and human health (Gilchrist et al., 2007; Hoffman et al., 2011; Gerber et al., 2013). It is estimated that livestock farms contribute to about 14% of the world's total greenhouse gas emission and about 64% of the total NH_3 emission (Gerber et al., 2013). The energy and exergy cost of livestock farms is affected by almost all the contributing factors of the processes such as the composition of the feed given to the cattle (Koknaroglu, 2008).

Figure 5.4 shows the main inputs and outputs of a livestock farm. The production method and composition and amount of the feed have an impact both on the energy and exergy consumption in the system and on the health, growth, and weight of the animals. There are also a large number of studies investigating the effect of animal diet on the CH_4, N, and ammonia emission (Hristov et al., 2013; Doreau et al., 2014).

The amount of diesel fuel and electricity consumed depends mainly on the employed technology and amount of the irreversibilities within the farm. Careful examinations of the energy and exergy flow for each operation within the farm can pinpoint the inefficient processes and sources of losses. Producing the fuel and electricity from renewable resources such as solar or wind energy reduces the CEnC and CExC.

Depending on the type of the farm, the main products include either meat or animal products (such as milk and cream). During livestock farming, there results also a considerable amount of waste, which has the potential of creating a number of environmental problems, if not handled with care. Contamination of rice with antibiotics originating from the urine of cattle reaching to the fields through the underground water is

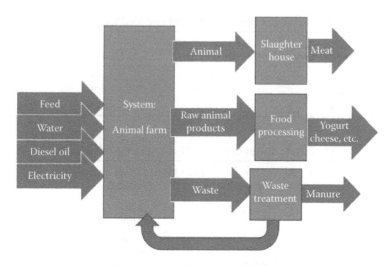

FIGURE 5.4   Schematic representation of the animal farm with its inputs and outputs.

reported by Hawker et al. (2013). In the past, livestock manure has caused serious water and air pollution problems (Cowling and Nilsson, 1995; Reda, 1996; Johnsen et al., 2001). Generally, 2–15 m^3 water/ton of live carcass is used in a slaughterhouse (Lattouf and Oliveira, 2003).

In their review paper, Loyon et al. (2016) discussed different farming techniques and outlined the best available technology for livestock farming in Europe. Based on the applied farming techniques, energy efficiency, exergy destruction rate, and greenhouse gas and methane emissions will change. The following two examples demonstrate the thermodynamic analysis of livestock farming.

Example 5.14: Meat Production

Example Figure 5.14.1 summarizes the steps in meat production. The first step is livestock farming, where cows are fed 9.4 kg of concentrate and 3.1 kg of roughage daily for 210 days (Senel, 1986; Coskun et al., 1997). An average 210-day-old cattle is assumed to weigh 500 kg with a body composition of 10% fat and 19.7% protein. We will calculate the exergy cost of producing 1 kg of beef.

Heavy-duty trucks with a 10 ton capacity, which consumed 0.25 L/km of fuel, are employed to transport the cows from the dairy farm to the slaughterhouse for 300 km. The density of diesel oil is about 0.832 kg/L with an energy equivalence of 57.45 MJ/L and its carbon dioxide emission factor is 0.94 $kg\,CO_2$/kg of diesel oil (Lal, 2004). Eleven cows are transported to the slaughterhouse per heavy-duty truck. An average slaughtering plant has a capacity to process 400 cows per day operated for 300 days in a year. The carcasses are kept in a refrigerated storehouse for 24–36 h to reduce the moisture loss and metabolic and enzymatic activities. Then the carcasses are cut with electric saws in refrigerated work environments at about 12°C. The capacity of meat cutting line

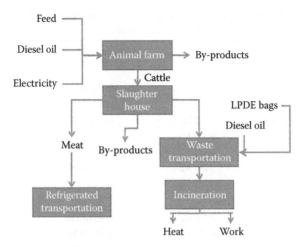

EXAMPLE FIGURE 5.14.1 Summary of the steps in meat production.

is 15 ton/h. The energy used for grinding and freezing varies between 1.3 and 1.8 MJ/kg meat, for storage between 5 and 25.5 MJ/kg meat, for transportation between 2.69 and 6.6 MJ/kg meat, and for slaughtering and cutting into pieces between 2.5 and 15.5 MJ/kg meat (Carlsson- Kanyama and Faist, 2000).

Example 5.15: Milk Production

We will calculate the exergy cost of producing 1 kg of cow milk from the annual energy requirements in a diary farm as summarized in Example Table 5.15.1. Feed and diesel are the inputs needed for the raw milk production process, and the by-products are calves and organic manure. The energy utilized for the transportation includes that consumed for the transportation of the milk to the laboratory for analysis and that used in the factory for processing and for the transportation of the feed from the manufacturer to the dairy farm. The energy utilization for dairy operations includes those of preparing the feed, feeding, inspection, veterinary care, waste removal, and milking and those of water and direct electric consumption. Each cow gives birth to one calf with an average mass of 40 kg during this period (Koknaroglu, 2010).

EXAMPLE TABLE 5.15.1 Annual Energy Utilization and Milk Yield in a Dairy Farm

Feed	13,596 MJ/cow
Dairy operations	2,322 MJ/cow
Transportation	1,035 MJ/cow
Machinery	1,017 MJ/cow
Average lactation	4,498.6 kg milk/cow

Source: Adapted from Table 1 of Sorgüven, E. and Özilgen, M., *Energy*, 40, 214, 2012.

EXAMPLE TABLE 5.15.2 Calculation of the Exergy of the Inputs (Feed) and the Outputs (Raw Milk) of the Raw Milk Production Process

	Flow Rate	Chemical Exergy	Total Exergy (MJ)
Inputs			
Concentrate	0.478 kg	18.7 MJ/kg	8.937
Roughage	0.493 kg	18.7 MJ/kg	9.225
Green chopped forage	0.106 kg	18.7 MJ/kg	1.987
Diesel oil	0.389 MJ	44.4 MJ/kg	17.268
Electricity	0.292 MJ	1.0 MJ/MJ	0.292
Natural gas	0.292 MJ	42.7 MJ/kg	12.455
Inputs total			50.164
Raw milk ingredients (main output)			
Protein	0.033 kg	25.35 MJ/kg	0.837
Fat	0.036 kg	37.14 MJ/kg	1.337
Carbohydrate	0.047 kg	17.49 MJ/kg	0.822
Water	0.884 kg	0.04 MJ/kg	0.038
Raw milk total	1.000 kg		3.034

EXAMPLE TABLE 5.15.3 Calculation of the Chemical Exergy of the Organic Manure

	Composition	Chemical Exergy (MJ/kg)
Hemicellulose	0.209	16.56
Cellulose	0.249	18.54
Lignin	0.129	25.00
Protein	0.119	25.35
Water	0.294	0.04
Organic manure		14.34

Source: Adapted from Table A1 of Sorgüven, E. and Özilgen, M., *Energy*, 40, 214, 2012.

The results show that 50.164 MJ of exergy is consumed to produce 1 kg of raw milk, which has a chemical exergy of 3.034 MJ (Example Table 5.15.2).

Note that one of the side products of this process is the organic manure. Assuming that organic manure is composed of hemicellulose, cellulose, lignin, protein, and water, its chemical exergy can be calculated as the sum of chemical exergies of the components, which is 14.34 MJ/kg (Example Table 5.15.3).

Cows produce 0.069 kg of organic manure per kg of raw milk. This means that during the production of 1 kg of raw milk, the side product organic manure carries 0.99 MJ of exergy out of the system.

Example 5.16: Egg Production

In Example Table 5.16.1, energy utilization data for the production of eggs by caged chicken are adapted from Wiedemann and McGahan (2011) and the data for the production of the eggcups are adapted from Table 3 of Zabaniotou and Kassidi (2003). The eggcup packages are made of recycled paper by using natural gas. An average egg weighs 60 g; six eggs are put in a 12 cm × 18 cm × 6 cm eggcup, and 120 eggcups are placed in a 50 cm × 56 cm × 62 cm carton. The energy cost of palletizing is negligible when compared to those of the production of eggs and packaging in the eggcups and cartons (Example Figure 5.16.1).

EXAMPLE TABLE 5.16.1 Energy Utilization for Packaged Egg Production

	Energy Utilization (MJ/Eggs)		
	Natural Gas	Electric Power	Diesel
Egg production	4,280	2140	4,280
Eggcups	13,779	158	7,333
Cardboard box production	51	6	—
Total	18,110	2304	11,613

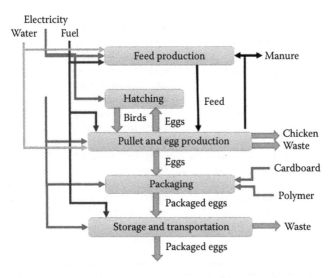

EXAMPLE FIGURE 5.16.1 Process flow diagram of the packaged egg production process.

5.4 Food Processing

Processing of each food may differ substantially from the others; the same food may also be processed through numerous steps. We will discuss the thermodynamic aspects of food processing with a few specific examples. The specific cumulative energy utilization (CEnC), specific cumulative exergy utilization (CExC), and the specific CCO_2E values for some of the inputs of the food industry are given in Table 5.5. We will discuss the

TABLE 5.5 Cumulative Energy Consumption, Cumulative Exergy Consumption, and Cumulative Carbon Dioxide Emission Values for the Inputs of the Food Industry

	Specific CEnC	Specific CExC	Specific CCO_2E
Animal feed	2.8 MJ/kg (Koknaroglu, 2010)	18.7 MJ/kg (Jorgensen et al., 2005)	0.23 kg/kg[a]
Diesel	57.5 MJ/kg (Banaeian et al., 2011)	53.2 MJ/kg (Szargut et al., 1988)	0.94 kg/kg (Lal, 2004)
Natural gas	50.0 MJ/kg (Szargut et al., 1988)	48.7 MJ/kg (Szargut et al., 1988)	0.06 kg/MJ (PAS, 2008)
Electricity	1.0 kJ/kJ (Szargut et al., 1988)	4.17 kJ/kJ (Szargut et al., 1988)	0.14 kg/MJ (PAS, 2008)
LDPE	48 MJ/kg (Saygin et al., 2009)	46.5 (Dewulf and van Langenhove, 2004)	2.1 (http://eco2gobox.com/pdfs/ECO2GO_Report_FINAL.pdf)
HDPE	3.2 MJ/kg (Gielen and Tom, 2010)	86.0 MJ/kg (Dewulf and van Langenhove, 2004)	0.45 kg/kg[b]
PP	1.8 MJ/kg (Gielen, 2010)	85.2 MJ/kg (Dewulf and van Langenhove, 2004)	0.25 kg/kg[b]
PS	0.9 MJ/kg (Gielen, 2010)	91.9 MJ/kg (Dewulf and van Langenhove, 2004)	0.12 kg/kg[b]
PLA	54.0 MJ/kg (Vink et al., 2003)	78.0 MJ/kg based on the data given by de Swaan Arons et al. (2004)	1.8 kg/kg (Vink et al., 2003)
PVC	57 (Gielen, 2010)	19.7 MJ/kg (Dewulf and van Langenhove, 2004)	1.8 kg/kg (http://www.pvc4pipes.com/sustainability/lci-data)
Corrugated board	43.3 (calculated based on the data from Chow et al., 2005)	16.5 MJ/kg (Lattouf and Oliveira, 2003)	1.17 kg/kg (calculated based on the data from Chow et al., 2005)
PET	83.8 MJ/kg (http://www.smart-ecofilms.com/index.php?nav_id=7&unav_id=7)	86.8 MJ/kg (Dewulf and van Langenhove, 2004)	2.33 kg/kg (Kirsen, 2008)
Organic manure	0.35 MJ/kg	5.33 MJ/kg	0.0462 kg/kg

[a] Calculated assuming that 45% of the energy for the feed production comes from electricity, 30% from diesel oil, and 25% from natural gas.

[b] Calculated assuming that only electrical energy is used for this process.

energy utilization in food processing in Example E5.17, energy and exergy utilization in Example 5.18, and energy savings in Example E5.19 and the impact of packaging materials on the environment in Example 5.20.

Example 5.17: Energy Utilization in Wine and Brandy Making

A schematic flow diagram of wine and brandy production process is described in Example Figure 5.17.1 and the energy utilization for the production of each input or in each production stage of wine and brandy is described in Example Tables 5.17.1 and 5.18.1 (adapted from Özilgen, submitted).

Wine bottle caps were made of 5 g of high-density polyethylene (HDPE). Each six-bottle beer package was wrapped with 0.22 m² HDPE (thickness 0.5 mm) after placing 0.07 m² of cardboard to the bottom. The pallets had a size of 120 cm × 100 cm; five layers of packages are placed on each pallet; there were 32 packages

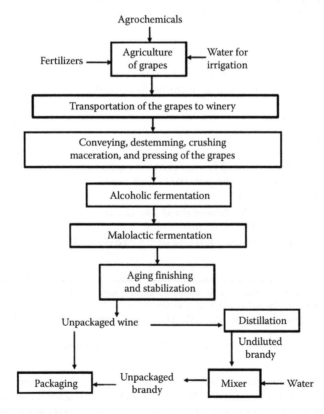

EXAMPLE FIGURE 5.17.1 Schematic description of the wine and brandy production processes.

EXAMPLE TABLE 5.17.1   Energy Utilization during Production of Wine

Agriculture of the grapes	MJ/Ton Wine
Chemical fertilizers	1,722
Agrochemicals	150
Diesel oil	2,426
Electricity	6,656
Agriculture total	10,954
Wine making	
Agriculture of the grapes (3.28 tons of grapes is needed to produce 1 ton of wine)	35,929
Transportation of 3.28 tons of grapes for 50 km	449
Wine making; energy utilization for fork trucks, sterilizing and cleaning, heating, cooling ventilation, bottling (including compressor and disgorging, pumping, filtration and mixing, and grape processing [3.5 kJ/L wine, 30% N, 70% E, ρ_{wine} = 0.975 kg/L])	1,458
Packaging and transportation	
Bottle making	5,342
Filling the bottles with Shanghai Jiadi (model BRl–UHT–CH–1SJ, China) filling machine, 2500 bottles/h, 4.5 kW	9
Labeling	3
Paper for labels	6
HDPE caps	21
Cardboard	847
Box making and printing with Shanghai Liu Xiang (China) carton maker, 250 cartons/min, 135 kJ/s (including dryer)	43
HDPE for palletizing	36
Palletizing with Dalian Jialin (model JT-1200, China) palletizing machine, 20 cartons/min, 9 kW	36
Transportation of the wine to the market in refrigerated trucks for 100 km	164
Wine total	44,373

Sources: Data for the agriculture of the grapes adapted from Ozkan, B. et al., *Energy*, 32, 1500, 2007; data for a winery producing 10,000–50,000 bottles of wine/year adapted from Smyth, M. and Nesbitt, A., *Energy Sustain. Dev.*, 23, 85, Figure 9, 2014.

Note: 3.28 tons of grapes are needed to produce 1 ton of wine.

in each layer. There was one layer of cardboard at the top and the bottom of each layer; the pallet is wrapped with HDPE. Nine bottles of wine are placed in a 34 cm × 25 cm × 25 cm carton, and then six layers of cartons were placed on a pallet; there are 16 cartons in one layer.

EXAMPLE TABLE 5.17.2 Energy Utilization during Production of Brandy

	MJ/Ton Brandy
Agriculture of grapes plus wine making (7 tons of wine after deducting the energy utilized for packaging and marketing is used to produce 1 ton of brandy)	265,062
Transportation of the grapes to the winery	3,143
Distillation (adapted from Cortella and Da Porto, 2003)	1,066
Undiluted brandy total	269,271
Mixing 1000 L of ingredients for 10 min in Shanghai Jiadi mixer (model JDGH-1000, China), 11 kW (springwater is used for bringing the alcohol content to 40%, at no energy cost)	6
Unpackaged brandy total	107,711
Packaging	
Same type of packaging being used with wine	6,343
Transportation of the brandy to the market	122
Brandy total	114,176

As we see from Example Tables 5.17.1 and 5.17.2 brandy production requires 2.25-fold energy to that of wine.

Example 5.18: Energy and Exergy Utilization and Carbon Dioxide Production during Olive Oil Production Process

Let us analyze the agriculture of soybean, sunflower, and olive oil. Globally, 220.9 million tons of soybean oil (56% of production), 9.0 million tons of sunflower oil (7% of production), and 2.9 million tons of olive oil (less than 1% of production) were produced in 2008 (Özilgen and Sorgüven, 2011). In addition to their role in nutrition, soybean and sunflower oils are raw materials for biodiesel production. Olive oil is a delicacy and healthy food. Each of these vegetable oil production processes involves both biochemical and mechanical processes (Example Table 5.18.1).

In the farm, the renewable input is solar power, and the nonrenewable inputs involve chemical fertilizers, agrochemicals (insecticides, herbicides, and fungicides), irrigation water, diesel, and electric power (Example Figure 5.18.1). The mechanical processes begin with the transportation of the harvested vegetable from the farm to the factory. In the factory, the common steps in the vegetable oil production are cleaning, grinding, and pressing; separation, refining, and packaging of the oil; and bottling, boxing, palletizing, and shrink-wrapping of the final product (Example Figure 5.18.2 and Example Tables 5.18.2 and 5.18.3).

EXAMPLE TABLE 5.18.1 Energy and Exergy Consumption and CO_2 Emission associated with Olive Oil Production Process

Processing Step and Equipment Details	Capacity (Ton of Olives/h)	Energy Consumption (MJ/h)	Energy Utilization for Processing 1 Ton of Olives (MJ/Ton)	CExC for Processing 1 Ton of Olives (MJ/Ton)	CO_2 Emission during Processing 1 Ton of Olives (kg/Ton)
Agriculture			8,533.2	7,904.7	164.9
Olive carrier conveyor (Polat Machinery, Turkey)	6	Motor—4.0	0.7	2.9	0.1
Olive washing unit (Polat Machinery, Turkey)	6	Aspirator—10.9 3.6	1.8 0.6	7.5 2.5	0.3 0.1
Olive feeding screw conveyor (Siemens)	4	5.4	1.4	5.8	0.2
Crushing machine (PMS 370-Polat Machinery, Turkey)	3	69.3	23.1	96.3	3.2
Modular mixing (Q4 100L 4A, Siemens)	0.3	23.8	79.2	330.3	11.1
Oil press machine (Anyang GEMCO Energy Machinery, China; Model YZS-120)	0.3	39.6	132	550.4	18.5
Decanter (centrifugation) (PX 40-Polat Machinery, Turkey)	5	69.9	14.0	58.4	2.0
Pumping (*Siemens, inventor*)	4	7.9	2.0	8.3	0.3
Screw conveying (PD-22-Polat Machinery, Turkey)	3	5.4	1.8	7.5	0.3
Separating (PMS 405-Polat Machinery)	2	19.8	9.9	41.3	1.4
Oil pumping (Polat Machinery, Turkey)	2 tons of olive oil/h	5.4	0.7	2.9	0.1

(Continued)

EXAMPLE TABLE 5.18.1 (*Continued*) Energy and Exergy Consumption and CO_2 Emission associated with Olive Oil Production Process

Processing Step and Equipment Details	Capacity (Ton of Olives/h)	Energy Consumption (MJ/h)	Energy Utilization for Processing 1 Ton of Olives (MJ/Ton)	CExC for Processing 1 Ton of Olives (MJ/Ton)	CO_2 Emission during Processing 1 Ton of Olives (kg/Ton)
Heat exchanger system (PMS DB 80 Polat Machinery, Turkey)	3 tons of olive oil/h	6.5	0.5	2.1	0
Olive oil filling machine (Jiangsu, Model ZP 7, China)	1000 bottles/h	11.2	1.4	5.8	0.2
Bottle pasteurizer (Zhangjiagang City Nanxin Technology, China)	Steam ($P = 0.6$ MPa) consumption rate, 600 kg/h		470	458.2	28.2
Labeler (Shanghai Peiyu Machinery, Ltd, China; *Model* SPC-SORL-TL)	5,000–30,000 bottles/h	24	0.1	0.4	0.0
Packaging			535.5	1,076.5	31.9
Transportation of olive oil to 550 km			222.1	206.6	3.0
Waste management	Batch process		−1.4	−5.8	57.3
	Continuous process		−2.8	−11.7	114.4
Total			10,028.6	10,762.6	323.1

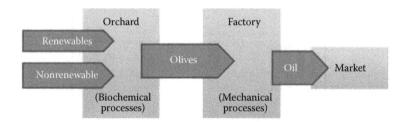

EXAMPLE FIGURE 5.18.1 Flow chart of the olive agriculture. Chemical fertilizers, water for irrigation, chemical pesticides, and diesel (including utilization for machine work) are the inputs of the processes.

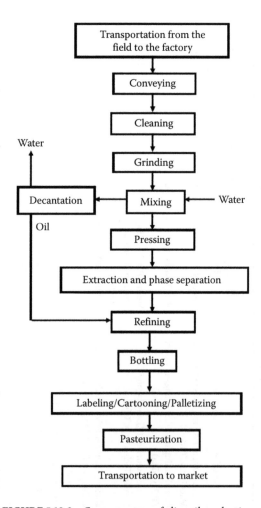

EXAMPLE FIGURE 5.18.2 Common steps of olive oil production processes.

EXAMPLE TABLE 5.18.2   Energy and Exergy Consumption and CO_2 Emission associated with Production of Olive Oil Packaging Materials

Packaging Material	Energy Utilization for Packaging Oil from 1 Ton of Olives (MJ/Ton)	CExC for Packaging Oil from 1 Ton of Olives (MJ/Ton)	CO_2 Emission during Packaging Oil from 1 Ton of Olives (kg/Ton)
Polylactic acid for making bottles	371.3	538.1	12.4
Paper labels for bottles	49.9	55.7	5.0
Cardboard for bottles	8.6	74.3	1.1
Stretch film for secondary packaging	11.9	17.2	0.4
Total	441.7	685.3	18.9

EXAMPLE TABLE 5.18.3   Energy and Exergy Consumption and CO_2 Emission associated with Primary and Secondary Packaging of Olive Oil

Processing Step and Equipment Details	Capacity (Ton of Olives/h)	Energy Consumption (MJ/h)	Energy Utilization for Processing 1 Ton of Olives (MJ/Ton)	CExC for Processing 1 Ton of Olives (MJ/Ton)	CO_2 Emission during Processing 1 Ton of Olives (kg/Ton)
Packaging material			441.7	685.3	18.9
Carton printing and box making machine (Shanghai Liu Xiang, China)	—	—	27.0	112.6	3.6
Carton filling (Shanghai Peifeng Electronics Co., Ltd., China)	15 box/min	36	33.4	139.3	4.7
Palletizing (Dalian Jialin Machine Manufacture Co., Ltd., China)	15–30 cartons/min	36	33.4	139.3	4.7
Total			535.5	1076.5	31.9

In the food industry, studies on energy efficiency began in the late 1970s, substantial improvement was achieved over the years, and the efficiency approached to the theoretical maximum in some sectors, including chemical fertilizer production (Kongshaug, 1998; Anundskas, 2000). The energy efficiency was improving by about 1% every year in the Dutch food industry in the early 2000s (Ramirez et al., 2006a). The ratio of the energy utilization in the Taiwanese food industry to the gross domestic production of the country showed a continuous decline from 2006 to 2010 (Ma et al., 2012). It is estimated that about 5%–15% of the energy utilized at each step of food processing is still being wasted (Klemes et al., 2008). Rigorous studies are being done in the food industry to enumerate the energy utilization (Masanet et al., 2008; Khan et al., 2009; Canning et al., 2010; Therkelsen et al., 2014).

Energy input directly for the production of the food itself offers the largest potential for savings. Technological progress required for higher energy efficiencies goes beyond the suggestions by Masanet et al. (2012) and implies fundamental changes like drying the foods in modern solar facilities, as reviewed by Sharma et al. (2009), instead of relying on the natural gas. In 1998, 1.2% of the world's energy demand and approximately 1.2% of the total greenhouse gas emissions were associated with fertilizer production; fortunately, energy consumption for chemical fertilizer production is decreasing over the years and approaching to the theoretical minimum in modern factories, for example, 40 MJ/kg of nitrogenous fertilizer (Kongshaug, 1998; Anundskas, 2000). Providing the chemical fertilizers from energy-efficient chemical plants may help to reduce energy utilization. Chemical fertilizer use in agriculture is generally not based on soil analysis and is much more than the actual need (Esengun et al., 2007). Introducing some microorganisms to the soil may stimulate nitrogen fixation, solubilize the insoluble minerals, and reduce the need to produce and transport large amounts of chemical fertilizers (Aslantas et al., 2007). Fertilizer management practices may reduce energy utilization up to 72%; consequently, herbicide utilization and pollution also decrease (Clements et al., 1995; Hülsbergen et al., 2001; Snyder et al., 2009; Ren et al., 2010). Pishgar-Komleh et al. (2012) argue that energy efficiency during agriculture of the potatoes varies with the farm size. Since traveling between the small fields to carry the equipment and the people may be among the reasons for the lower energy efficiency, the same trend may also be valid during agriculture of the other commodities; therefore, employing the optimum-size fields may help to reduce the energy expenditure and subsequent carbon dioxide emissions. Energy efficiency is also reported to be lower in small breweries (Sturm et al., 2013), since small enterprises may not be willing to take the risk of investing for a newer and more efficient technology. Lower energy efficiency is also reported in smaller dairies, when compared to the larger ones (Xu and Flapper, 2009).

Only the light that is in the wavelength range of 400–700 nm (which constitutes about 45% of the total solar energy) may be used in photosynthesis. About 25% of the referred 45% may be absorbed by the plants; the rest is lost by various reasons including reflection. Receiving nonoptimal levels of radiation reduces the efficiency of photosynthesis further; therefore, only 3%–6% of total solar radiation may be used in photosynthesis (Miyamato, 1997). Ways to increase the photosynthetic efficiency of the plants is being actively researched to improve their yields, including those of the grain crops (Long et al., 2006). Kaltsas et al. (2007) reported use of insect traps instead of pesticides during cultivation of organic olives. Improving the efficiency of these traps and extending their use to other crops may reduce the environmental impact and energy cost of agrochemicals. Technological advancement of the same proportion as achieved in the farmland should be achieved in the factories too to make our food more environmentally friendly. The recent publications by Xu and Flapper (2011), Wu et al. (2013), and

Rodriguez-Gonzales et al. (2015) offer serious recommendations regarding substitution of the less-energy-efficient process steps with the more-energy-efficient ones and may be regarded as among the major publications pointing the direction to the food industry toward increasing the energy efficiency and decreasing the subsequent environmental pollution caused by energy utilization.

There are numerous studies reporting savings in energy utilization by making changes in packaging or the design of the processes by replacing some sets of equipment with the less-energy-demanding ones. Galitsky et al. (2003) commented that up to 44 MJ/barrel, corresponding to 220 MJ/ton beer of energy savings, may be achieved in the individual steps of brewing. Barley malt, a key ingredient in beer, is produced in three steps. First, it is immersed and drained in water for 40 h, to increase its water content, allowing kernels to germinate for 4–6 days, and then the kernels are dried from 85% to 4% moisture content (Kribs and Spolek, 1997), which requires thermal energy of 3.9 MJ per kg of dried malt (Bala, 1984). Kribs and Spolek (1997) developed a drying schedule and reduced the energy requirement by about 20%.

Data presented by De Monte et al. (2005) while discussing the mass–energy balance with respect to the various processes involved in coffee packaging imply that packaging of coffee when marketed in 3 kg white latten cans requires an electric power input of 345 MJ per ton of coffee for the production of the can plus packaging. When the same product is marketed in 125 g white latten cans, 2,894 MJ per ton of coffee plus 233 MJ per ton of coffee of energy from natural gas and electric power, respectively, are used with the same purpose, implying that nine times more energy is utilized with the smaller-size containers. In Example 5.19, energy savings in orange juice production process is reported when we use different technologies. Example 5.20 describes a different case; by using four different packaging materials, we end up with the same CEnC, CExC, and CCO_2E values, and in this case the decision is based on the environmental concern.

Recycling or eliminating the waste of the food plants is a challenge for the food industry. Active research is being carried out to achieve this purpose. Musee et al. (2007) presented 90 strategies for minimization of the winery waste. Composting appears to be a feasible way for recycling the solid winery waste. Ruggieri et al. (2012) reported that energy expenditure for carrying the waste with tractors with this purpose was actually less than that of the savings from the chemical fertilizers. The same procedure may be applied also with some other agricultural waste. Dong et al. (2015) constructed a microbial fuel cell, where brewery wastewater was treated on energy self-sufficient way.

Example 5.19: Energy Savings in Concentrated Orange Juice Production Process

Example Table 5.19.1 describes energy utilization during production of orange juice with pasteurization (Case 1) and then 4.7-fold concentration in a multiple-effect evaporator system (Case 2), 4.7-fold concentration in a multiple-effect

EXAMPLE TABLE 5.19.1   Energy Utilization during Production of Concentrated Orange Juice

Agriculture of the Oranges (Adapted from Özkan et al., 2004)	Energy Utilization MJ/ Ton of Oranges
Chemical fertilizers and manure	244
Agrochemicals	67
Diesel oil	402
Electricity	1
Agriculture total	2,091
Orange Juice Production Processes	*MJ/ton orange juice*
Case 1—single-strength pasteurized orange juice	
Agriculture of the fruit (1 ton of juice is produced from 3.1 tons of oranges; density of the juice, 1.014 kg/m³)	6,482
Transportation of the oranges from orchards to factory for 100 km	426
Processing of the juice, 1,239 MJ/ton (adapted from Waheed et al., 2008 after excluding packaging)	1,232
Energy utilization for the packaging (same as that of milk; adapted from Table 3 of Williams and Wikström, 2011)	2,697
Distribution of the packed product by traveling for 500 km in refrigerated trucks	164
Pasteurized single-strength packaged orange juice total	10,997
Case 2—4.7-fold concentration in a multiple-effect evaporator system	
Agriculture of the fruit (1 ton of juice is produced from 3.1 tons of oranges and concentrated 4.7 times, having a density of the concentrated juice of 1.066 kg/m³)	31,365
Transportation of the oranges from orchards to factory for 100 km	2,002
Concentration (adapted from Gul and Harasek, 2012)	335
Energy utilization for the packaging (same as that of milk; adapted from Table 3 of Williams and Wikström, 2011)	2,697
Freezing of the product in tunnel freezer (GEA technologies, model SMA-4), 500 kg/h, power 211 kW	1,519
Distribution of the packed product by traveling for 100 km in refrigerated trucks	164
Total	38,082
Reconstituted (4.7 times diluted) juice	8,103
Case 3—4.7-fold concentration in a multiple-effect evaporator system in combination with a membrane system	
Agriculture of the fruit (1 ton of juice is produced from 3.1 tons of oranges and concentrated 4.7 times, having a density of the concentrated juice of 1.066 kg/m³)	31,365
Transportation of the oranges from orchards to factory for 100 km	2,002
Concentration (adapted from Gul and Harasek, 2012)	54
Energy utilization for the packaging (same as that of milk; adapted from Table 3 of Williams and Wikström, 2011)	2,697
Freezing of the product in tunnel freezer (GEA technologies, model SMA-4), 500 kg/h, power 211 kW	1,519

(Continued)

EXAMPLE TABLE 5.19.1 (*Continued*) Energy Utilization during Production of Concentrated Orange Juice

Agriculture of the Oranges (Adapted from Özkan et al., 2004)	Energy Utilization MJ/ Ton of Oranges
Distribution of the packed product by traveling for 100 km in refrigerated trucks	164
Total	37,801
Reconstituted (4.7 times diluted) juice	8,043
Case 4—4.7-fold concentration with freeze concentration	
Agriculture of the fruit (1 ton of juice is produced from 3.1 tons of oranges and concentrated 4.7 times, having a density of the concentrated juice of 1.066 kg/m³)	31,365
Transportation of the oranges from orchards to factory for 100 km	2,002
Processing of the juice, 1,239 MJ/ton (adapted from Waheed et al., 2008, after excluding packaging)	5,790
3,700 kg of water being removed to produce 1 ton of concentrate (freeze concentration of the orange juice is achieved with the HybridICE® technology, with energy utilization of 100 kW h/m³ of ice removed; adapted from Zvinowanda et al., 2014)	1,332
Energy utilization for the packaging (same as that of milk; adapted from Table 3 of Williams and Wikström, 2011)	2,697
Distribution of the packed product by traveling for 100 km in refrigerated trucks	164
Total	43,350
Reconstituted (4.7 times diluted) juice	9,223

evaporator system in combination with a membrane system (Case 3), and 4.7-fold concentration with freeze concentration (Case 4). Calculations have shown that among all of these alternatives, Case 3 requires the lowest energy utilization.

Example 5.20: Energy and Exergy Utilization and Carbon Dioxide Emission when Different Materials Are Used for Packaging Flavored Yogurt

Sorgüven and Özilgen (2012) calculated the values of the CEnC, CExC, and the CCO_2E for the flavored yogurt production process. It is possible to market the yogurt in HDPE, polypropylene (PP), polystyrene (PS), and PLA packaging materials. Example Table 5.20.1 presents the values of the thermodynamic parameters CEnC, CExC, and CCO_2E for the packaged product in each case.

Example Table 5.20.1 gives almost the same CEnC, CExC, and CCO_2E values for the packaged flavored yogurt. Since PLA is biodegradable, it may be the preferred packaging material.

EXAMPLE TABLE 5.20.1 Energy and Exergy Consumed and CO_2 Emitted to Produce 1 Ton of Strawberry-Flavored Yogurt

Energy Utilization for		CEnC (MJ/ Ton Flavored Yogurt)	CExC (MJ/ Ton Flavored Yogurt)	CCO_2E (kg CO_2/ Ton Flavored Yogurt)
Unpackaged flavored yogurt production		28,836.9	75,616.5	1229.5
Primary packaging	HDPE	19.2	516.0	2.7
	PP	10.8	511.2	1.5
	PS	5.4	551.4	0.8
	PLA	324.0	468.0	45.4
Container making and filling machinery		1,851.4	7,720.3	259.2
Waste transport		32.4	30.0	0.4
Storage		5.4	22.5	0.8
Transport to store		636.2	589.1	8.9
Retail air conditioning		5.4	22.5	0.8
Total (flavored yogurt packaged in HDPE containers)		31,387	84,517	1503
Total (flavored yogurt packaged in PLA containers)		31,692	84,469	1545
Total (flavored yogurt packaged in PP containers)		31,379	84,512	1502
Total (flavored yogurt packaged in PS containers)		31,373	84,552	1501

Source: Adapted from Sorgüven, E. and Özilgen, M., *Energy*, 40, 214, 2012.

Nomenclature

In general, the following characters are employed to mean the following properties throughout the text; some characters that are used only once or twice are defined at the same place as they are used.

c	Concentration (kg/m^3 or mol/m^3)
CDP	Cumulative degree of perfection
$CECO_2$	Cumulative emission of CO_2 (kg/kg)
CEnC	Cumulative consumption of energy (J/kg)
CExC	Cumulative consumption of exergy (J/kg)
c_p	Specific heat determined under constant pressure (J/kg °C)
c_v	Specific heat determined under constant volume (J/kg °C)
e	Specific energy (J/kg)
E	Energy (J)
e_k	Specific kinetic energy (J/kg)
E_k	Kinetic energy (J)
e_p	Specific potential energy (J/kg)
E_p	Potential energy (J)
ex	Specific exergy (J/kg)
ex_{ch}	Specific chemical exergy (J/kg)
$ex_{element}$	Specific chemical exergy of an element (J/kg)
ex_{th}	Specific thermomechanical exergy (J/kg)
Ex	Exergy (J)
F	Force (N)
F_d	Drag force (N)
g	Specific Gibbs free energy (J/kg K)
G	Gibbs free energy (J/kg K)
g_a	Gravitational acceleration (9.81 m/s^2)
g_f	Specific Gibbs free energy of formation (J/kg K)
h	Specific enthalpy (J/kg K)
H	Enthalpy (J/kg K)
h_l	Height (m)
h_f	Specific enthalpy of formation (J/kg K)

n_i	Number of moles of the ith compound in a mixture
LHV	Lower heating value (J/mol)
N_A	6.02×10^{-23} (1/mol)
p	Pressure (Pa)
p_c	Critical pressure (Pa)
p_r	Reduced pressure
q	Heat per unit mass (J/kg)
Q	Heat (J)
q_c	Charge (C)
r	Radius (m)
R	Gas constant (8.314 J/mol K)
s	Specific entropy (J/kg K)
S	Entropy (J/kg K)
s_f	Specific entropy of formation (J/kg K)
t	Time (s, min, or h)
T	Temperature (°C or K)
T_c	Critical temperature (K)
T_r	Reduced temperature
u	Specific internal energy (m³/kg)
U	Internal energy (m³/kg)
v	Specific volume (m³/kg)
V	Volume (m³/kg)
V_e	Electric potential difference (J/C)
w	Work per unit mass (J/m)
W	Work (J)
y_i	Mole fraction of the ith compound in the gas or vapor phase of a mixture
x_i	Mole fraction of the ith compound in a mixture
w_i	Mass fraction of the ith compound in a mixture
z	Compressibility factor
θ	Angle of rotation (radians)
μ	Chemical potential (J/kg mol K)
η	Efficiency
ρ	Density (kg/m³)
σ	Surface tension (J/m²)
τ	Torque (N m)
υ	Velocity (m/s)

References

Ackermann M, Lin C, Bergstrom CT, Doebeli M (2007). On the evolutionary origin of aging. *Cell* 6:235–244.

Albert M (1995). Physiologic and clinical principles of eccentrics. In: Albert M (ed.), *Eccentric Muscle Training in Sports and Orthopedics*, 2nd ed. Churchill Livingstone, New York.

Alberty RA (2003). *Thermodynamics of Biochemical Reactions*. John Wiley & Sons, Hoboken, NJ.

Aljundi HI (2009). Energy and exergy analysis of a steam power plant in Jordan. *Applied Thermal Engineering* 29:324–328.

Alti M, Özilgen M (1994). Statistical process analysis in Broiler feed formulation. *Journal of the Science of Food and Agriculture* 66:13–20.

Amat JA, Aguilera E, Visser GH (2007). Energetic and developmental costs of mounting an immune response in greenfinches (*Carduelis chloris*). *Ecological Research* 22:282–287.

Amour DS. Artificial daylight lamp. United States Patent US2725461 A. 1955. https://www.google.com/patents/US2725461?hl=tr.

Annamalai K, Silva CA (2012). Entropy stress and scaling of vital organs over life span based on allometric laws. *Entropy* 14:2550–2577.

Anundskas A (2000). Technical improvements in mineral nitrogen fertilizer production. In: *Harvesting Energy with Fertilizers*. European Fertilizer Manufacturers Association, Brussels, Belgium.

Aoki L (1989). Entropy flow and entropy production in the human body in basal conditions. *Journal of Theoretical Biology* 141:11–21.

Aoki L (1994). Entropy production in human life span: A thermodynamical measure of aging. *Age* 17:29–31.

Apicella MA, Robinson JA (1970). Physicochemical properties of *Neisseria meningitidis* group C and Y polysaccharide antigens. *Infection and Immunity* 2:392–397.

Appleton B, Koci J, Harris R, Sevebeck K, Alleman D, Swanson L (2000). Trees for problem landscape sites—Air pollution. Virginia Cooperative Extension and Virginia Tech University Publication Number 430-022, Blacksburg, VA. Available at: http://www.waynesboronurseries.com/lists/trees%20for%20problem%20landscape%20sites%20-%20air%20pollution.htm (accessed November 11, 2014).

Arias-Gonzalez JR (2012). Entropy involved in fidelity of DNA replication. *PLoS ONE* 7(8): e42272.doi:10.1371/journal.pone.0042272.

Ariji Y, Kimura Y, Gotoh M, Sakuma S, Zhao YP, Ariji E (2001). Blood flow in and around the masseter muscle: Normal and pathologic features demonstrated by color Doppler sonography. *Oral Surgery, Oral Medicine, Oral Pathology, Oral Radiology and Endodontics* 91:472–482.

Aslantas R, Cakmakci R, Sahin F (2007). Effect of plant growth promoting rhizobacteria on young apple tree growth and fruit yield under orchard conditions. *Scientia Horticulturae* 111:371–377.

Attwell D, Laughlin SB (2001). An energy budget for signaling in the grey matter of the brain. *Journal of Cerebral Blood Flow & Metabolism* 21:1133–1145.

Aubert A, Costalat R (2005). Interaction between astrocytes and neurons studied using a mathematical model of compartmentalized energy metabolism. *Journal of Cerebral Blood Flow and Metabolism* 25:1476–1490.

Avraamides M, Fatta D (2008). Resource consumption and emissions from olive oil production: A life cycle inventory case study in Cyprus. *Journal of Cleaner Production* 16:809–821.

Ayres RU (1998). Eco-thermodynamics: Economics and the second law. *Ecological Economics* 26:189–209.

Ayres RU (1999). The second law, the fourth law, recycling and limits to growth. *Ecological Economics* 29:473–483.

Bailey JE, Ollis DF (1986). *Biochemical Engineering Fundamentals*, 2nd ed. McGraw-Hill, New York.

Bala BK (1984). Simulation of deep bed malt drying. *Journal of Agricultural Engineering Research* 30:235–244.

Baldwin WW, Kubitschek EH (1984). Buoyant density variation during the cell cycle of *Saccharomyces cerevisiae*. *Journal of Bacteriology* 158:701–704.

Balmer RT (1982). Entropy and aging in biological systems. *Chemical Engineering Communications* 17:171–181.

Balzhiser RE, Samuels MR, Eliassen JD (2004). *Chemical Engineering Thermodynamics, the Study of Energy, Entropy and Equilibrium*. Prentice Hall, Englewood Cliffs, NJ, 1972.

Banaeian N, Omid M, Ahmadi H (2011). Energy and economic analysis of greenhouse strawberry production in Tehran province of Iran. *Energy Conversion and Management* 5:1020–1025.

Bao Y, Mukai K, Hishiki T, Kubo A, Ohmura M, Sugiura Y, Matsuura T et al. (2013). Energy management by enhanced glycolysis in G1-phase in human colon cancer cells in vitro and in vivo. *Molecular Cancer Research* 11:973–985.

Barclay CJ (1996). Mechanical efficiency and fatigue of fast and slow muscles of the mouse. *The Journal of Physiology* 497:781–794.

Barclay CJ (2008). Getting energy to where it is needed is the problem in the failing heart. *The Journal of Physiology* 586(21): 5037–5038.

Barja G. Free radicals and aging. *Trends in Neurosciences* 27:595–600.

Barnett MW, Larkman PM (2007). The action potential. *Practical Neurology* 7(3):192–197.

Baruque-Ramos J, Hiss H, Vicentin MA, da Paz MF, Peixoto A, Leal MBB, Gonçalves PC, Raw I (1996). Batch cultivation kinetics of *Neisseria meningitidis* (serogroup c) in Frantz medium: Growth and polysaccharide production. *Brazilian Archives of Biology and Technology* 39:215–220.

Battino R, Strong LE, Wood SE (1997). A brief history of thermodynamics notation. *Journal of Chemical Education* 74:304–305.

Battley EH (1993). Calculation of entropy change accompanying growth of *Escherichia coli* K-12 on succinic acid. *Biotechnology and Bioengineering* 41:422–428.

Battley EH (1998). The development of direct and indirect methods for the study of the thermodynamics microbial growth. *Thermochimica Acta* 309:17–37.

Battley EH (1999). An empirical method for estimating the entropy of formation and the absolute entropy of dried microbial biomass for use in studies on thermodynamics of microbial growth. *Thermochimica Acta* 326:7–15.

Battley EH, Putham RL, Boerio-Goates J (1997). Specific heat measurements from 10 to 300 K and derived thermodynamic functions of lyophilized cells of *Saccharomyces cerevisiae* including the absolute entropy and entropy formation at 298.15 K. *Thermochimica Acta* 298:37–46.

Battley EH, Stone JR (2000). A comparison of values for the entropy and the entropy of formation of selected organic substances of biological importance in the state, as determined experimentally or calculated empirically. *Thermochimica Acta* 349:153–161.

Bayer E. Process for producing solid, liquid and gaseous fuels from organic starting material. United States Patent, 5,114,541, 1992.

Bejan A. *Entropy Generation Minimization*. CRC Press, New York, 1996.

Bejan A. *Advanced Engineering Thermodynamics*, 3rd ed. Wiley, Hoboken, NJ, 2006.

Beloukas AI, Magiorkinis E, Tsoumakas TL, Kosma AG, Diamantis A (2013). Milestones in the history of research on cardiac energy metabolism. *Canadian Journal of Cardiology* 29:1504–1511.

Bensinger SJ, Christofk HR (2012). New aspects of the Warburg effect in cancer cell biology. *Seminars in Cell & Developmental Biology* 23:352–361.

Benson SW (1965). Bond energies. *Journal of Chemical Education* 42:502–518.

Berg JM, Tymoczko JL, Stryer L (2002). *Biochemistry*, 5th ed. WH Freeman, New York.

Berry JD, Boese DJ, Law DKS, Zollinger WD, Tsang RSW (2005). Molecular analysis of monoclonal antibodies to group variant capsular polysaccharide of *Neisseria meningitidis*: Recurrent heavy. *Molecular Immunology* 42:335–344.

Berthiaume R, Bouchard C, Rosen MA (2001). Exergetic evaluation of the renewability of a biofuel. *Exergy an International Journal* 1:256–268.

Black NH, Davis HN (1914). *Practical Physics: Fundamental Principles and Applications to Daily Life*. Macmillan, New York.

Boardman B, Palmer J (2003). Consumer choice and carbon consciousness for electricity. European Commission Altener Program Final Report, Environmental Challenge Institute. Available at: http://www.electricitylabels.com/downloads/4CE_Final_Report.pdf (accessed May 7, 2016).

Boregowda SC, Choate RE, Handy R (2009). Entropy generation analysis of human thermal stress responses. *ISRN Thermodynamics* 1–11:Article ID 830103.

Borowitzka MA (1999). Commercial production of microalgae: Ponds, tanks, tubes and fermenters. *Journal of Biotechnology* 70:313–321.

Bowden J (November 2011). Blood sugar battles, how to prevent, treat, and even cure type 2 diabetes. *Better Nutrition*, 44–48; http://www.betternutrition.com/prevent-treat-cure-type-2-diabetes/features/featurearticles/1125 (accessed November 7, 2011).

Branchini BR, Rosenberg JC, Ablamsky DM, Taylor KP, Southworth TL, Linder SJ (2011). Sequential bioluminescence resonance energy transfer–fluorescence resonance energy transfer-based ratiometric protease assays with fusion proteins of firefly luciferase and red fluorescent protein. *Analytical Biochemistry* 414:239–245.

Brehmer B (2008). Chemical biorefinery perspectives: The valorisation of functionalised chemicals from biomass resources compared to the conventional fossil fuel production route. Wageningen University, Wageningen, the Netherlands. Available at: http://www.lcacenter.org/InLCA2006/Brehmer-abstract.pdf (accessed December 9, 2010).

British Petroleum Company (June 2011). BP statistical review of world energy, London, UK. Available at: http://www.bp.com/content/dam/bp-country/de_de/PDFs/brochures/statistical_review_of_world_energy_full_report_2011.pdf (accessed November 2, 2015).

Buchholz AC, Schoeller DA (2004). Is calorie a calorie? *American Journal of Clinical Nutrition* 79(suppl):899S–906S.

Canning P, Charles A, Huang S, Polenske KR, Waters A (March 2010). *Energy Use in the U.S. Food System*. ERR-94, U.S. Department of Agriculture Economic Research Service, Washington, DC. Available at: http://www.ers.usda.gov/media/136418/err94_1_.pdf (accessed November 12, 2015).

Cardona TD, Driscoll RH, Paterson JL, Srzednicki GS, Kim WS (2002). Optimization conditions for heat pump dehydration of lactic acid bacteria. *Drying Technology* 20:1611–1632.

Carlsson-Kanyama A, Faist M. Energy use in the food sector: A data survey. Available at: http://mmm.comuv.com/wordpress/wp-content/uploads/2010/06/Energy-use-in-the-food-sector-Carlsson-Kanyama-and-Fiest.pdf, 2000 (accessed September 8, 2013).

Carmeli E, Coleman R, Zeznick AZ (2002). The biochemistry of aging muscle. *Experimental Gerontology* 37:477–489.

Carnot (1824). Réflexions sur la puissance motrice du feu et sur les machines propres à développer cette puissance. Bachelier, Paris, France.

Castelnovo M, Bowles RK, Reiss H, Gelbart WM (2003). Osmotic force resisting chain insertion in a colloidal suspension. *The European Physical Journal* E10:191–197.

Casten TR, Schewe PF (2009). Getting the most from energy recycling waste heat can keep carbon from going sky high. *American Scientist* 97:26–33.

Çatak J, Develi AÇ, Sorgüven E, Özilgen M, Serap İnal H (2015). Entropy generated by the masseter muscles during chewing: An indicator of life expectancy? *International Journal of Exergy* 18:46–66.

Caton JA (2000a). A review of investigations using the second law of thermodynamics to study internal combustion engines. SAE Paper No: 2000-01-1081, Society of Automotive Engineers Inc, Warrendale, PA.

Caton JA (2000b). On destruction of availability (exergy) due to combustion process—With specific application to internal combustion engines. *Energy* 25:1097–1117.

Çengel Y, Boles MA (2007). *Thermodynamics an Engineering Approach*, 6th ed. McGraw-Hill, Singapore.

Cengel Y, Boles M (2011). *Thermodynamics: An Engineering Approach*. McGraw-Hill, Singapore.

Çengel YA, Cimbala JM (2006). *Fluid Mechanics Fundamentals and Applications*. McGraw-Hill, Boston, MA

Cesari M, Vellas B, Gambassi G (2013). The stress of aging. *Experimental Gerontology* 48:451–456.

Chang R (2005). *Chemistry*. McGraw-Hill, New York.

ChemSpider, 2013. *N-Acetylneuraminic Acid*, Royal Society of Chemistry, London, UK. http://www.chemspider.com/Chemical-Structure.10292217.html (accessed August 16, 2013).

Chinnery PF, Samuels DC, Elson J, Turnbull DM (2002). Accumulation of mitochondrial DNA mutations in aging, cancer, and mitochondrial disease: Is there a common mechanism? *The Lancet* 360:1323–1325.

Cho YI, Kensey KR (1991). Effects of the non-Newtonian viscosity of blood on flows in a diseased arterial vessel. Part 1: Steady flows. *Biorheology* 28:241–262.

Choresh O, Bayarmagnae B, Lewis RV (2009). Spider web glue: Two proteins expressed from opposite strands of the same DNA sequence. *Biomacromolecules* 10(10):2852–2856.

Chow S, Ganji AR, Hackett B (2005). Opportunities for energy efficiency and demand response in corrugated cardboard manufacturing facilities. In: *Industrial Energy Technology Conference*, New Orleans, LA.

Chueh EF, Swanson C (1973a). Estimating liquid heat capacity. *Chemical Engineering Progress* 69:83–85.

Chueh EF, Swanson C (1973b). Estimation of liquid heat capacities. *The Canadian Journal of Chemical Engineering* 51:596–600.

Clauss M, Hofmann RR, Streich WJ, Fickel J, Hummel J (2008). Higher masseter muscle mass in grazing than in browsing ruminants. *Oecologia* 157:377–385.

Clements DR, Weise SF, Brown R, Stonehouse DP, Hume DJ, Swanton CJ (1995). Energy analysis of tillage and herbicide inputs in alternative weed management systems. *Agriculture, Ecosystems & Environment* 52:119–128.

Clemson Cooperative Extension (2014). Small grains production. Available at: https://www.clemson.edu/extension/rowcrops/small_grains/pdfs/disease_control.pdf (accessed July 24, 2014).

Constable JK, Barclay CJ, Gibbs CL (1997). Energetics of lengthening in mouse and toad skeletal muscles. *The Journal of Physiology* 505.1:205–215.

Cordier J-L, Butsch BM, Birou B, Stokar U (1987). The relationship between elemental composition and heat of combustion of microbial biomass. *Applied Microbiology and Biotechnology* 25:305–312.

Cortella G, Da Porto C (2003). Design of a continuous distillation plant for the production of spirits originating from fermented grape. *Journal of Food Engineering* 58:379–385.

Cortossa S, Aon MA, Iglesias AA, Lloyd D (2002). *An Introduction to Metabolic and Cellular Engineering*. Word Scientific Pub. Co., London, U.K.

Coskun B, Seker E, Inal F (1997). Hayvan Beslenme. Selcuk Universitesi, Veteriner Fakültesi Yayin Unitesi, Konya, sayfa 75–90.

Cowling E, Nilsson J (1995). Acidification research: Lessons from history and visions of environmental futures. *Water, Air, and Soil Pollution* 85:279–292.

Cramer J, Wissema E, Lammers E, Faaij A, Hamelinck C, Bergsma G, van den Heuvel E, Junginger M, Smeets E (2006). Criteria for sustainable biomass production, Final report from the project group, Enschede, the Netherlands. Available at: http://www.globalproblems-globalsolutions-files.org/unf_website/PDF/criteria_ sustainable_biomass_prod.pdf (accessed July 15, 2016).

da Paz MF, Baruque-Ramos J, Hiss H, Vicentin MA, Leal MBB, Raw I (2003). Polysaccharide production in batch processes of *Neisseria meningitidis* serogroup C comparing Frantz, modified Frantz and Catlin 6 cultivation media. *Brazilian Journal of Microbiology* 34:27–32.

Damman K, Navis G, Voors AA, Asselbergs FW, Smilde TDJ, Cleland JGF, Veldhuisen DJV, Hilledge HL (2007). Worsening renal function and prognosis in heart failure: Systematic review and meta-analysis. *Journal of Cardiac Failure* 13:599–608.

Daniels F, Alberty RA (1975). *Physical Chemistry*, 4th ed., Wiley, New York.

Darnell JE, Lodish HF, Baltimore D (1986). *Molecular Cell Biology*. Scientific American Books Inc., New York.

Davey M, Jan M (2010). Sunflower (*Heliant annulus L.*): Genetic improvement using conventional and in vitro technologies. *Journal of Crop Improvement* 24:349–391.

Davies PCW, Rieper E, Tuszynski JA (2013). Self-organization and entropy generation in a living cell. *Biosystems* 111:1–10.

De D, Singh RS, Chandra H (2001). Technological impact on energy consumption in rainfed soybean cultivation in Madhya Pradesh. *Applied Energy* 70:193–213.

Değerli B, Küçük K, Sorgüven E, Özilgen M (2015). Thermodynamic analysis of serogroup C antigen production by *Neisseria meningitides*. *International Journal of Exergy* 16:1–21.

del Castillo LF, Vera-Cruz P (2011). Thermodynamic formulation of living systems and their evolution. *Journal of Modern Physics* 2:379–391.

Demetrius L, Legendre S, Harremöes P (2009). Evolutionary entropy: A predictor of body size, metabolic rate and maximal life span. *Bulletin of Mathematical Biology* 71:800–818.

Demirel Y (2010). Nonequilibrium thermodynamics modeling of coupled biochemical cycles in living cells. *Journal of Non-Newtonian Fluid Mechanics* 165:953–972.

Demirel Y, Sandler SI (2004). Nonequilibrium thermodynamics in engineering and science. *Journal of Physical Chemistry* 108:31–43.

De D, Singh RS, Chandra H (2001). Technological impact on energy consumption in rainfed soybean cultivation in Madhya Pradesh. *Applied Energy* 70:193–213.

De Monte M, Padoano E, Pozzetto D (2005). Alternative coffee packaging: An analysis from a life cycle point of view. *Journal of Food Engineering* 66:405–411.

de Pedro MA, Quintela JC, Holtje JV, Schwartz H (1997). Murein segregation in *Escherichia coli*. *Journal of Bacteriology* 179:2823–2834.

de Swaan Arons J, van de Kooi H, Sankaranarayanan K (2004). *Efficiency and Sustainability in the Energy and Chemical Industries*. Marcel Dekker, New York.

Dewulf J, van Langenhove H (2004). Thermodynamic optimization of the life cycle of plastics by exergy analysis. *International Journal of Energy Research* 28:969–976.

Diaz-Ruiz R, Rigoulet M, Devin A (2011). The Warburg and Crabtree effects: On the origin of cancer cell energy metabolism and of yeast glucose repression. *Biochica et Biophysica Acta* 1807:568–576.

Dietz, M.W (1995). Development of metabolism and thermoregulation in galliforms. PhD thesis, University of Utrecht, Utrecht, the Netherlands.

Dinçer I, Çengel Y (2001). Energy, entropy and exergy concepts and their roles in thermal engineering. *Entropy* 3:116–149.

Domalski ES (1972). Selected values of heats of combustion and heats of formation of organic compounds containing the elements C, H, N, O, P, and S. *Journal of Physical and Chemical Reference Data* 1:221–277.

Domalski ES, Evans WH, Jobe Jr TL, Milne TA. Thermodynamic data for biomass conversion and waste incineration. Solar Technical Information Program, U.S. Department of Energy, September 1986. Available at: http://www.nrel.gov/biomass/pdfs/2839.pdf (accessed August 16, 2014).

Domalski ES, Hearing ED (1993). Estimation of the thermodynamic properties of C-H-N-O-S-halogen compounds at 298.15 K. *Journal of Physical and Chemical Reference Data* 22(4):805–1159.

Dong Y, Qu Y, He W, Du Y, Liu J, Han X, Feng Y (2015). A 90-liter stackable baffled microbial fuel cell for brewery wastewater treatment based on energy self-sufficient mode. *Bioresource Technology* 195:66–72.

Doreau M, Ferlay A, Rochette Y, Martin C (2014). Effects of dehydrated lucerne and soya bean meal on milk production and composition, nutrient digestion, and methane and nitrogen losses in dairy cows receiving two different forages. *Animal: An International Journal of Animal Bioscience* 8(3):420–430.

Dortmund Data Bank (2008). *Some Mixture Properties of Ethanol and Water*, DDBST Dortmund Data Bank Software & Separation Technology GmbH, Oldenburg, Germany. Available at: http://www.ddbst.com/ddb.html (accessed February 3, 2013).

Eide MH (2002). Life cycle assessment (LCA) of industrial milk production. *International Journal of Life Cycle Assessment* 7:115–126.

Eisenstein EM, Eisenstein DL, Sarma JSM, Knapp H, Smith JC (2012). Some new speculative ideas about the "behavioral homeostasis theory" as to how the simple learned behaviors of habituation and sensitization improve organism survival throughout phylogeny. *Communicative & Integrative Biology* 5(3): 233–239.

Elson JL, Samuels DC, Turnbull DM, Chinnery PF (2001). Random intracellular drift explains the clonal expansion of mitochondrial DNA mutations with age. *American Journal of Human Genetics* 68:802–806.

Emanuel KA (1991). The theory of hurricanes. *Annual Reviews of Fluid Mechanics* 32:176–196.

Erdal G, Esengun K, Erdal H, Gunduz O (2007). Energy use and economic analysis of sugar beet production in Tokat province of Turkey. *Energy* 32:35–41.

Erzurum Valiligi Il Gida Tarim ve Hayvancilik Müdürlügü, Cavdar Yetistiriciliği 2013.

Esengun K, Erdal G, Gunduz O, Erdal H (2007). An economic analysis and energy use in stake-tomato production in Tokat province of Turkey. *Renewable Energy* 32:1873–1881.

Everman WJ. Chemical weed control in wheat, barley, oats, rye, and triticale. In: *North Carolina Agricultural Chemicals Manual, 2015.* Available at: http://www.smallgrains.ncsu.edu/_Pubs/Xtrn/AgChemHerbicides.pdf (accessed July 24, 2015).

Evilevitch A, Castelnovo M, Knobler CM, Gelbart WM (2004). Measuring the force ejecting DNA from phage. *Journal of Physical Chemistry B* 108:6838–6843.

Exergoecology Portal Exergy Calculator. Available at: http://www.exergoecology.com/excalc/exergo/refenv/#equation1 (accessed August 16, 2013).

Fadere DA, Bamiro OA, Oni AO (2010). Energy and cost analysis of organic fertilizer production in Nigeria. *Energy* 35:332–340.

Fahraeus R, Lindqvist T (1931). The viscosity of blood in narrow capillary tubes. *American Journal of Physiology* 96:562–568.

Faure G (1991). *Principles and Applications of Inorganic Chemistry.* MacMillan, New York.

Feinman RD, Fine EJ (2004). "A calorie is a calorie" violates the second law of thermodynamics. *Nutrition Journal* 1–5, 3:doi:10.1186/1475-2891-3-9.

Fell DA, Small JR (1986). Fat synthesis in adipose tissue. An examination of stoichiometric constraints. *Biochemical Journal* 238:781–786.

Fine EJ, Feinman RD (2004). Thermodynamics of weight loss diets. *Nutrition and Metabolism* 1:15.

Finke E, Pörtner H-O, Lee P, Webber D (1996). Squid (*Lolliguncula brevis*) life in shallow waters: oxygen limitation of metabolism and swimming performance. *The Journal of Experimental Biology* 199:911–921.

Flatt JP (1970). Conversion of carbohydrate to fat in adipose tissue: An energy-yielding, and, therefore, self-limiting process. *Journal of Lipid Research* 11:131–143.

Fonseca JC, Miguel A, Pardal MA, Azeiteiro UM, Marques JC (2002). Estimation of ecological exergy using weighing parameters determined from DNA contents of organisms—A case study. *Hydrobiologia* 475–476:79–90.

Fore SR, Porter P, Lazarus W (2011). Net energy balance of small-scale on-farm biodiesel production from canola and soybean. *Biomass and Bioenergy* 35:2234–2244.

Fox RF (2004). Origin of life and energy. In: Cleveland CJ (ed.), *Encyclopedia of Energy*, Vol. 4, Elsevier, Amsterdam, the Netherlands, pp. 781–792.

French NA (1997) Modeling incubation temperature: The effects of incubator design, embryonic development, and egg size. *Poultry Science* 76:124–133.

Furukawa H, Kuroiwa TK, Mizushima S (1983). DNA injection during bacteriophage T4 infection of *Escherichia coli. Journal of Bacteriology* 154:938–945.

Galitsky C, Martin N, Worrell E, Lehman B (September 2003). Energy efficiency improvement and cost saving opportunities for breweries. Report No: LBNL-50934, Ernest Orlando Lawrence Berkeley National Laboratory Berkeley, CA.

Garland J (2013). Energy management—A critical role in cancer induction? *Critical Reviews in Oncology/Hematology* 88:198–217.

Garrels RM, Christ CL (1965). *Solutions, Minerals, Equilibria.* Harper & Row, New York.

Gaviao MBD, Engelen L, van der Bilt A (2004). Chewing behavior and salivary secretion. *European Journal of Oral Sciences* 112:19–24.

Genç S, Aksan-Kurnaz I, Özilgen M (2011). Astrocyte-neuron lactate shuttle may boost ATP supply to the neurons under hypoxic conditions- an in silico study. *BMC Systems Biology* 5:162–175.

Genç S, Sorgüven E, Aksan-Kurnaz I, Özilgen M (2013a). Exergetic efficiency of ATP production in neuronal glucose metabolism. *International Journal of Exergy* 13:60–84.

Genç S, Sorgüven E, Özilgen M, Aksan-Kurnaz I (2013b). Unsteady exergy destruction of the neuron under dynamic stress conditions. *Energy* 59:422–431.

Gerber PJ, Steinfeld H, Henderson B, Mottet A, Opio C, Dijkman J, Falcucci A, Tempio G (2013). *Tackling Climate Change through Livestock—A Global Assessment of Emissions and Mitigation Opportunities*. Food and Agriculture Organization of the United Nations, FAO, Rome, Italy.

Getu HM, Bansal PK (2008). Thermodynamic analysis of an R744–R717 cascade refrigeration system. *International Journal of Refrigeration* 31:45–54.

Geyer H, Seyfarth A, Reinhard Blickhan R (2005). Spring-mass running: Simple approximate solution and application to gait stability. *Journal of Theoretical Biology* 232:315–328.

Gharagheizi F, Ilani-Kashkouli P, Mohammadi AH (2014). A group contribution method for determination of the standard molar chemical exergy of organic compounds. *Energy* 70:288–297.

Gibbs CL, Chapman JB (1974). Effects of stimulus conditions, temperature, and length on energy output of and toad sartorious. *American Journal of Physiology* 227:964–971.

Gielen D, Tom C (2010). Proposal for energy and CO_2 emission indicators in the petrochemical sector. Available at: https://www.researchgate.net/profile/Cecilia_Tam/publication/237752174_Proposal_for_Energy_and_CO_2_Emission_Indicators_in_the_Petrochemical_Sector/links/0f31753ba82c23b3ec000000.pdf (accessed July 12, 2016).

Giese H, Snyder WK, van Oostrom C, van Steeg H, Dolle MET, Vijg J (2002). Age related mutation accumulation at lacZ reporter locus in normal and tumor tissues of Trp53-deficient mice. *Mutation Research* 514:153–163.

Gilchrist M, Greko C, Wallinga DB, Beran GW, Riley DG, Thorne PS (2007). The potential role of concentrated animal feeding operations in infectious disease epidemics and antibiotic resistance. *Environmental Health Perspectives* 115:313–316.

Goldemberg J, Coelho ST, Guardabasii P (2008). The sustainability of ethanol from sugarcane. *Energy Policy* 36:2086–2097.

Gomi K, Kajiyama N (2001). Oxyluciferin, a luminescence product of firefly luciferase, is enzymatically regenerated into luciferin. *Journal of Biological Chemistry* 276:36508–36513.

Gosline JM, DeMont ME (1985). Jet-propelled swimming in squids. *Scientific American* 252:96–103.

Gougeon R, Harrigan K, Tremblay J-F, Hedrei P, Lamarche M, Morais JA (2005). Increase in the thermic effect of food in women by adrenergic amines extracted from *Citrus Aurantium*. *Obesity Research* 13:1187–1194.

Granata GP, Brandon LJ (2002). The thermic effect of food and obesity: Discrepant results and methodological variations. *Nutrition Reviews* 60:223–233.

Green DW, Perry RH (1997). *Perry's Chemical Engineers' Handbook*, 7th ed. McGraw-Hill, New York.

Griffin TM, Main PP, Farley CT (2004). Biomechanics of quadrupedal walking: How do four-legged animals achieve inverted pendulum-like movements? *Journal of Experimental Biology* 207:3545–3558.

Grohe A, Hoelsch K (2012). Mechanistic model for the synthesis of N-acetylneuraminic acid using N-acetylneuraminate lyase from *Escherichia coli* K12. *Journal of Molecular Catalysis B: Enzymatic* 83:1–7.

Gronert S (2006). An alternative interpretation of the C–H bond strengths of alkanes. *Journal of Organic Chemistry* 71:1209–1219.

Gul S, Harasek M (2012). Energy saving in sugar manufacturing through the integration of environmental friendly new membrane process for thin juice pre-concentration. *Applied Thermal Engineering* 43:128–133.

Gungor A, Erbay Z, Hepbasli A (2011a). Exergetic analysis and evaluation of a new application of gas engine heat pumps (GEHPs) for food drying processes. *Applied Energy* 88:882–891.

Gungor A, Erbay Z, Hepbasli A (2011b). Exergoeconomic analyses of a gas engine driven heat pump drier and food drying process. *Applied Energy* 88:2677–2684.

Gunsolus JL (1999). Herbicide resistant weeds. North Central Regional Extension Publication No. 468. Available at: http://www.extension.umn.edu/agriculture/crops/weed-management/herbicide-resistant-weeds/ (accessed September 11, 2014).

Guo X, Honggui L, Xu H, Woo S, Dong H, Lu F, Lange AJ, Wu C (2012). Glycolysis in the control of blood glucose homeostasis. *Acta Pharmaceutica Sinica B* 2:358–367.

Guyton AC, Hall JE (2016). *Textbook of Medical Physiology*, 12th ed. Elsevier Saunders, New York.

Hahn E, Wild P, Hermanns U, Sebbel P, Glockshuber R, Häner M, Taschner N, Burkland P, Aebi U, Müller SA (2002). Exploring the 3D molecular architecture of *Escherichia coli* type 1 pili. *Journal of Molecular Biology* 323:845–857.

Hamalainen M (2005). Thermodynamics and information in aging: Why aging is not mystery and how we will be able to make rational interventions. *Rejuvenation Research* 8(1):29–36.

Hannam AG, Stavness I, Lloyd JE, Fels S (2008). A dynamic model of jaw and hyoid biomechanics during chewing. *Journal of Biomechanics* 41:1069–1076.

Hansson SE (2007). What is technological science? *Studies in History and Philosophy of Science-A* 38:523–527

Harman D (1992). Free radical theory of aging. *Mutation Research/DNA Aging* 275:257–266.

Harrison J, Fewell JH (2002). Environmental and genetic influences on metabolic rate in the honey bee, *Apis mellifera*. *Comparative Biochemistry and Physiology, Part A*, 133:323–333.

Hasselquist D, Nilsson J-A (2012). Physiological mechanisms mediating costs of immune responses: What can we learn from studies of birds? *Animal Behaviour* 83:1303–1312

Hawker DW, Cropp R, Boonsaner M (2013). Uptake of zwitterionic antibiotics by rice (*Oryza sativa L.*) in contaminated soil. *Journal of Hazardous Materials* 263:458–466.

Hayflick L (2007). Entropy explains aging, genetic determinism explains longevity, and undefined terminology explains misunderstanding both. *PLoS Genetics* 3:2351–2354.

Hayman DM, Xiao Y, Yao Q, Jiang Z, Lindsey ML, Han H-C (2012). Alterations in pulse pressure affect artery function. *Cellular and Molecular Bioengineering* 5:474–487.

Hayne DT (2008). *Biological Thermodynamics*. Cambridge University Press, Cambridge, U.K.

He Z-H, Chillingworth RK, Brune M, Corrie JET, Webb MR, Ferenczi MA (1999). The efficiency of contraction in rabbit skeletal muscle fibres, determined from the rate of release of inorganic phosphate. *The Journal of Physiology* 517.3:839–854.

Heglund NC, Cavagna GA (1987). Mechanical work, oxygen consumption, and efficiency in isolated frog and rat muscle. *American Journal of Physiology* 253: C22–C29.

Hellevang KJ (March 1995). Grain moisture content effects and management. North Dakota State University, Fargo, AE-905. Available at: https://www.ag.ndsu.edu/extension-aben/documents/ae905.pdf (accessed April 24, 2015).

Helsel ZR (1992). Energy and alternatives for fertilizer and pesticide use. In: Fluck RC (ed.), *Energy in Farm Production*, Vol. 6. Elsevier, New York, pp. 177–201.

Henriques AWS, Jessouroun E, Lima EL, Alves TLM (2006). Capsular polysaccharide production by *Neisseria meningitidis* serogroup C: Optimization of process variables using response surface methodology. *Process Biochemistry* 41:1822–1828.

Hepbasli A, Colak N, Hancioglu E, Icier F, Erbay Z (2010). Exergoeconomic analysis of plumb drying in a heat pump conveyor dryer. *Drying Technology* 28:1385–1395.

Hershey D (1974). *Lifespan and Factors Affecting It*. Charles C. Thomas Publisher, Springfield, MA.

Hershey D, Wang H (1980). *A New Age-Scale for Humans*. Lexington Books, Lexington, KY.

Hill AV (1938). The heat of shortening and dynamic constants of muscle. *Proceedings of the Royal Society of London B, Biological Sciences* 126:136–195.

Hinderink AP, Kerkof FPJM, Lie ABK, de Swaan Arons J, van der Kooi HJ (1996). Exergy analysis with flow sheet simulator. *Chemical Engineering Science* 51:4693–4700.

Hockachka P, Hartline P, Fields J (1977). Octopine as an end product of anaerobic glycolysis in the chambered nautilus. *Science* 195(4773):72–74.

Hodgkin AL, Huxley AF (1952). A quantitative description of membrane current and its application to conduction and excitation in nerve. *Journal of Physiology* 117(4):500–544.

Hodgkin AL, Rushton AH (1946). The electrical constants of a crustacean nerve fibre. *Proceedings of the Royal Society London* 133:444–479.

Hoffmann M, Hilton-Taylor C, Angulo A, Boehm M, Brooks TM, Butchart SHM, Carpenter KE et al. (2010). The impact of conservation on the status of the world's vertebrates. *Science* 330(6010):1503–1509.

Hollinger HB, Zenzen MJ (1982). An interpretation of macroscopic irreversibility within the Newtonian framework. *Philosophy of Science* 49:309–354.

Holmberg H, Ruohonen P, Ahtila P (2009). Determination of the real loss of power for a condensing and a backpressure turbine by means of second law analysis. *Entropy* 11(4):702–712.

Holmes JW (2006). Teaching from classic papers: Hill's model of muscle contraction. *Advances in Physiology Education* 30:67–72.

Holmgren M. Xsteam for MATLAB. www.x-eng.com. Available at: http://assets.openstudy.com/updates/attachments/4e134b420b8b56e555996beb-slnkktn-1310012884358-xsteamformatlab.pdf (accessed June 14, 2013).

Hopkins WG, Hüner NPA (2009). *Introduction to Plant Physiology*, 4th ed. John Wiley & Sons Hoboken, NJ.

Houshyar E, Sheikh Davoodi MJ, Nassiri SM (2010). Energy efficiency for wheat production using data envelopment analysis (DEA) technique. *Journal of Agricultural Technology* 6(4):663–672.

Hovelius K, Hansson P-A (1999). Energy- and exergy analysis of rape seed oil methyl ester (RME) production under Swedish conditions. *Biomass and Bioenergy* 17:279–291.

Hristov AN, Oh J, Firkins JL, Dijkstra J, Kebreab E, Waghorn G, Makkar HPS et al. (2013). Mitigation of methane and nitrous oxide emissions from animal operations: I. A review of enteric methane mitigation options. *Journal of Animal Science* 91:5045–5069.

Hrubec Z, Ryder RA (1980). Traumatic limb amputations and subsequent mortality from cardiovascular disease and other causes. *Journal of Chronic Diseases* 1233:239–250.

Hülsbergen KJ, Feil B, Biermann S, Rathke, GW, Kalk WD, Diepenbrock W (2001). A method of energy balancing in crop production and its application in a long-term fertilizer trial. *Agriculture, Ecosystems & Environment* 86:303–321.

Hürriyet Daily News, February 27, 2012, Anatolia News Agency, Istanbul, Turkey.

Hurst T (2010). Could high-tech plastic pallets mean the end for disposable wooden pallets? Earth and Industry. Available at: http://earthandindustry.com/2010/09/is-the-pallet-of-the-future-made-out-of-plastic (accessed January 17, 2016).

Hursting SD, Sarah M Dunlap SM, Ford NA, Hursting MJ, Laura M Lashinger LM (2013). Calorie restriction and cancer prevention: A mechanistic perspective. *Cancer & Metabolism* 1–10, doi:10.1186/2049-3002-1-10.

Huxley AF (1957). Muscle structure and theories of construction. *Progress in Biophysics and Biophysical Chemistry* 7:255–318.

Huxley AF, Niedergerke R (1954). Structural changes in muscle during contraction: Interference microscopy of living muscle Fibres. *Nature* 173:971–973.

Huxley H, Hanson J (1954). Changes in the cross-striations of muscle during contraction and stretch and their structural interpretation. *Nature* 173:973–976.

Huxley HE (2008). Memories of early work on muscle contraction and regulation in the 1950's and 1960's. *Biochemical and Biophysical Research Communications* 369:34–42.

Institute of Medicine (US) (2005). Dietary reference intakes for energy, carbohydrate, fiber, fat, fatty acids, cholesterol, protein, and amino acids, Vol. 1. National Academies Press Washington, DC.

Ishay JS (2004). Hornet flight is generated by solar energy: UV irradiation counteracts anaesthetic effects. *Journal of Electron Microscopy* 53:623–633.

Ishay JS, Sverdlov A, Pertsis V, Gavrilov Y, Steinberg D (2004). Light modulates electric phenomena in hornet cuticle. *Photochemistry and Photobiology* 80:115–118.

Jamison D, Jamison K (1968). A note on the entropy of partially known languages. *Information and Control* 12:164–167.

Janas T, Janas T (2011). Membrane oligo- and polysialic acids. *Biochimica et Biophysica Acta* 1808:2923–2932.

Jin J, Hassanzadeh P, Perotto G, Sun W, Brenckle MA, Kaplan D, Omenetto FG, Rolandi M (2013). A biomimetic composite from solution self-assembly of chitin nanofibers in a silk fibroin matrix. *Advanced Materials* 25:1–7, doi: 10.1002/adma.201301429.

Joback KG, Reid RC (1987). Estimation of pure-component properties from group-contributions. *Chemical Engineering Communications* 57:233–243.

Johnsen G, Wasteson Y, Heir E, Berget OI, Herikstad H (2001). *Escherichia coli* O157:H7 in faeces from cattle, sheep and pigs in the southwest part of Norway during 1998 and 1999. *International Journal of Food Microbiology* 65:193–200.

Johnson WH, Ambrose JR (1963). Heat of oxidation of aqueous sulfur dioxide with gaseous chlorine. *Journal of Research of the National Bureau of Standards* 67A:427–430.

Jorfeldt L, Wahren J (1971). Leg blood flow during exercise in man. *Clinical Science* 41:459–473.

Jorgensen SE (2015). New method to calculate the work energy of information and organisms. *Ecological Modeling* 295:18–20.

Jorgensen SE, Ladegaard N, Debeljak M, Marques JC (2005). Calculations of exergy for organisms. *Ecological Modeling* 185:165–175

Jorgensen SE, Ludovisi A, Nielsen SN (2010). The free energy and information embodied in the amino acid chains of organisms. *Ecological Modeling* 221:2388–2392.

Jorgensen SE, Nielsen SN (2007). Application of exergy as thermodynamic indicator in ecology. *Energy* 32:673–685.

Jorgensen SE, Fath B, Bastiononi S, Marques M, Müller F, Nielsen SN, Patten BC, Tiezzi E, Ulanowicz R (2007). *A New Ecology*. Systems Perspectives, Elsevier, Amsterdam.

Joshi VS, Bajaj IB, Survase SA, Singhal RS, Kennedy JF (2009). Meningococcal polysaccharide vaccines: A review. *Carbohydrate Polymers* 75:553–565.

Joubert F, Mazet JL, Mateo P, Hoerter JA (2002). 31P NMR detection of subcellular creatine kinase fluxes in the perfused rat heart: Contractility modifies energy transfer pathways. *Journal of Biological Chemistry* 277:18469–18476.

Joubert F, Wilding JR, Fortin, D, Domergue-Dupont, V, Novotova M, Ventura-Clapier R, Veksler V (2008). Local energetic regulation of sarcoplasmic and myosin ATPase is differently impaired in rats with heart failure. *The Journal of Physiology* 586:5181–5192.

Jubrias SA, Vollestad NK, Gronka RK, Martin J, Kushmerick MJ (2008). Contraction coupling efficiency of human first dorsal interosseous muscle. *The Journal of Physiology* 586.7:1993–2002.

Kallivroussis L, Natsis A, Papadakis G (2002). The Energy balance of sunflower production for biodiesel in Greece. *Biosystems Engineering* 81:347–354.

Kaltsas AM, Mamolos AP, Tsatsarelis CA, Nanos GD, Kalburtji KL (2007). Energy budget inorganic and conversion olive groves. *Agriculture, Ecosystems & Environment* 122:243–251.

Kapustenko PO, Ulyev LM, Boldyryev SA, Garev AO (2008). Integration of a heat pump into the heat supply system of a cheese production plant. *Energy* 33:882–889.

Kaweckyj N, Frye W, Hilling L, Lovering L, Schmitt L, Leeuw W (2011). The business of dentistry: Patient records and records management. Available at: http://dentalcare.com/media/en-US/education/ce390/ce390.pdf (accessed August 12, 2013).

Keeling AE. *Great Britain and Her Queen*. Project Gutenberg eBook. Available at http://www.gutenberg.org/etext/13103 (accessed August 15, 2016).

Keenan JH, Keyes FG (1936). *Thermodynamic Properties of Steam Including Data for the Liquid and Solid Phases*. John Wiley & Sons, New York.

Keenan JH, Shapiro AH (1947). History and exposition of the laws of thermodynamics. *Mechanical Engineering* 69:915–917.

Kennedy BK, Austriaco NR Jr., Guarente L (1994). Daughter cells of *Saccharomyces cerevisiae* from old mothers display a reduced life span. *The Journal of Cell Biology* 127:1985–1993.

Kernegger L, Carstensen J, Zaldívar J-M (2008). Application of specific eco-exergy to FAO fisheries data. *The Open Fish Science Journal* 1:11–18.

Ketzer LA, de Meis L (2008). Heat production by skeletal muscles of rats and rabbits and utilization of glucose 6-phosphate as ATP regenerative system by rats and rabbits heart Ca^{2+}-ATPase. *Biochemical and Biophysical Research Communications* 369:265–269.

Khan S, Khan MA, Hanjra MA, Mu J (2009). Pathways to reduce the environmental footprints of water and energy inputs in food production. *Food Policy* 34: 141–149.

Kirsen SD (2008). Life cycle assessment of PET bottle. Dokuz Eylul University Graduate School of Natural Applied Sciences. Available at: http://www.fbe.deu.edu.tr/all_files/tez_arsivi/2008/yl_t2427.pdf (accessed January 17, 2016).

Kita Y, Shimizu M, Shibayama S, Yoshio H, Ino H, Mabuchi H (1996). Correlation between mycardical dysfunction and changes in myosin isoenzymes in diabetic rat hearts. *Journal of Diabetes and Its Complications* 10:38–42.

Kita Y, Shimizu M, Sugihara N, Shimizu K, Yoshio H, Shibayama S, Tekada R (1991). Correlation between histopathological changes and mechanical dysfunction in diabetic rat hearts. *Diabetes Research and Clinical Practice* 11:177–188.

Kitagaki H, Takagi H (2014). Mitochondrial metabolism and stress response of yeast: Applications in fermentation technologies. *Journal of Bioscience and Bioengineering* 117:383–393.

Kitani K (2007). Pharmacological interventions in aging and age-associated disorders. *Geriatrics & Gerontology International* 7:97–103.

Kitani K, Minami C, Yamamoto T, Kanai S, Ivy GO, Carrillo M-C (2002). Pharmacological interventions in aging and age-associated disorders: Potential of propargylamines for human use. *Annals of the New York Academy of Sciences* 959:295–307.

Kitchin J. Interacting with the steam Entropy-temperature chart. MATLAB in ChemE@CMU, Posted on November 21, 2011. Available at: http://matlab.cheme.cmu.edu/2011/11/21/interacting-with-the-steam-entropy-temperature-chart/ (accessed May 13, 2015).

Klasing KC (1998). Nutritional modulation of resistance to infectious diseases. *Poultry Science* 77:1119–1125.

Klass DL (1998). Biomass for renewable energy and fuels. In: *Encyclopedia of Energy*. Elsevier, Oxford, U.K.

Klemes J, Smith R, Kim J (eds.) (2008). *Handbook of Water and Energy Management in Food Processing*, Woodhead, Cambridge, UK.

Kobayashi T, Honma K, Shingaki S, Nakajima T (2001). Changes in masticatory function after orthognathic treatment in patients with mandibular prognathism. *British Journal of Oral and Maxillofacial Surgery* 39:260–265.

Koch Al, Higgins ML, Doyle RJ (1981). Surface tension-like forces determine bacterial shapes: *Streptococcus faecium. Journal of General Microbiology* 123:151–161.

Koknaroglu H (2010). Cultural energy analyses of dairy cattle receiving different concentrate levels. *Energy Conversion and Management* 51:955–958.

Kolovos N, Georgakopoulos A, Filippidis A, Kavouridis C (2002). Utilization of lignite reserves and simultaneous improvement of dust emissions and operation efficiency of a power plant by controlling the calcium (total and free) content of the fed lignite. Application on the Agios Dimitrios power plant, Ptolemanis, Greece. *Energy & Fuels* 16:1516–1522.

Kongshaug G (1998). Energy consumption and greenhouse gas emissions in fertilizer production. In: *IFA Technical Conference*, Marrakech, Morocco.

Kosmadaki MG, Gilhrest BA (2004). The role of telomeres in skin aging/photoaging. *Micron* 35:155–159.

Kotas TJ (1996). *The Exergy Method of Thermal Plant Analysis*. Krieger, Melbourne, FL.

Koubassova NA, Tsaturyan AK (2011). Molecular mechanism of actin–myosin motor in muscle. *Biochemistry (Moscow)* 76:1484–1506.

Kribs JD, Spolek GA (1997). Drying energy conservation for deep-bed barley-malt kilns. *Journal of Agricultural Engineering Research* 68:367–373.

Küçük K, Tehavita R, Sorgüven E, Demirel Y, Özilgen M (2015). Bioenergetics of growth and lipid production in *Chlamydomonas reinhardtii*. *Energy* 83:503–510.

Lafel L (1999). Ketone bodies: A review of physiology, pathophysiology and application of monitoring to diabetes. *Diabetes Metabolic Research Reviews* 15:412–426.

Lai CY, Jaruga E, Borghouts C, Jazwinski SM (2002). A mutation in the ATP2 gene abrogates the age asymmetry between mother and daughter cells of the yeast *Saccharomyces cerevisiae*. *Genetics* 162:73–87.

Laidler KJ (1980). *Physical Chemistry with Biological Applications*. Benjamin/Cummings Publishing Co., Menlo Park, CA, p. 167.

Lal R (2004). Carbon emission from farm operations. *Environment International* 30:981–990.

Lattouf R, Oliveira S, Exergy analysis of environmental impact mitigation processes. In: *17th International Congress of Mechanical Engineering*, Sao Paulo, Brazil, November 10–14 2003. Available at: http://www.abcm.org.br/anais/cobem/2003/html/pdf/COB03-1900.pdf (accessed January 28, 2016).

Legramandi MA, Schepens B, Cavagna GA (2013). Running humans attain optimal elastic bounce in their teens. *Scientific Reports* 3:1310, doi:10.1038/srep01310.

Lems S (2009). Thermodynamic explorations into sustainable energy conversion, learning from living systems. PhD dissertation, Technical University of Delft, Rotterdam, the Netherlands.

Lems S, van der Kooi HJ, Arons JD (2009). The second-law implications of biochemical energy conversion: Exergy analysis of glucose and fatty-acid breakdown in the living cell. *International Journal of Exergy* 6:228–248.

Lewins J (2011). Entropy pollution of the environment: A teaching approach to the second law. *International Journal of Mechanical Engineering Education* 39(1): 60–67.

Li X, Dash RK, Pradhan RK, Qi F, Thompson M, Vinnakota KC, Wu F, Yang F, Beard DA (2010). A database of thermodynamic quantities for the reactions of glycolysis and the tricarboxylic acid cycle. *The Journal of Physical Chemistry* B 114:16068–16082.

Li X, Dash RK, Pradhan RK, Qi F, Thompson M, Vinnakota KC, Wu F, Yang F, Beard DA (2010). A database of thermodynamic quantities for the reactions of glycolysis and the tricarboxylic acid cycle. *The Journal of Physical Chemistry* B 114:16068–16082.

Li H, Sharma LK, Li YP, Idowu A, Liu D, Lu J, Bai Y (2013). Comparative bioenergetic study of neuronal and muscle mitochondria during aging. *Free Radical Biology and Medicine* 63:30–40.

Lights of America (2014). Fluorex replacement bulb, model 9166B, 65W, 3M, China specification sheet . https://www.amazon.com/LIGHTS-AMERICA-9166B-Fluorex-Replacement/dp/B000BO6G94 (accessed July 13, 2016).

Lin ECC, Hirota Y, Jacob F (1971). On the process of cellular division in *Escherichia coli*. VI. Use of a methocel-autoradiographic method for the study of cellular division in *Escherichia coli*. *Journal of Bacteriology* 108:375–385.

Linari M, Woledge RC, Curtin NA (2003). Energy storage during stretch of active single fibers from frog skeletal muscle. *The Journal of Physiology* 548.2:461–474.

Linko M, Haikara A, Ritala A, Penttila M (1998). Recent advances in the malting and brewing industry. *Journal of Biotechnology* 65:85–98.

Long SP, Zhu X-G, Naidu SL, Ort DR (2006). Can improvement in photosynthesis increase crop yields? *Plant Cell and Environment* 29:315–330.

Loyon L, Burton CH, Misselbrook T, Webb J, Philippe FX, Aguilar M, Doreau M et al. (2016). Best available technology for European livestock farms: Availability, effectiveness and uptake. *Journal of Environmental Management* 166:1–11.

Luo LF (2009). Entropy production in cancer and reversal of entropy flow as an anticancer therapy. *Frontiers of Physics in China* (4):122–139.

Luria SE, in Gunsalus I, Stainer RY (eds.) (1960). *The Bacteria*, Vol. 1, Chapter 1. Academic Press, New York.

Lusting HC (2006). Childhood obesity: Behavioral aberration or biochemical drive? Reinterpreting the first law of thermodynamics. *Nature of Clinical Practice Endocrinology & Metabolism* 2:447–458.

Ma C-M, Chen M-H, Hong G-B (2012). Energy conservation status in Taiwanese food industry. *Energy Policy* 50:458–463.

Machado MF, Saraiva JM (2005). Thermal stability and activity regain of horseradish peroxidase in aqueous mixtures of imidazolium-based ionic liquids. *Biotechnology Letters* 27:1233–1239.

Macklem PT (2008). Emergent phenomena and the secrets of life. *Journal of Applied Physiology* 104:1844–1846.

Mady CEK, de Olivera S (2013). Human body exergy metabolism. *International Journal of Thermodynamics* 16:73–80.

Mady CEK, Ferreira MS, Yanagihara JI, Saldiva PHN, de Olivera S (2012). Modeling the exergy behavior of human body. *Energy* 45:546–553.

Magid J, Granstedt A, Dýrmundsson Ó, Kahiluoto H, Ruissen T. Urban areas—Rural areas and recycling—The organic way forward? *Proceedings from NJF*, Seminar No. 327, Copenhagen, Denmark, August 20–21, 2001.

Mahan B (1969). *University Chemistry*, 2nd ed. Addison-Wesley Pub. Co., Manila, Philippines.

Mariunas M, Daunoravicien K, Griskevicius J, Andrasiut J (2008). Research of relation between muscle biosignal and systolic blood pressure, and application of its characteristics for evaluation of efficiency. *Journal of Vibroengineering* 10:329–334.

Markham MR (2013). Electrocyte physiology: 50 years later. *The Journal of Experimental Biology* 216:2451–2458.

Marrero J, Gani R (2001). Group-contribution based estimation of pure component properties. *Fluid Phase Equilibria* 183–184:183–208.

Martinez-Salas E, Martin JA, Vicente M (1981). Relationship of *Escherichia coli* density to growth rate and cell age. *Journal of Bacteriology* 147:97–100.

Masanet E, Worrel E, Graus W, Galitsky C (2008). Energy efficiency improvement and cost saving opportunities for the fruit and vegetable processing industries: An energy star guide for energy and plant managers. LBNL-59289-Revision, Lawrence Berkeley National Laboratory, Berkeley, CA. Available at http://www.energystar. gov/ia/business/industry/Food-Guide.pdf (accessed October 19, 2014).

Masanet ER, Therkelsen PR, Worrell E (2012). Energy efficiency improvement and cost saving opportunities for the baking industry. Energy Star Guide for Plant and Energy Managers. LBNL-6112E-Revision, Lawrence Berkeley National Laboratory, Berkeley, CA. Available at: http://eetd.lbl.gov/sites/all/files/baking_ guide_final_28dec2012c.pdf (accessed October 19, 2014).

Matsumota T, Akutsu S, Wakana N, Morito M, Shimada S, Yamane A (2006). The expression of insulin-like growth factors, their receptors, and binding proteins are related to the mechanism regulating masseter muscle mass in the rat. *Archives of Oral Biology* 51:603–611.

Matsushita S, Matsushita M, Itoh H, Hagiwara K, Takahashi R, Ozawa T, Kuramota K (2003). Multiple pathology and tails of disability: Space–time structure of disability in longevity. *Geriatrics & Gerontology International* 3:189–199.

Metting FB (1996). Biodiversity and application of microalgae. *Journal of Industrial Microbiology and Biotechnology* 17:477–489.

Minea V (2013a). Drying heat pumps—Part I: System integration. *International Journal of Refrigeration* 36:643–658.

Minea V (2013b). Drying heat pumps—Part II: Agro-food, biological and wood products. *International Journal of Refrigeration* 36:659–673.

Miyamato K (ed.) (1997). Chapter 1: Biological Energy production. In: *Renewable Biological Systems for Alternative Sustainable Energy Production*. Agricultural Services Bulletin 128, FAO, Rome, Italy, p. 2.

Modan M, Peles E, Halkin H, Nitzan H, Azaria M, Sanford Gitel S, Dolfin D, Modan B (1998). Increased cardiovascular disease mortality rates in traumatic lower limb amputees. *The American Journal of Cardiology* 82:1242–1247.

Moradi H, Grzymala-Busse JW, Roberts JA (1998). Entropy of English text: Experiments with humans and a machine learning system based on rough sets. *Information Sciences* 104:31–47.

Moran MJ (1982). *Availability Analysis: A Guide to Efficient Energy Use.* Prentice-Hall, Englewood Cliffs, NJ.

Morán-Zorzano M, Alonso-Casajus N, Munoz FJ, Viale AM, Baroja-Fernandez E, Eydallin G, Pozueta-Romero J (2007). Occurrence of more than one important source of ADPglucose linked to glycogen biosynthesis in *Escherichia coli* and Salmonella. *FEBS Letters* 581:4423–4429.

Morgenthaler FD, Koski DM, Rudolf Kraftsik, Henry P-G, Gruetter R (2006). Biochemical quantification of total brain glycogen concentration in rats under different glycemic states. *Neurochemistry International* 48:616–622.

Morley D, Litwak K, Ferber P, Spence P, Dowling R, Meyns B, Griffith B, Burkhoff D (2007). Hemodynamic effects of partial ventricular support in chronic heart failure: Results of simulation validated with in vivo data. *Journal of Thoracic and Cardiovascular Surgery* 133:21–28.

Murphy H, Oddy DJ (2013). Some clarifications but yet more questions regarding the early days of the New Zealand frozen-meat trade. *The Mariner's Mirror* 99:212–219.

Murr R, Thieriot H, Zoughaib A, Clodic D (2011). Multi-objective optimization of a multi water-to-water heat pump system using evolutionary algorithm. *Applied Energy* 88:3580–3591.

Musee N, Lorenzen L, Aldrich C (2007). Cellar waste minimization in the wine industry: A systematic approach. *Journal of Cleaner Production* 15:417–431.

Nakamura M, Maki S, Amano Y, Ohkita Y, Niwa K, Hirano T, Ohmiya Y, Niwa H (2005). Firefly luciferase exhibits bimodal action depending on the luciferin chirality. *Biochemical and Biophysical Research Communications* 331:471–475.

The National Institute of Standards and Technology (2014). *NIST Chemistry WebBook*, Gaithersburg, MD. Available at http://webbook.nist.gov/chemistry/ (accessed November 8, 2014).

Nelson DL, Cox MM (2013). *Lehninger, Principles of Biochemistry*, 6th ed. Freeman, New York.

Neptune RR, Kautz SA (2001). Muscle activation and deactivation dynamics: The governing properties in fast cyclical human movement performance. *Exercise and Sport Sciences Review* 29:76–81.

Neptune RR, van der Bogert AJ (1998). Standard mechanical energy analyses do not correlate with the muscle work in cycling. *Journal of Biomechanics* 31:239–245.

Niaousnakis M, Halvadakis CP (2006). *Olive Processing Waste Management Literature Review and Patent Survey*, 2nd ed. Pergamon, Athens, Greece.

Nilsson J-A (2002). Metabolic consequences of hard work. *Proceedings of the Royal Society of London, Part B* 269:1735–1739.

Nilsson K, Flysjö A, Davis J, Sim S, Unger N, Bell S (2010). Comparative life cycle assessment of margarine and butter consumed in the UK, Germany and France. *International Journal of Life Cycle Assessment* 15:916–926.

Nuibe T (2007). *Energy Intensity in Industrial Sub Sectors*. The Energy Conservation Center, Tokyo, Japan. Available at: http://www.asiaeec-col.eccj.or.jp/eas/01/pdf/02.pdf.

Nylund N-O, Erkkila K (2005). Heavy-duty truck emissions and fuel consumption simulating real-world driving laboratory conditions. Paper presented in the *DEER Conference*, Chicago, IL, August 21—25, 2005. Available at: http://www1.eere.energy.gov/vehiclesandfuels/pdfs/deer_2005/session5/2005_deer_erkkila.pdf (accessed August 5, 2015).

Nyström T (2007). A bacterial kind of aging. *PLoS Genetics* 3:2355–2357.

O'Neill MC, Schmitt D (2012). The gaits of primates: Center of mass mechanics in walking, cantering and galloping ring-tailed lemurs, *Lemur catta*. *Journal of Experimental Biology* 215:1728–1739.

Oren MN, Ozturk HH (2006). An analysis of energy utilization for sustainable wheat and cotton production in Southeastern Anatolia region of Turkey. *Journal of Sustainable Agriculture* 29:119–130.

Ormerod JOM, Ashrafian H, Frenneaux MP (2008). Impaired energetics in heart failure—A new therapeutic agent. *Pharmacology & Therapeutics* 119:264–274.

Özilgen M (1988). Kinetics of amino acid production by over-producer mutant microorganisms. *Enzyme and Microbial Technology* 10:110–114.

Özilgen M (2011). *Endüstrileşme Sürecinde Bilgi Birikiminin Öyküsü (Build-up of Knowledge During the Progression of Industrialization)*, 2nd ed., Arkadaş Yayınevi, Ankara (In Turkish).

Özilgen M (2014). *Artistic Narrative of Technology*. Yeditepe University Press, Istanbul, Turkey.

Özilgen M. Energy utilization and carbon dioxide emission in beverage industry. Submitted for publication.

Özilgen M, Ollis DF, Ogrydziak D (1988). Kinetics of batch fermentations with *Kluyveromyces fragilis*. *Enzyme and Microbial Technology* 10:165–172.

Özilgen M, Sorgüven E (2011). Energy and exergy utilization and carbon dioxide emission in vegetable oil production. *Energy* 36:5954–5967.

Özkan B, Akcaöz H, Karadeniz F (2004). Energy requirement and economic analysis of citrus production in Turkey. *Energy Conversion and Management* 45(11–12): 1821–1830.

Özkan B, Fert C, Karadeniz CF (2007). Energy and cost analysis for greenhouse and open-field grape production. *Energy* 32:1500–1504.

Paphangkorakit J, Chaiyapanya N, Sriladlao P, Pimsupa S (2008). Determination of chewing efficiency using muscle work. *Archives of Oral Biology* 53:533–537.

Park S-H (1992). Falling oil prices and exchange rate fluctuation. In: Shojai S, Katz BS (eds.), *The Oil Market in the 1980s*. Praeger, New York.

PAS (2008). *Guide to PAS 2050, How to Assess the Carbon Footprint of Goods and Services*. Carbon Trust, London, UK.

Patzek TW (2004). Thermodynamics of the corn—Ethanol biofuel cycle. *Critical Reviews in Plant Sciences*, 23:519–567.

Pavy-Le Traon A, Heer M, Narici MV, Rittweger J, Vernikos J (2007). From space to Earth: Advances in human physiology from 20 years of bed rest studies (1986–2006). *European Journal of Applied Physiology* 101:143–194.

Peakall DB, Witt PN (1976). The energy budget of an orb web building spider. *Comparative Biochemistry and Physiology* 54:187–190.

Pearce JD, Edwards MS, Craven TE, English WP, Mondi MM, Reavis SW, Hansen KJ (2005). Renal duplex parameters, blood pressure, and renal function in elderly people. *American Journal of Kidney Diseases* 45:842–850.

Peiro LT, Lombardi L, Méndez GV, Durany XG (2010a). Life cycle assessment (LCA) and exergetic life cycle assessment (ELCA) of the production of biodiesel from used cooking oil (UCO). *Energy* 35:889–893.

Peiro LT, Méndez GV, Sciubba E, Durany XG (2010b). Extended exergy accounting applied to biodiesel production. *Energy* 35:2861–2869.

Pellegrini LF, de Oliveira S Jr (2011). Combined production of sugar, ethanol and electricity: Thermoeconomic and environmental analysis and optimization. *Energy* 36:3704–3715.

Penn State College of Agricultural Sciences Extension. Suggested insecticides, rates, and restrictions for small-grain insect control. Available at: http://extension.psu.edu/agronomy-guide/pm/tables/table-2-5-8 (accessed July 24, 2014).

Perry RH, Chilton CH (1973). *Chemical Engineer's Handbook*, 5th ed. McGraw Hill, New York.

Perry RH, Green DW, Maloney JO (1997). *Perry's Chemical Engineers Handbook*, 7th ed. RR Donnelley & Sons Company Chicago, IL.

Petela R (2008). An approach to the exergy analysis of photosynthesis. *Solar Energy* 82:311–328.

Petala R (2010) *Engineering Thermodynamics of Thermal Radiation for Solar Power Utilization*. McGraw Hill, New York.

Phillips S, Norton R (2012). Global wheat production and fertilizer use. *Better Crops* 96(3):4–6.

Pimentel D (1980). *Handbook of Energy Utilization in Agriculture*. CRC Press, Boca Raton, FL.

Pimentel D (1991). Ethanol fuels: Energy, security, economics, and the environment. *Journal of Agricultural and Environmental Ethics* 4:1–13.

Pimentel D, Patzek TW (2005). Ethanol production using corn, switchgrass, and wood; biodiesel production using soybean and sunflower. *Natural Resources Research* 14:65–76.

Pishgar-Komleh SH, Ghahderrijani M, Sefeedpari P (2012). Energy consumption and CO_2 emissions of potato production based on different farm size levels in Iran. *Journal of Cleaner Production* 33:183–191.

Plotkin M, Hod I, Zaban A, Boden SA, Bagnall DM, Galushko D, Bergman DJ (2010). Solar energy harvesting in the epicutile of the oriental hornet (*Vespa orientalis*). *Naturwissenschaften* 97:1067–1076.

Polyak K, Li Y, Zhu H et al. (1998). Somatic mutations of the mitochondrial genome in human colorectal tumors. *Nature Genetics* 20:291–293.

Polychchronaki EA, Douma AC, Dantsism TH, Giourga C, Lomou A (2007). Energy analysis as indicator of agricultural sustainability: The case of Western Macedonia and Epirus, Greece. In: *Proceedings of the 10th International Conference on Environmental Science and Technology*, Kos Island, Greece, 2007, pp. A1190–A1198.

Poritosh R, Nei D, Okadome H, Nakamura N, Orikasa T, Shiina T (2007). Life cycle inventory analysis of fresh tomato distribution systems in Japan considering the quality aspect. *Journal of Food Engineering* 86:225–233.

Priambodo A, Kumar S (2001). Energy use and carbon dioxide emission of Indonesian small and medium scale industries. *Energy Conversion and Management* 42:1335–1348.

Prigogine I (December 8, 1977). Time structure and fluctuations. Nobel Lecture. Available at: http://pchen.ccer.edu.cn/homepage/Time,%20Structure,%20and%20 Fluctuations.pdf (accessed April 7, 2014).

Prigogine I, Wiame JM (1946). Biologie et thermodynamique des phénomènes irréversibles. *Experientia* 2:451–453.

Pringle AT, Forsdyke J, Rose AH (1979). Scanning electron microscope of *Saccharomyces cerevisiae* spheroplast formation. *Journal of Bacteriology* 140:289–293.

Ptasinski KJ, Prinsa MJ, Pierika A (2007). Exergetic evaluation of biomass gasification. *Energy* 32:568–574.

Qin Z, Compton BG, Lewis JA, Markus J, Buehler MJ (2015). Structural optimization of 3D-printed synthetic spider webs for high strength. *Nature Communications* 6:7038, doi: 10.1038/ncomms8038.

Raberg L, Nillson JA, Ilmonen P, Stjernman M, Hasselquist D (2000). The cost of an immune response: Vaccination reduces parental effort. *Ecology Letters* 3:382–386.

Radegran G, Saltin B (1999). Human femoral artery diameter in relation to knee extensor muscle mass, peak blood flow, and oxygen uptake. *AJP-Heart* 278: H162–H167.

Rafai S, Jibuti L, Peyla P (2010). Effective viscosity of microswimmer suspensions. *Physics Review Letters* 104:098102.

Rakopoulos CD, Giakoumis EG (2006). Second-law analyses applied to internal combustion engines operation. *Progress in Energy and Combustion Science* 32:2–47.

Rall W (1977). Core conductor theory and cable properties of neurons. In: Kandel ER (ed.), *Handbook of Physiology, the Nervous System, Cellular Biology of Neurons.* American Physiological Society, Bethesda, MD.

Ramirez CA, Blok K, Neelis M, Patel M (2006a). Adding apples and oranges: The monitoring of energy efficiency in the Dutch food industry. *Energy Policy* 34: 1720–1735.

Ramirez CA, Patel M, Blok K (2006b). From fluid milk to milk powder: Energy use and energy efficiency in the European dairy industry. *Energy* 31:1984–2004.

Ramirez CA, Patel M, Blok K (2006c). How much energy to process one pound of meat? A comparison of energy use and specific energy consumption in the meat industry of four European countries. *Energy* 31:2047–2063.

Rastogi SC (2007). *BioChemistry*, 2nd ed. McGraw-Hill New Delhi, IN.

Reda ALDL (1996). *Simulation and Control of Stormwater Impacts on River Water Quality.* University of London, London, U.K.

Ren T, Christie P, Wang J, Chen Q, Zhang F (2010). Root zone soil nitrogen management to maintain high tomato yields and minimum nitrogen losses to the environment. *Scientia Horticulturae* 125:25–33.

Reggiani C, Potma EJ, Bottinelli R, Canepari M, Pellegrino MA, Stienen GJM (1997). Chemo-mechanical energy transduction in relation to myosin isoform composition in skeletal muscle fibers of the rat. *The Journal of Physiology* 502.2:449–460.

Ribeiro JP, Chiappa GR, Callegaro CC (2012). The contribution of inspiratory muscles function to exercise limitation in heart failure: Pathophysiological mechanisms. *Brazilian Journal of Physical Therapy* 16:261–267.

Richardson RB (2009). Ionizing radiation and aging: Rejuvenating an old idea. *Aging* 1:887–902.

Rikimaru H, Kikuchi M, Itoh M, Tashiro M, Watanabe M (2001). Mapping energy metabolism in jaw and tongue muscles during chewing. *Journal of Dental Research* 8:1849–1853.

Rinzel J, Rall W (1974). Transient response in a dendritic neuron model for current injected at one branch. *Biophysical Journal* 14:759–790.

Rittié L, Fisher GJ (2002). UV-light-induced signal cascades and skin aging. *Ageing Research Reviews* 1:705–720.

Rittman BE, McCarty PL (2001). *Environmental Biotechnology: Principles and Applications.* McGraw-Hill Higher Education, New York.

Rivero R, Garfias M (2006). Standard chemical exergy of elements updated. *Energy* 31:3310–3326.

Rodriguez-Gonzales O, Buckow R, Koutchma T, Balasubramaniam VM (2015). Energy requirements for alternative processing technologies—Principles, assumptions, and evaluation of efficiency. *Comparative Reviews in Food Science* 14:536–554.

Rose GJ (December 2004). Insights into neural mechanisms and evolution of behaviour from electric fish. *Nature Reviews Neuroscience* 5:943–951.

Rossmann MG, Mesyanzhinov VV, Arisaka F, Leiman PG (2004). The bacteriophage T4 DNA injection machine. *Current Opinion in Structural Biology* 14:171–180.

Roy P, Nei D, Okadome H, Nakamura N, Orikasa T, Shiina T (2008). Life cycle inventory analysis of fresh tomato distribution systems in Japan considering the quality aspects. *Journal of Food Engineering* 86:225–233.

Rubenstein D, Yin W, Frame M (2011). *Biofluid Mechanics: An Introduction to Fluid Mechanics, Macrocirculation, and Microcirculation*, Elsevier, Boston, MA.

Ruggieri L, Cadena E, Martínez-Blanco J, Gasol CM, Rieradevall J, Gabarrell X, Gea T, Sort X, Sánchez A (2012). Recovery of organic wastes in the Spanish wine industry. Technical, economic and environmental analyses of the composting process. *Journal of Cleaner Production* 17:830–838.

Salminen A, Kaarniranta K (2010). Genetics vs. entropy: Longevity factors suppress the NF-κB-driven entropic aging process. *Ageing Research Reviews* 9:298–314.

Sanchez IC (2011). Entropy of living versus non-living systems. *Journal of Modern Physics* 2:654–657.

SantaLucia J Jr., Hicks D (2004). The thermodynamics of DNA structural motifs. *Annual Review of Biophysics and Biomolecular Structure* 33:415–440.

Savery T (1702). The Miners Friend or, an engine to raise water by fire described: and of the manner of fixing it in mines: with an account of the several other uses it is applicable unto, and an answer to the objections made against it. Printed for S. Crouch, London, U.K. (accessed January 7, 2014).

Schmolz E, Hoffmeister D, Lamprecht I (2002). Calorimetric investigations on metabolic rates and thermoregulation of sleeping honeybees (*Apis mellifera carnica*). *Thermochimica Acta* 282:221–227.

Schrödinger E (1944). *What Is Life? The Physical Aspects of a Living Cell*. Cambridge University Press, Cambridge, U.K.

Schulz GE, Schirmer RH (1979). *Principles of Protein Structure*. Springer-Verlag, New York.

Scottish Shipbuilding Innovations, http://www.martinfrost.ws/.

Senel HS (1986). Hayvan Beslenme. Istanbul Universitesi, Veteriner Fakültesi Yayınları, Istanbul, sayfa 141–171.

Sengupta B, Stemmler M, Laughlin SB, Niven JE (2010). Action potential energy efficiency varies among neuron types in vertebrates and invertebrates. *PLoS Computational Biology* 6(7): e1000840.

Seyfried TN, Kiebish MA, Marsh J, Shelton LM, Huysentruyt LC, Mukherjee P (2011). Metabolic management of brain cancer. *Biochimica et Biophysica Acta* 1807:577–594.

Shamir L, Wolkow CA, Ilya G (2009). Goldberg. Quantitative measurement of aging using image texture entropy. *Bioinformatics* 25:3060–3063.

Shannon CE (July/October, 1948). A mathematical theory of communication. *Bell System Technical Journal* 27: 379–423.

Shannon CE (January 1951). Prediction and entropy of printed English. *Bell System Technical Journal* 30 50–64.

Sharma A, Chen CR, Lan NV (2009). Solar-energy drying systems: A review. *Renewable and Sustainable Energy Reviews* 13:1185–1210.

Sheehan J, Camobreco V, Duffield J, Grabowski M, Shapouri H (1998a). Life cycle inventory of biodiesel and petroleum diesel for use in an urban bus. Final Report, USDA, USDE, National Renewable Energy Laboratory, Golden, CO. Available at: http://www.nrel.gov/docs/legosti/fy98/24089.pdf.

Sheehan J, Dunahay T, Benemann J, Roessler P (1998b). *A Look Back at the US Department of Energy's Aquatic Species Program—Biodiesel from Algae*. National Renewable Energy Laboratory, Golden, CO.

Shieh JA, Fan LT (1982). Estimation of energy (enthalpy) and exergy (availability) contents in structurally complicated materials. *Energy Sources* 6:1–46.

Shier D, Buttler J, Lewis R (2015). *Hole's Essentials of Human Anatomy and Physiology*, 12th ed. McGraw-Hill, International ed, New York.

Shoemaker JK, MacDonald MJ, Hughson RL (1997). Time course of brachial artery diameter responses to rhythmic handgrip exercise in humans. *Cardiovasclar Research* 35:125–131.

Siclair DA, Guarente L (1997). Extrachromosomal rDNA circles–A cause of aging in yeast. *Cell* 91:1033–1042.

Silflow CD, Lefebvre PA (2001). Assembly and motility of eukaryotic clia and flagella. Lessons from *Chlamydomonas reinhardtii*. *Plant Physiology* 127:1500–1507.

Silva CA, Annamalai K (2008). Entropy generation and human aging: Lifespan entropy and effect of physical activity level. *Entropy* 10:100–123.

Silva CA, Annamalai K (2009). Entropy generation and human aging: Lifespan entropy and effect of diet composition and caloric restriction diets. *Journal of Thermodynamics* 10:100–123. doi:10.1155/2009/186723

Singh JM (2002). On farm energy use pattern in different cropping systems in Haryana India. MSc thesis, International Institute of Management University of Flensburg, Flensburg, Germany.

Skhiri M, Hunt SA, Denault AY, Haddad F (2010). Evidence-based management of right heart failure: A systematic review of an empiric field. *Revista Española de Cardiología* 63:451–471.

Smith E (2008). Thermodynamics of natural selection II: Chemical Carnot cycles. *Journal of Theoretical Biology*, 252:198–212.

Smith NP, Barclay CJ, Loiselle DS (2008). The efficiency of muscle contraction. *Progress in Biophysics & Molecular Biology* 88:1–58.

Smith JM, van Ness HC, Abbott MM (2005). *Introduction to Chemical Engineering Thermodynamics*, 7th ed. McGraw-Hill, Singapore.

Smyth M, Nesbitt A (2014). Energy and English wine production: A review of energy use and benchmarking. *Energy for Sustainable Development* 23:85–91.

Snyder CS, Bruulsema TW, Jensen TL, Fixen PE (2009). Review of greenhouse gas emissions from crop production systems and fertilizer management effects. *Agriculture, Ecosystems & Environment* 133:247–266.

Solomon BD, Krishna K (2011). The coming sustainable energy transition: History, strategies, and outlook. *Energy Policy* 39:7422–7431.

Sorgüven E, Özilgen M (2010). Thermodynamic assessment of algal biodiesel utilization. *Renewable Energy* 35:1956–1966.

Sorgüven E, Özilgen M (2011). Energy and exergy utilization and carbon dioxide emission in vegetable oil production. *Energy* 36:5954–5967.

Sorgüven E, Özilgen M (2012). Energy utilization, carbon dioxide emission, and exergy loss in flavored yogurt production process. *Energy* 40:214–225.

Sorgüven E, Özilgen M (2013). Thermodynamic efficiency of synthesis, storage and breakdown of the high-energy metabolites by photosynthetic microalgae. *Energy* 58:679–687.

Sorgüven E, Özilgen M (2015). First and second law work production efficiency of a muscle cell. *International Journal of Exergy* 18:142–156.

Soy Stats (2009) World Statistics, 2009. Available at: http://www.soystats.com/2009/Default-frames.htm.

Spakowitz AJ, Wang Z-G (2005). DNA packaging in Bacteriophage: Is twist important? *Biophysical Journal* 88:3912–3925.

Spolaore P, Joannis-Cassan C, Duran E, Isambert A (2006). Commercial application of microalgae. *Journal of Bioscience and Bioengineering* 101:87–96.

Stabentheiner A, Kovac H, Hetz SK, Kafer H, Stabentheiner G (2012). Assessing honeybee and wasp thermoregulation and energetics-New insights by combination of flow-through resprometry with infrared thermography. *Thermochimica Acta* 534:77–86.

Steinsaltz D, Mohan G, Kolb M (2012). Markov models of aging: Theory and practice. *Experimental Gerontology* 47:792–802.

Steven A (1993). Conformational change—An alternative energy source? *Biophysical Journal* 65:5–6.

Stewart TJ, Jackson DR, Smith DS, Shannon SF, Cremo CR, Baker JE (2013). Actin sliding velocities are influenced by the driving forces of actin-myosin binding. *Cellular and Molecular Bioengineering* 6:26–37.

Stewart EJ, Madden R, Paul G, Taddei F (2005). Aging and death in an organism that reproduces by morphologically symmetric division. *PLoS Biol* 3:295–300.

Stock JB, Rauch B, Roseman S (1977). Periplasmic space in *Salmonella typhimirium* and *Escherichia coli*. *Journal of Biological Chemistry* 252:7850–7861.

Struchtruo H, Rosen M (2002). How much work is lost in an irreversible turbine? *Exergy, an International Journal* 2:152–158.

Stryer L. (1981). *Biochemistry*, 2nd ed. WH Freeman, San Francisco, CA.

Sturm B, Hugenschmidt S, Joyce S, Hofacker W, Roskilly AP (2013). Opportunities and barriers for efficient energy use in a medium-sized brewery. *Applied Thermal Engineering* 53:397–404.

Sumithran P, Proietto J (2008). Ketogenic diets for weight loss: A review of their principles, safety and efficacy. *Obesity Research & Clinical Practice* 2:1–13.

Szargut J, Morris DR, Steward FR (1988). *Exergy Analysis of Thermal, Chemical, and Metallurgical Processes*. Hemisphere Publishing, New York.

Szargut J (2007). *Egzergia poradnik obliczania i stosowania*. Editorial of Silesian University of Technology, Gliwice, Poland.

Szargut J, Styrylska T (1964). Approximate evaluation of the exergy of fuels. *Brennst Waerme Kraft* 16:589–596 (In German).

Szargut J, Valero A, Stanek W, Valero A (2005). Towards an international reference environment of chemical exergy. Elsevier Science. Available at: http://www.exergoecology.com/papers/towards_int_re.pdf.

Takahashi T, Cho RY, Murata T, Mizuno T, Kikuchi M, Mizukami K, Kosaka H, Takahashi K, Wada Y (2009). Age-related variation in EEG complexity to photic stimulation: A multiscale entropy analysis. *Clinical Neurophysiology* 120:476–483.

Talens L, Villalba G, Gabarrell X (2007). Exergy analysis applied to biodiesel production. *Resources Conservation and Recycling* 51:397–407.

Tataranni PA, Larson DE, Snitker S, Ravussin E (1995). Thermic effect of food in humans: Methods and results from use of a respiratory chamber. *American Journal of Clinical Nutrition* 61:1013–1019.

Technology 2005. *National Institute of Standards NIST Chemistry WebBook* (cited July 2007. Available at: http://webbook.nist.gov/chemistry/(accessed June 12, 2013).

Teh KY, Lutz AE (2011). Thermodynamic analysis of fermentation and anaerobic growth of baker's yeast for ethanol production. *Journal of Biotechnology* 147:80–87.

Tevatia R, Demirel Y, Blum P (2012). Kinetic modeling of photoautotropic growth and neutral lipid accumulation in terms of ammonium concentration in *Chlamydomonas reinhardtii. Bioresource Technology* 119:419–424.

Thamsiriroj T, Murphy JD (2008). Is it better to import palm oil from Thailand to produce biodiesel than to produce biodiesel from indigenous Irish rape seed? *Applied Energy* 86:595–604. doi:10.1016/j.apenergy.2008.07.010.

Therkelsen P, Masanet E, Worrell E (2014). Energy efficiency opportunities in the U.S. commercial baking industry. *Journal of Food Engineering* 130:14–22.

Thurston RH (1878). A History of the Growth of the Steam Engine. *Journal of Food Engineering* 130:14–22.

Tinoco Jr. I, Sauer K, Wang JC, Puglisi JD, Harbison G, Rovnyak D (2013). *Physical Chemistry: Principles and Applications in Biological Sciences*, 5th ed. Prentice-Hall, Englewood Cliffs, NJ.

Tipi T, Cetin B, Vardar A (2009). An analysis of energy use and input costs for wheat production in Turkey. *Journal of Food, Agriculture & Environment* 7(2):352–356.

Tobin AJ, Morel RE (1997). *Asking about Cells.* Harcourt Brace & Company, Orlando, FL.

Todd GW (1982). Photosynthesis and respiration of vegetative and reproductive parts of wheat and barley plants in response to increasing temperature. *Proceedings of the Oklahoma Academy of Science* 62:57–62.

Tokunaga K, Shukuya M (2011). Human-body exergy balance calculation under un-steady state conditions. *Building and Environment* 46:2220–2229.

Toledo FGS, Goodpaster BH (2013). The role of weight loss and exercise in correcting skeletal muscle mitochondrial abnormalities in obesity, diabetes and aging. *Molecular and Cellular Endocrinology* 379:30–34.

Toprak Mahsulleri Ofisi 2014. *Turkey Rye Area-Production-Yield TMO Purchase.* TMO, Ankara, Turkey. http://www.tmo.gov.tr/Upload/Document/istatistikler/2012_ing_tables/3_Ryb.pdf (accessed July 24, 2014).

Tortola GJ, Grabowski SR (1993). *Principles of Anatomy and Physiology*, 7th ed. Harper Collins College Publishers, New York.

Toussaint O, Michiels C, Raes M, Remacle J (1995). Cellular aging and the importance of energetic factors. *Experimental Gerontology* 30:1–22.

Uche J, Martinez A, Carrasquer B (2013). Exergy as a guide to allocate environmental costs for implementing the Water Framework Directive in the Ebro River. *Desalination and Water Treatment* 51:4207–4217.

ULS (2007). A study of packaging efficiency as it relates to waste prevention. Report, American Chemistry Council. Available at: http://www.americanchemistry.com/plastics/doc.asp?CID=1593&DID=6072.

United States Department of Agriculture 2014. *World Agricultural Production*. United States Department of Agriculture, Washington, DC. http://apps.fas.usda.gov/psdonline/circulars/production.pdf (accessed July 24, 2014).

U.S. Census Bureau (2012). The National Data Book, 2012. http://www.census.gov/compendia/statab/cats/births_deaths_marriages_divorces/life_expectancy.html (accessed November 12, 2013).

Vendelin M, Eimre M, Seppet E, Peet N, Andrienko T, Lembra M, Engelbrecht J, Seppet EK, Saks VA (2004). Intracellular diffusion of adenosine phosphates is locally restricted in cardiac muscle. *Molecular and Cellular Biochemistry* 256–257(1–2):229–241.

Ventura-Clapier R, Garnier A, Veksler V (2003). Energy metabolism in heart failure. *Journal of Physiology* 555:1–13.

Vieira J, da Silva LP, Esteves da Silva JCG (2012). Advances in the knowledge of light emission by firefly luciferin and oxyluciferin. *Journal of Photochemistry and Photobiology B: Biology* 117:33–39.

Villa RF, Gorini A, Ferrari F, Hoyer S (2013). Energy metabolism of cerebral mitochondria during aging, ischemia and post-ischemic recovery assessed by functional proteomics. *Neurochemistry International* 63:765–781.

Vink ETH, Rabago KR, Glassner DA, Gruben PR (2003). Applications of life cycle assessment to Nature Works™ polylactide (PLA) production. *Polymer Degradation and Stability* 80:403–419.

Vinyard CJ, Taylor AB (2010). A preliminary analysis of the relationship between jaw-muscle architecture and jaw-muscle electromyography during chewing across primates. *The Anatomical Record (Hoboken)* 293:572–582.

Vitolo S, Petarca L, Bresci B (1999). Treatment of olive oil industry wastes. *Bioresource Technology* 67:129–137.

Voliantis S, Secher NH (2002). Arm blood flow and metabolism during arm and combined arm and leg exercise in humans. *The Journal of Physiology* 544:977–984.

von Stockar U, Liu J (1999). Does microbial life always feed on negative entropy? Thermodynamic analysis of microbial growth. *Biochimica et Biophysica Acta* 1412:191–211.

von Stockar U, Maskow T, Liu J, Marison IW, Patino R (2006). Thermodynamics of microbial growth and metabolism: An analysis of current situation. *Journal of Biotechnology* 121:517–533

Wagman DD, Evans WH, Parker VB, Schumm RH, Halow I, Bailey SM, Churney KL, Nuttall RL (1982). The NBS Tables of chemical thermodynamic properties: Selected values for inorganic and C1 and C2 organic substances in SI units. *Journal of Physical Chemistry* 11, supplement 2.

Wagner A (2005). Energy constraints on the evolution of gene expression. *Molecular Biology and Evolution* 22:1365–1374.

Waheed MA, Jekayinfa SO, Ojediran JO, Imeokparia OE (2008). Energetic analysis of fruit juice processing operations in Nigeria. *Energy* 33:35–45.

Wang JF, Dai YP, Gao L (2008). Exergy analyses and parametric optimizations for different cogeneration power plants in cement industry. *Applied Energy* 86:941–948.

Wang Y, Kubota H, Yamada N, Irie T, Akiyama H (2011). Quantum yields and quantitative spectra of firefly bioluminescence with various bivalent metal ions. *Photochemistry and Photobiology* 87:846–852.

Weast RC, Astle WJ, Beyer WH (1986). *CRC Handbook of Chemistry and Physics*, 66th ed. CRC Press, Boca Raton, FL.

West J, Bianconi G, Severini S, Teschendorff AE (2012). Differential network entropy reveals cancer system hallmarks. *Scientific Reports* 2:802:1–8.

Whitwell JC, Toner RK (1973). *Conservation of Mass and Energy*. McGraw-Hill, New York.

Wiedemann SG, McGahan EJ (2011). Environmental assessment of an egg production supply chain using life cycle assessment. Final Project Report, Australian Egg Corporation Limited. Available at: http://www.fsaconsulting.net/fsa/docs/Environmental_Assessment_Egg_Production.pdf (accessed July 31, 2014).

Willems PA, Cavagna GA, Heglund NC (1995). External, internal and total work in human locomotion. *The Journal of Experimental Biology* 198:379–393.

Williams H, Wikström F (2011). Environmental impact of packaging and food losses in the life cycle perspective: A comparative analysis of five food items. *Journal of Cleaner Production* 19:43–48.

Williams PL (1995). *Gray's Anatomy*, 38th ed., Vol. 58. Churchill Livingstone, New York.

Wittmus H, Olson L, Lane D (1975). Energy requirements for conventional versus minimum tillage. *Journal of Soil and Water Conservation* 3:72–75.

Woods TL, Garrels RM (1987). *Thermodynamic Values at Low Temperature for Natural Inorganic Materials: An Uncritical Summary*. Oxford University Press, New York.

Worldwatch Institute (2007). *Biofuels for Transport—Global Potential and Implications for Energy and Agriculture*, Earthscan, London, U.K.

Wright SE, Rosen MA (2004). Exergetic efficiencies and the exergy content of terrestrial solar radiation. *ASME Journal of Solar Energy Engineering* 126:673–676.

Wright SE, Rosen MA, Scott DS, Haddow JB (2002a). The exergy flux of radiative heat transfer for the special case of the blackbody radiation. *Exergy, an International Journal* 2:24–33.

Wright SE, Rosen MA, Scott DS, Haddow JB (2002b). The exergy flux of radiative heat transfer with an arbitrary spectrum. *Exergy, an International Journal* 2:69–77.

Wu H, Tassou SA, Karayiannis TG (2013). Modelling and control approaches for energy reduction in continuous frying systems. *Applied Energy* 112:939–948.

Wu W, Liu Y (2010). Radiation entropy flux and entropy production of the earth system. *Reviews of Geophysics* 48:RG2003. Available at: http://www.bnl.gov/envsci/pubs/pdf/2010/BNL-81482-2008-JA.pdf.

Xu T, Flapper J (2011). Reduce energy use and greenhouse gas emissions from global dairy processing facilities. *Energy Policy* 39:234–247.

Xu T, Flapper J, Kramer KJ (2009). Characterization of energy use and performance of global cheese processing. *Energy* 34:1993–2000.

Yang Q, Chen B, Ji X, He YF, Chen GQ (2009). Exergetic evaluation of corn-ethanol production in China. *Communications in Nonlinear Science and Numerical Simulation* 14:2450–2461.

Yaniv Y, Juhaszova M, Sollott SJ (2013). Age-related changes of myocardial ATP supply and demand mechanisms. *Trends in Endocrinology and Metabolism* 24:495–505.

Yoo C, Jun SY, Lee JY, Ahn CY, Oh HM (2010). Selection of microalgae for lipid production under high levels of carbon dioxide. *Bioresource Technology* 101:571–4.

Zabaniotou A, Kassidi E (2003). Life cycle assessment applied to egg packaging made from polystyrene and recycled paper. *Journal of Cleaner Production* 11:549–559.

Zakon HH (2003). Insight into the mechanisms of neuronal processing from electric fish. *Current Opinion in Neurobiology* 13:744–750.

Zhang F, Tong C, Xie Z, Lu J (2007). Exergy as an ecological application used in the recovery process of benthic communities. *Integrative Zoology* 1:1–9.

Zhang M, Li G, Mu HL, Ning YD (2011). Energy and exergy efficiencies in the Chinese transportation sector, 1980–2009. *Energy* 36:770–776.

Zhou L, Salem JE, Saidel GM, Stanley WC, Cabrera ME (2005). Mechanistic model of cardiac energy metabolism predicts localization of glycolysis to cytosolic subdomain during ischemia. *American Journal of Physiology: Heart and Circulatory Physiology* 288:2400–2411.

Zotin AA, Zotin AI (1997). Phenomenological theory of ontogenesis. *International Journal of Developmental Biology* 41:917–921.

Zvinowanda CM, Zikalala NN, Tabani Mtombeni T, Maree JP, Louw WJ, Oosthuizen F (2014). Process optimization of freeze desalination of brine using HybridICE™ pilot plant. *Desalination and Water Treatment* 52:6315–6324.

Index

Milton Keynes UK
Ingram Content Group UK Ltd.
UKHW021825071024
449327UK00021B/1438